Coding Theory

Coding Theory

Algorithms, Architectures, and Applications

André Neubauer
Münster University of Applied Sciences, Germany

Jürgen Freudenberger
HTWG Konstanz, University of Applied Sciences, Germany

Volker Kühn
University of Rostock, Germany

John Wiley & Sons, Ltd

Copyright © 2007   John Wiley & Sons Ltd, The Atrium, Southern Gate, Chichester,
                      West Sussex PO19 8SQ, England

                      Telephone (+44) 1243 779777

Email (for orders and customer service enquiries): cs-books@wiley.co.uk
Visit our Home Page on www.wileyeurope.com or www.wiley.com

All Rights Reserved. No part of this publication may be reproduced, stored in a retrieval system or transmitted in any form or by any means, electronic, mechanical, photocopying, recording, scanning or otherwise, except under the terms of the Copyright, Designs and Patents Act 1988 or under the terms of a licence issued by the Copyright Licensing Agency Ltd, 90 Tottenham Court Road, London W1T 4LP, UK, without the permission in writing of the Publisher. Requests to the Publisher should be addressed to the Permissions Department, John Wiley & Sons Ltd, The Atrium, Southern Gate, Chichester, West Sussex PO19 8SQ, England, or emailed to permreq@wiley.co.uk, or faxed to (+44) 1243 770620.

Designations used by companies to distinguish their products are often claimed as trademarks. All brand names and product names used in this book are trade names, service marks, trademarks or registered trademarks of their respective owners. The Publisher is not associated with any product or vendor mentioned in this book. All trademarks referred to in the text of this publication are the property of their respective owners.

This publication is designed to provide accurate and authoritative information in regard to the subject matter covered. It is sold on the understanding that the Publisher is not engaged in rendering professional services. If professional advice or other expert assistance is required, the services of a competent professional should be sought.

Other Wiley Editorial Offices

John Wiley & Sons Inc., 111 River Street, Hoboken, NJ 07030, USA

Jossey-Bass, 989 Market Street, San Francisco, CA 94103-1741, USA

Wiley-VCH Verlag GmbH, Boschstr. 12, D-69469 Weinheim, Germany

John Wiley & Sons Australia Ltd, 42 McDougall Street, Milton, Queensland 4064, Australia

John Wiley & Sons (Asia) Pte Ltd, 2 Clementi Loop #02-01, Jin Xing Distripark, Singapore 129809

John Wiley & Sons Canada Ltd, 6045 Freemont Blvd, Mississauga, Ontario, L5R 4J3, Canada

Wiley also publishes its books in a variety of electronic formats. Some content that appears in print may not be available in electronic books.

Anniversary Logo Design: Richard J. Pacifico

Library of Congress Cataloging-in-Publication Data

Neubauer, Andre.
  Coding theory : algorithms, architectures and applications / Andre Neubauer, Jürgen Freudenberger, Volker Kühn.
      p. cm.
  ISBN 978-0-470-02861-2 (cloth)
  1. Coding theory. I Freudenberger, Jrgen. II. Kühn, Volker. III. Title
  QA268.N48 2007
  003′.54–dc22

British Library Cataloguing in Publication Data

A catalogue record for this book is available from the British Library

ISBN   978-0-470-02861-2 (HB)

Typeset in 10/12pt Times by Laserwords Private Limited, Chennai, India

This book is printed on acid-free paper responsibly manufactured from sustainable forestry in which at least two trees are planted for each one used for paper production.

Contents

Preface ix

1 Introduction 1
   1.1 Communication Systems 1
   1.2 Information Theory 3
       1.2.1 Entropy 3
       1.2.2 Channel Capacity 4
       1.2.3 Binary Symmetric Channel 5
       1.2.4 AWGN Channel 6
   1.3 A Simple Channel Code 8

2 Algebraic Coding Theory 13
   2.1 Fundamentals of Block Codes 14
       2.1.1 Code Parameters 16
       2.1.2 Maximum Likelihood Decoding 19
       2.1.3 Binary Symmetric Channel 23
       2.1.4 Error Detection and Error Correction 25
   2.2 Linear Block Codes 27
       2.2.1 Definition of Linear Block Codes 27
       2.2.2 Generator Matrix 27
       2.2.3 Parity-Check Matrix 30
       2.2.4 Syndrome and Cosets 31
       2.2.5 Dual Code 36
       2.2.6 Bounds for Linear Block Codes 37
       2.2.7 Code Constructions 41
       2.2.8 Examples of Linear Block Codes 46
   2.3 Cyclic Codes 62
       2.3.1 Definition of Cyclic Codes 62
       2.3.2 Generator Polynomial 63
       2.3.3 Parity-Check Polynomial 67
       2.3.4 Dual Codes 70
       2.3.5 Linear Feedback Shift Registers 71
       2.3.6 BCH Codes 74
       2.3.7 Reed–Solomon Codes 81

	2.3.8	Algebraic Decoding Algorithm	84
2.4	Summary		93

3 Convolutional Codes 97
- 3.1 Encoding of Convolutional Codes 98
  - 3.1.1 Convolutional Encoder 98
  - 3.1.2 Generator Matrix in the Time Domain 101
  - 3.1.3 State Diagram of a Convolutional Encoder 103
  - 3.1.4 Code Termination 104
  - 3.1.5 Puncturing 106
  - 3.1.6 Generator Matrix in the D-Domain 108
  - 3.1.7 Encoder Properties 110
- 3.2 Trellis Diagram and the Viterbi Algorithm 112
  - 3.2.1 Minimum Distance Decoding 113
  - 3.2.2 Trellises 115
  - 3.2.3 Viterbi Algorithm 116
- 3.3 Distance Properties and Error Bounds 121
  - 3.3.1 Free Distance 121
  - 3.3.2 Active Distances 122
  - 3.3.3 Weight Enumerators for Terminated Codes 126
  - 3.3.4 Path Enumerators 129
  - 3.3.5 Pairwise Error Probability 131
  - 3.3.6 Viterbi Bound 134
- 3.4 Soft-input Decoding 136
  - 3.4.1 Euclidean Metric 136
  - 3.4.2 Support of Punctured Codes 137
  - 3.4.3 Implementation Issues 138
- 3.5 Soft-output Decoding 140
  - 3.5.1 Derivation of APP Decoding 141
  - 3.5.2 APP Decoding in the Log Domain 145
- 3.6 Convolutional Coding in Mobile Communications 147
  - 3.6.1 Coding of Speech Data 147
  - 3.6.2 Hybrid ARQ 150
  - 3.6.3 EGPRS Modulation and Coding 152
  - 3.6.4 Retransmission Mechanism 155
  - 3.6.5 Link Adaptation 156
  - 3.6.6 Incremental Redundancy 157
- 3.7 Summary 160

4 Turbo Codes 163
- 4.1 LDPC Codes 165
  - 4.1.1 Codes Based on Sparse Graphs 165
  - 4.1.2 Decoding for the Binary Erasure Channel 168
  - 4.1.3 Log-Likelihood Algebra 169
  - 4.1.4 Belief Propagation 174
- 4.2 A First Encounter with Code Concatenation 177
  - 4.2.1 Product Codes 177

		4.2.2	Iterative Decoding of Product Codes	180
	4.3	Concatenated Convolutional Codes		182
		4.3.1	Parallel Concatenation	182
		4.3.2	The UMTS Turbo Code	183
		4.3.3	Serial Concatenation	184
		4.3.4	Partial Concatenation	185
		4.3.5	Turbo Decoding	186
	4.4	EXIT Charts		188
		4.4.1	Calculating an EXIT Chart	189
		4.4.2	Interpretation	191
	4.5	Weight Distribution		196
		4.5.1	Partial Weights	196
		4.5.2	Expected Weight Distribution	197
	4.6	Woven Convolutional Codes		198
		4.6.1	Encoding Schemes	200
		4.6.2	Distance Properties of Woven Codes	202
		4.6.3	Woven Turbo Codes	205
		4.6.4	Interleaver Design	208
	4.7	Summary		212

5 Space–Time Codes 215

	5.1	Introduction		215
		5.1.1	Digital Modulation Schemes	216
		5.1.2	Diversity	223
	5.2	Spatial Channels		229
		5.2.1	Basic Description	229
		5.2.2	Spatial Channel Models	234
		5.2.3	Channel Estimation	239
	5.3	Performance Measures		241
		5.3.1	Channel Capacity	241
		5.3.2	Outage Probability and Outage Capacity	250
		5.3.3	Ergodic Error Probability	252
	5.4	Orthogonal Space–Time Block Codes		257
		5.4.1	Alamouti's Scheme	257
		5.4.2	Extension to More than Two Transmit Antennas	260
		5.4.3	Simulation Results	263
	5.5	Spatial Multiplexing		265
		5.5.1	General Concept	265
		5.5.2	Iterative APP Preprocessing and Per-layer Decoding	267
		5.5.3	Linear Multilayer Detection	272
		5.5.4	Original BLAST Detection	275
		5.5.5	QL Decomposition and Interference Cancellation	278
		5.5.6	Performance of Multi-Layer Detection Schemes	287
		5.5.7	Unified Description by Linear Dispersion Codes	291
	5.6	Summary		294

A Algebraic Structures — 295
- A.1 Groups, Rings and Finite Fields — 295
  - A.1.1 Groups — 295
  - A.1.2 Rings — 296
  - A.1.3 Finite Fields — 298
- A.2 Vector Spaces — 299
- A.3 Polynomials and Extension Fields — 300
- A.4 Discrete Fourier Transform — 305

B Linear Algebra — 311

C Acronyms — 319

Bibliography — 325

Index — 335

Preface

Modern information and communication systems are based on the reliable and efficient transmission of information. Channels encountered in practical applications are usually disturbed regardless of whether they correspond to information transmission over noisy and time-variant mobile radio channels or to information transmission on optical discs that might be damaged by scratches. Owing to these disturbances, appropriate channel coding schemes have to be employed such that errors within the transmitted information can be detected or even corrected. To this end, channel coding theory provides suitable coding schemes for error detection and error correction. Besides good code characteristics with respect to the number of errors that can be detected or corrected, the complexity of the architectures used for implementing the encoding and decoding algorithms is important for practical applications.

The present book provides a concise overview of channel coding theory and practice as well as the accompanying algorithms, architectures and applications. The selection of the topics presented in this book is oriented towards those subjects that are relevant for information and communication systems in use today or in the near future. The focus is on those aspects of coding theory that are important for the understanding of these systems. This book places emphasis on the algorithms for encoding and decoding and their architectures, as well as the applications of the corresponding coding schemes in a unified framework.

The idea for this book originated from a two-day seminar on coding theory in the industrial context. We have tried to keep this seminar style in the book by highlighting the most important facts within the figures and by restricting the scope to the most important topics with respect to the applications of coding theory, especially within communication systems. This also means that many important and interesting topics could not be covered in order to be as concise as possible.

The target audience for the book are students of communication and information engineering as well as computer science at universities and also applied mathematicians who are interested in a presentation that subsumes theory and practice of coding theory without sacrificing exactness or relevance with regard to real-world practical applications. Therefore, this book is well suited for engineers in industry who want to know about the theoretical basics of coding theory and their application in currently relevant communication systems.

The book is organised as follows. In Chapter 1 a brief overview of the principle architecture of a communication system is given and the information theory fundamentals underlying coding theory are summarised. The most important concepts of information theory, such as entropy and channel capacity as well as simple channel models, are described.

Chapter 2 presents the classical, i.e. algebraic, coding theory. The fundamentals of the encoding and decoding of block codes are explained, and the maximum likelihood decoding rule is derived as the optimum decoding strategy for minimising the word error probability after decoding a received word. Linear block codes and their definition based on generator and parity-check matrices are discussed. General performance measures and bounds relating important code characteristics such as the minimum Hamming distance and the code rate are presented, illustrating the compromises necessary between error detection and error correction capabilities and transmission efficiency. It is explained how new codes can be constructed from already known codes. Repetition codes, parity-check-codes, Hamming codes, simplex codes and Reed–Muller codes are presented as examples. Since the task of decoding linear block codes is difficult in general, the algebraic properties of cyclic codes are exploited for efficient decoding algorithms. These cyclic codes, together with their generator and parity-check polynomials, are discussed, as well as efficient encoding and decoding architectures based on linear feedback shift registers. Important cyclic codes such as BCH codes and Reed–Solomon codes are presented, and an efficient algebraic decoding algorithm for the decoding of these cyclic codes is derived.

Chapter 3 deals with the fundamentals of convolutional coding. Convolutional codes can be found in many applications, for instance in dial-up modems, satellite communications and digital cellular systems. The major reason for this popularity is the existence of efficient decoding algorithms that can utilise soft input values from the demodulator. This so-called soft-input decoding leads to significant performance gains. Two famous examples for a soft-input decoding algorithm are the Viterbi algorithm and the Bahl, Cocke, Jelinek, Raviv (BCJR) algorithm which also provides a reliability output. Both algorithms are based on the trellis representation of the convolutional code. This highly repetitive structure makes trellis-based decoding very suitable for hardware implementations.

We start our discussion with the encoding of convolutional codes and some of their basic properties. It follows a presentation of the Viterbi algorithm and an analysis of the error correction performance with this maximum likelihood decoding procedure. The concept of soft-output decoding and the BCJR algorithm are considered in Section 3.5. Soft-output decoding is a prerequisite for the iterative decoding of concatenated convolutional codes as introduced in Chapter 4. Finally, we consider an application of convolutional codes for mobile communication channels as defined in the Global System for Mobile communications (GSM) standard. In particular, the considered hybrid ARQ protocols are excellent examples of the adaptive coding systems that are required for strongly time-variant mobile channels.

As mentioned above, Chapter 4 is dedicated to the construction of long powerful codes based on the concatenation of simple convolutional component codes. These concatenated convolutional codes, for example the famous turbo codes, are capable of achieving low bit error rates at signal-to-noise ratios close to the theoretical Shannon limit. The term *turbo* reflects a property of the employed iterative decoding algorithm, where the decoder output of one iteration is used as the decoder input of the next iteration. This concept of iterative decoding was first introduced for the class of low-density parity-check codes. Therefore, we first introduce low-density parity-check codes in Section 4.1 and discuss the relation between these codes and concatenated code constructions. Then, we introduce some popular encoding schemes for concatenated convolutional codes and present three methods to analyse the performance of the corresponding codes. The EXIT chart method

PREFACE

in Section 4.4 makes it possible to predict the behaviour of the iterative decoder by looking at the input/output relations of the individual constituent soft-output decoders. Next, we present a common approach in coding theory. We estimate the code performance with maximum likelihood decoding for an ensemble of concatenated codes. This method explains why many concatenated code constructions lead to a low minimum Hamming distance and therefore to a relatively poor performance for high signal-to-noise ratios. In Section 4.6 we consider code designs that lead to a higher minimum Hamming distance owing to a special encoder construction, called the woven encoder, or the application of designed interleavers.

The fifth chapter addresses space–time coding concepts, a still rather new topic in the area of radio communications. Although these techniques do not represent error-correcting codes in the classical sense, they can also be used to improve the reliability of a data link. Since space–time coding became popular only a decade ago, only a few concepts have found their way into current standards hitherto. However, many other approaches are currently being discussed. As already mentioned before, we restrict this book to the most important and promising concepts.

While classical encoders and decoders are separated from the physical channel by modulators, equalisers, etc., and experience rather simple hyperchannels, this is not true for space–time coding schemes. They directly work on the physical channel. Therefore, Chapter 5 starts with a short survey of linear modulation schemes and explains the principle of diversity. Next, spatial channel models are described and different performance measures for their quantitative evaluation are discussed. Sections 5.4 and 5.5 introduce two space–time coding concepts with the highest practical relevance, namely orthogonal space–time block codes increasing the diversity degree and spatial multiplexing techniques boosting the achievable data rate. For the latter approach, sophisticated signal processing algorithms are required at the receiver in order to separate superimposed data streams again.

In the appendices a brief summary of algebraic structures such as finite fields and polynomial rings is given, which are needed for the treatment especially of classical algebraic codes, and the basics of linear algebra are briefly reviewed.

Finally, we would like to thank the Wiley team, especially Sarah Hinton as the responsible editor, for their support during the completion of the manuscript. We also thank Dr. rer. nat. Jens Schlembach who was involved from the beginning of this book project and who gave encouragement when needed. Last but not least, we would like to give special thanks to our families – Fabian, Heike, Jana and Erik, Claudia, Hannah and Jakob and Christiane – for their emotional support during the writing of the manuscript.

ANDRÉ NEUBAUER
Münster University of Applied Sciences

JÜRGEN FREUDENBERGER
HTWG Konstanz University of Applied Sciences

VOLKER KÜHN
University of Rostock

1

Introduction

The reliable transmission of information over noisy channels is one of the basic requirements of digital information and communication systems. Here, transmission is understood both as transmission in space, e.g. over mobile radio channels, and as transmission in time by storing information in appropriate storage media. Because of this requirement, modern communication systems rely heavily on powerful channel coding methodologies. For practical applications these coding schemes do not only need to have good coding characteristics with respect to the capability of detecting or correcting errors introduced on the channel. They also have to be efficiently implementable, e.g. in digital hardware within integrated circuits. Practical applications of channel codes include space and satellite communications, data transmission, digital audio and video broadcasting and mobile communications, as well as storage systems such as computer memories or the compact disc (Costello *et al.*, 1998).

In this introductory chapter we will give a brief introduction into the field of channel coding. To this end, we will describe the information theory fundamentals of channel coding. Simple channel models will be presented that will be used throughout the text. Furthermore, we will present the binary triple repetition code as an illustrative example of a simple channel code.

1.1 Communication Systems

In Figure 1.1 the basic structure of a digital communication system is shown which represents the architecture of the communication systems in use today. Within the transmitter of such a communication system the following tasks are carried out:

- source encoding,
- channel encoding,
- modulation.

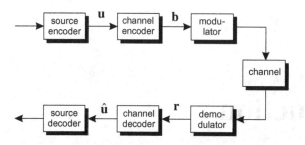

Principal structure of digital communication systems

- The sequence of information symbols **u** is encoded into the sequence of code symbols **b** which are transmitted across the channel after modulation.

- The sequence of received symbols **r** is decoded into the sequence of information symbols **û** which are estimates of the originally transmitted information symbols.

Figure 1.1: Basic structure of digital communication systems

In the receiver the corresponding inverse operations are implemented:

- demodulation,
- channel decoding,
- source decoding.

According to Figure 1.1 the *modulator* generates the signal that is used to transmit the sequence of symbols **b** across the channel (Benedetto and Biglieri, 1999; Neubauer, 2007; Proakis, 2001). Due to the noisy nature of the channel, the transmitted signal is disturbed. The noisy received signal is demodulated by the *demodulator* in the receiver, leading to the sequence of received symbols **r**. Since the received symbol sequence **r** usually differs from the transmitted symbol sequence **b**, a *channel code* is used such that the receiver is able to detect or even correct errors (Bossert, 1999; Lin and Costello, 2004; Neubauer, 2006b). To this end, the channel encoder introduces redundancy into the information sequence **u**. This redundancy can be exploited by the channel decoder for error detection or error correction by estimating the transmitted symbol sequence **û**.

In his fundamental work, Shannon showed that it is theoretically possible to realise an information transmission system with as small an error probability as required (Shannon, 1948). The prerequisite for this is that the information rate of the information source be smaller than the so-called channel capacity. In order to reduce the information rate, *source coding* schemes are used which are implemented by the source encoder in the transmitter and the source decoder in the receiver (McEliece, 2002; Neubauer, 2006a).

INTRODUCTION

Further information about source coding can be found elsewhere (Gibson *et al.*, 1998; Sayood, 2000, 2003).

In order better to understand the theoretical basics of information transmission as well as channel coding, we now give a brief overview of information theory as introduced by Shannon in his seminal paper (Shannon, 1948). In this context we will also introduce the simple channel models that will be used throughout the text.

1.2 Information Theory

An important result of information theory is the finding that error-free transmission across a noisy channel is theoretically possible – as long as the information rate does not exceed the so-called channel capacity. In order to quantify this result, we need to measure information. Within Shannon's information theory this is done by considering the statistics of symbols emitted by information sources.

1.2.1 Entropy

Let us consider the discrete memoryless *information source* shown in Figure 1.2. At a given time instant, this discrete information source emits the random discrete symbol $\mathcal{X} = x_i$ which assumes one out of M possible symbol values $x_1, x_2, \ldots, x_M$. The rate at which these symbol values appear are given by the probabilities $P_\mathcal{X}(x_1), P_\mathcal{X}(x_2), \ldots, P_\mathcal{X}(x_M)$ with

$$P_\mathcal{X}(x_i) = \Pr\{\mathcal{X} = x_i\}.$$

Discrete information source

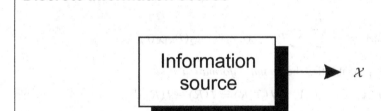

- The discrete information source emits the random discrete symbol $\mathcal{X}$.
- The symbol values $x_1, x_2, \ldots, x_M$ appear with probabilities $P_\mathcal{X}(x_1), P_\mathcal{X}(x_2), \ldots, P_\mathcal{X}(x_M)$.
- Entropy

$$I(\mathcal{X}) = -\sum_{i=1}^{M} P_\mathcal{X}(x_i) \cdot \log_2(P_\mathcal{X}(x_i)) \qquad (1.1)$$

Figure 1.2: Discrete information source emitting discrete symbols $\mathcal{X}$

The average information associated with the random discrete symbol $\mathcal{X}$ is given by the so-called *entropy* measured in the unit 'bit'

$$I(\mathcal{X}) = -\sum_{i=1}^{M} P_{\mathcal{X}}(x_i) \cdot \log_2\left(P_{\mathcal{X}}(x_i)\right).$$

For a binary information source that emits the binary symbols $\mathcal{X} = 0$ and $\mathcal{X} = 1$ with probabilities $\Pr\{\mathcal{X} = 0\} = p_0$ and $\Pr\{\mathcal{X} = 1\} = 1 - \Pr\{\mathcal{X} = 0\} = 1 - p_0$, the entropy is given by the so-called *Shannon function* or binary entropy function

$$I(\mathcal{X}) = -p_0 \log_2(p_0) - (1 - p_0) \log_2(1 - p_0).$$

1.2.2 Channel Capacity

With the help of the entropy concept we can model a channel according to Berger's channel diagram shown in Figure 1.3 (Neubauer, 2006a). Here, $\mathcal{X}$ refers to the input symbol and $\mathcal{R}$ denotes the output symbol or received symbol. We now assume that M input symbol values $x_1, x_2, \ldots, x_M$ and N output symbol values $r_1, r_2, \ldots, r_N$ are possible. With the help of the conditional probabilities

$$P_{\mathcal{X}|\mathcal{R}}(x_i|r_j) = \Pr\{\mathcal{X} = x_i | \mathcal{R} = r_j\}$$

and

$$P_{\mathcal{R}|\mathcal{X}}(r_j|x_i) = \Pr\{\mathcal{R} = r_j | \mathcal{X} = x_i\}$$

the conditional entropies are given by

$$I(\mathcal{X}|\mathcal{R}) = -\sum_{i=1}^{M} \sum_{j=1}^{N} P_{\mathcal{X},\mathcal{R}}(x_i, r_j) \cdot \log_2\left(P_{\mathcal{X}|\mathcal{R}}(x_i|r_j)\right)$$

and

$$I(\mathcal{R}|\mathcal{X}) = -\sum_{i=1}^{M} \sum_{j=1}^{N} P_{\mathcal{X},\mathcal{R}}(x_i, r_j) \cdot \log_2(P_{\mathcal{R}|\mathcal{X}}(r_j|x_i)).$$

With these conditional probabilities the *mutual information*

$$I(\mathcal{X}; \mathcal{R}) = I(\mathcal{X}) - I(\mathcal{X}|\mathcal{R}) = I(\mathcal{R}) - I(\mathcal{R}|\mathcal{X})$$

can be derived which measures the amount of information that is transmitted across the channel from the input to the output for a given information source.

The so-called *channel capacity* C is obtained by maximising the mutual information $I(\mathcal{X}; \mathcal{R})$ with respect to the statistical properties of the input $\mathcal{X}$, i.e. by appropriately choosing the probabilities $\{P_{\mathcal{X}}(x_i)\}_{1 \leq i \leq M}$. This leads to

$$C = \max_{\{P_{\mathcal{X}}(x_i)\}_{1 \leq i \leq M}} I(\mathcal{X}; \mathcal{R}).$$

If the input entropy $I(\mathcal{X})$ is smaller than the channel capacity C

$$I(\mathcal{X}) \stackrel{!}{<} C,$$

then information can be transmitted across the noisy channel with arbitrarily small error probability. Thus, the channel capacity C in fact quantifies the information transmission capacity of the channel.

INTRODUCTION

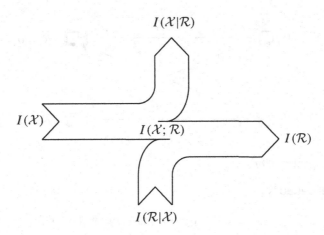

Figure 1.3: Berger's channel diagram

1.2.3 Binary Symmetric Channel

As an important example of a memoryless channel we turn to the *binary symmetric channel* or BSC. Figure 1.4 shows the channel diagram of the binary symmetric channel with bit error probability ε. This channel transmits the binary symbol $\mathcal{X} = 0$ or $\mathcal{X} = 1$ correctly with probability $1 - \varepsilon$, whereas the incorrect binary symbol $\mathcal{R} = 1$ or $\mathcal{R} = 0$ is emitted with probability ε.

By maximising the mutual information $I(\mathcal{X}; \mathcal{R})$, the channel capacity of a binary symmetric channel is obtained according to

$$C = 1 + \varepsilon \log_2(\varepsilon) + (1 - \varepsilon) \log_2(1 - \varepsilon).$$

This channel capacity is equal to 1 if $\varepsilon = 0$ or $\varepsilon = 1$; for $\varepsilon = \frac{1}{2}$ the channel capacity is 0. In contrast to the binary symmetric channel, which has discrete input and output symbols taken from binary alphabets, the so-called AWGN channel is defined on the basis of continuous real-valued random variables.[1]

---

[1] In Chapter 5 we will also consider complex-valued random variables.

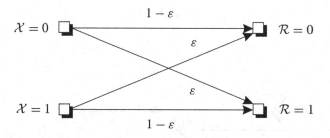

Figure 1.4: Binary symmetric channel with bit error probability ε

1.2.4 AWGN Channel

Up to now we have exclusively considered discrete-valued symbols. The concept of entropy can be transferred to continuous real-valued random variables by introducing the so-called differential entropy. It turns out that a channel with real-valued input and output symbols can again be characterised with the help of the mutual information $I(\mathcal{X}; \mathcal{R})$ and its maximum, the channel capacity C. In Figure 1.5 the so-called *AWGN channel* is illustrated which is described by the additive white Gaussian noise term $\mathcal{Z}$.

With the help of the signal power

$$S = \mathrm{E}\left\{\mathcal{X}^2\right\}$$

and the noise power

$$N = \mathrm{E}\left\{\mathcal{Z}^2\right\}$$

the channel capacity of the AWGN channel is given by

$$C = \frac{1}{2} \log_2\left(1 + \frac{S}{N}\right).$$

The channel capacity exclusively depends on the signal-to-noise ratio S/N.

In order to compare the channel capacities of the binary symmetric channel and the AWGN channel, we assume a digital transmission scheme using binary phase shift keying (BPSK) and optimal reception with the help of a matched filter (Benedetto and Biglieri, 1999; Neubauer, 2007; Proakis, 2001). The signal-to-noise ratio of the real-valued output

INTRODUCTION 7

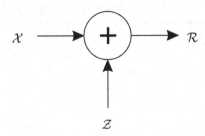

AWGN channel

- Signal-to-noise ratio $\frac{S}{N}$
- Channel capacity

$$C = \frac{1}{2} \log_2 \left(1 + \frac{S}{N}\right) \qquad (1.5)$$

Figure 1.5: AWGN channel with signal-to-noise ratio S/N

$\mathcal{R}$ of the matched filter is then given by

$$\frac{S}{N} = \frac{E_b}{N_0/2}$$

with bit energy E_b and noise power spectral density N_0. If the output $\mathcal{R}$ of the matched filter is compared with the threshold 0, we obtain the binary symmetric channel with bit error probability

$$\varepsilon = \frac{1}{2} \operatorname{erfc}\left(\sqrt{\frac{E_b}{N_0}}\right).$$

Here, $\operatorname{erfc}(\cdot)$ denotes the complementary error function. In Figure 1.6 the channel capacities of the binary symmetric channel and the AWGN channel are compared as a function of E_b/N_0. The signal-to-noise ratio S/N or the ratio E_b/N_0 must be higher for the binary symmetric channel compared with the AWGN channel in order to achieve the same channel capacity. This gain also translates to the coding gain achievable by soft-decision decoding as opposed to hard-decision decoding of channel codes, as we will see later (e.g. in Section 2.2.8).

Although information theory tells us that it is theoretically possible to find a channel code that for a given channel leads to as small an error probability as required, the design of good channel codes is generally difficult. Therefore, in the next chapters several classes of channel codes will be described. Here, we start with a simple example.

8 INTRODUCTION

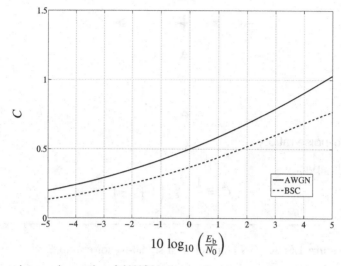

- Signal-to-noise ratio of AWGN channel

$$\frac{S}{N} = \frac{E_b}{N_0/2} \tag{1.6}$$

- Bit error probability of binary symmetric channel

$$\varepsilon = \frac{1}{2} \, \text{erfc}\left(\sqrt{\frac{E_b}{N_0}}\right) \tag{1.7}$$

Figure 1.6: Channel capacity of the binary symmetric channel vs the channel capacity of the AWGN channel

1.3 A Simple Channel Code

As an introductory example of a simple channel code we consider the transmission of the binary information sequence

$$00\underline{1}0\underline{1}110$$

over a binary symmetric channel with bit error probability $\varepsilon = 0.25$ (Neubauer, 2006b). On average, every fourth binary symbol will be received incorrectly. In this example we assume that the binary sequence

$$00\underline{0}0\underline{0}110$$

is received at the output of the binary symmetric channel (see Figure 1.7).

INTRODUCTION

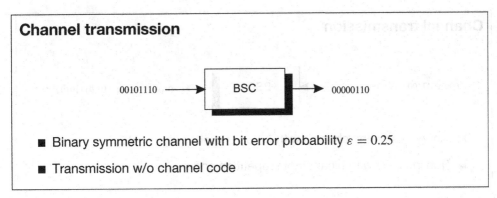

Figure 1.7: Channel transmission without channel code

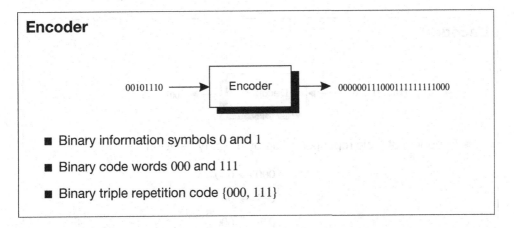

Figure 1.8: Encoder of a triple repetition code

In order to implement a simple error correction scheme we make use of the so-called binary *triple repetition code*. This simple channel code is used for the encoding of binary data. If the binary symbol 0 is to be transmitted, the encoder emits the code word 000. Alternatively, the code word 111 is issued by the encoder when the binary symbol 1 is to be transmitted. The encoder of a triple repetition code is illustrated in Figure 1.8.

For the binary information sequence given above we obtain the binary code sequence

$$000\,000\,111\,000\,111\,111\,111\,000$$

at the output of the encoder. If we again assume that on average every fourth binary symbol is incorrectly transmitted by the binary symmetric channel, we may obtain the received sequence

$$0\underline{1}0\,000\,\underline{0}11\,0\underline{1}0\,111\,0\underline{1}0\,111\,0\underline{1}0.$$

This is illustrated in Figure 1.9.

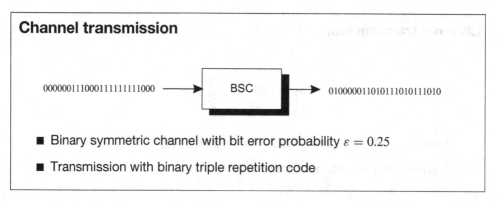

Figure 1.9: Channel transmission of a binary triple repetition code

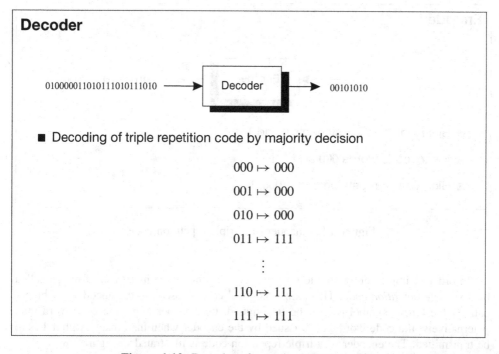

Figure 1.10: Decoder of a triple repetition code

The decoder in Figure 1.10 tries to estimate the original information sequence with the help of a *majority decision*. If the number of 0s within a received 3-bit word is larger than the number of 1s, the decoder emits the binary symbol 0; otherwise a 1 is decoded. With this decoding algorithm we obtain the decoded information sequence

$$00101\underline{0}10.$$

INTRODUCTION

As can be seen from this example, the binary triple repetition code is able to correct a single error within a code word. More errors cannot be corrected. With the help of this simple channel code we are able to reduce the number of errors. Compared with the unprotected transmission without a channel code, the number of errors has been reduced from two to one. However, this is achieved by a significant reduction in the transmission bandwidth because, for a given symbol rate on the channel, it takes 3 times longer to transmit an information symbol with the help of the triple repetition code. It is one of the main topics of the following chapters to present more efficient coding schemes.

2

Algebraic Coding Theory

In this chapter we will introduce the basic concepts of algebraic coding theory. To this end, the fundamental properties of block codes are first discussed. We will define important code parameters and describe how these codes can be used for the purpose of error detection and error correction. The optimal maximum likelihood decoding strategy will be derived and applied to the binary symmetric channel.

With these fundamentals at hand we will then introduce linear block codes. These channel codes can be generated with the help of so-called generator matrices owing to their special algebraic properties. Based on the closely related parity-check matrix and the syndrome, the decoding of linear block codes can be carried out. We will also introduce dual codes and several techniques for the construction of new block codes based on known ones, as well as bounds for the respective code parameters and the accompanying code characteristics. As examples of linear block codes we will treat the repetition code, parity-check code, Hamming code, simplex code and Reed–Muller code.

Although code generation can be carried out efficiently for linear block codes, the decoding problem for general linear block codes is difficult to solve. By introducing further algebraic structures, cyclic codes can be derived as a subclass of linear block codes for which efficient algebraic decoding algorithms exist. Similar to general linear block codes, which are defined using the generator matrix or the parity-check matrix, cyclic codes are defined with the help of the so-called generator polynomial or parity-check polynomial. Based on linear feedback shift registers, the respective encoding and decoding architectures for cyclic codes can be efficiently implemented. As important examples of cyclic codes we will discuss BCH codes and Reed–Solomon codes. Furthermore, an algebraic decoding algorithm is presented that can be used for the decoding of BCH and Reed–Solomon codes.

In this chapter the classical algebraic coding theory is presented. In particular, we will follow work (Berlekamp, 1984; Bossert, 1999; Hamming, 1986; Jungnickel, 1995; Lin and Costello, 2004; Ling and Xing, 2004; MacWilliams and Sloane, 1998; McEliece, 2002; Neubauer, 2006b; van Lint, 1999) that contains further details about algebraic coding theory.

Coding Theory – Algorithms, Architectures, and Applications André Neubauer, Jürgen Freudenberger, Volker Kühn
© 2007 John Wiley & Sons, Ltd

2.1 Fundamentals of Block Codes

In Section 1.3, the binary triple repetition code was given as an introductory example of a simple channel code. This specific channel code consists of two code words 000 and 111 of length $n = 3$, which represent $k = 1$ binary information symbol 0 or 1 respectively. Each symbol of a binary information sequence is encoded separately. The respective code word of length 3 is then transmitted across the channel. The potentially erroneously received word is finally decoded into a valid code word 000 or 111 – or equivalently into the respective information symbol 0 or 1. As we have seen, this simple code is merely able to correct one error by transmitting three code symbols instead of just one information symbol across the channel.

In order to generalise this simple channel coding scheme and to come up with more efficient and powerful channel codes, we now consider an information sequence $u_0 \, u_1 \, u_2 \, u_3 \, u_4 \, u_5 \, u_6 \, u_7 \ldots$ of discrete information symbols u_i. This information sequence is grouped into blocks of length k according to

$$\underbrace{u_0 \, u_1 \, \cdots \, u_{k-1}}_{\text{block}} \, \underbrace{u_k \, u_{k+1} \, \cdots \, u_{2k-1}}_{\text{block}} \, \underbrace{u_{2k} \, u_{2k+1} \, \cdots \, u_{3k-1}}_{\text{block}} \, \cdots .$$

In so-called q-nary (n, k) *block codes* the information words

$$u_0 \, u_1 \, \cdots \, u_{k-1},$$
$$u_k \, u_{k+1} \, \cdots \, u_{2k-1},$$
$$u_{2k} \, u_{2k+1} \, \cdots \, u_{3k-1},$$
$$\vdots$$

of length k with $u_i \in \{0, 1, \ldots, q-1\}$ are encoded separately from each other into the corresponding code words

$$b_0 \, b_1 \, \cdots \, b_{n-1},$$
$$b_n \, b_{n+1} \, \cdots \, b_{2n-1},$$
$$b_{2n} \, b_{2n+1} \, \cdots \, b_{3n-1},$$
$$\vdots$$

of length n with $b_i \in \{0, 1, \ldots, q-1\}$ (see Figure 2.1).[1] These code words are transmitted across the channel and the received words are appropriately decoded, as shown in Figure 2.2. In the following, we will write the information word $u_0 \, u_1 \, \cdots \, u_{k-1}$ and the code word $b_0 \, b_1 \, \cdots \, b_{n-1}$ as vectors $\mathbf{u} = (u_0, u_1, \ldots, u_{k-1})$ and $\mathbf{b} = (b_0, b_1, \ldots, b_{n-1})$ respectively. Accordingly, the received word is denoted by $\mathbf{r} = (r_0, r_1, \ldots, r_{n-1})$, whereas

---
[1] For $q = 2$ we obtain the important class of binary channel codes.

ALGEBRAIC CODING THEORY

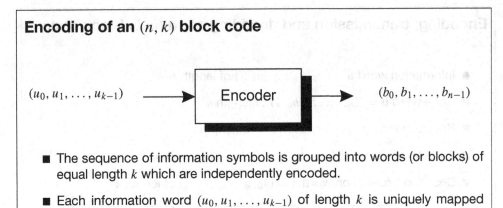

Figure 2.1: Encoding of an (n,k) block code

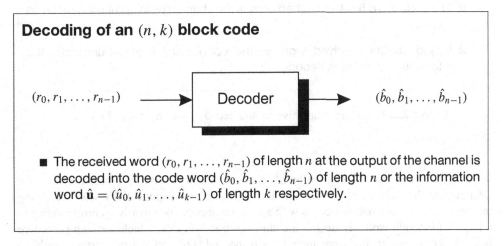

Figure 2.2: Decoding of an (n,k) block code

the decoded code word and decoded information word are given by $\hat{\mathbf{b}} = (\hat{b}_0, \hat{b}_1, \ldots, \hat{b}_{n-1})$ and $\hat{\mathbf{u}} = (\hat{u}_0, \hat{u}_1, \ldots, \hat{u}_{k-1})$ respectively. In general, we obtain the transmission scheme for an (n,k) block code as shown in Figure 2.3.

Without further algebraic properties of the (n,k) block code, the encoding can be carried out by a table look-up procedure. The information word **u** to be encoded is used to address a table that for each information word **u** contains the corresponding code word **b** at the respective address. If each information symbol can assume one out of q possible values,

> **Encoding, transmission and decoding of an (n, k) block code**
>
> - Information word $\mathbf{u} = (u_0, u_1, \ldots, u_{k-1})$ of length k
> - Code word $\mathbf{b} = (b_0, b_1, \ldots, b_{n-1})$ of length n
> - Received word $\mathbf{r} = (r_0, r_1, \ldots, r_{n-1})$ of length n
> - Decoded code word $\hat{\mathbf{b}} = (\hat{b}_0, \hat{b}_1, \ldots, \hat{b}_{n-1})$ of length n
> - Decoded information word $\hat{\mathbf{u}} = (\hat{u}_0, \hat{u}_1, \ldots, \hat{u}_{k-1})$ of length k
>
>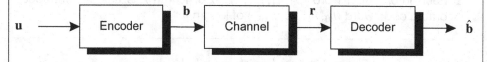
>
> - The information word $\mathbf{u}$ is encoded into the code word $\mathbf{b}$.
> - The code word $\mathbf{b}$ is transmitted across the channel which emits the received word $\mathbf{r}$.
> - Based on the received word $\mathbf{r}$, the code word $\hat{\mathbf{b}}$ (or equivalently the information word $\hat{\mathbf{u}}$) is decoded.

Figure 2.3: Encoding, transmission and decoding of an (n, k) block code

the number of code words is given by q^k. Since each entry carries n q-nary symbols, the total size of the table is $n q^k$. The size of the table grows exponentially with increasing information word length k. For codes with a large number of code words – corresponding to a large information word length k – a coding scheme based on a table look-up procedure is inefficient owing to the large memory size needed. For that reason, further algebraic properties are introduced in order to allow for a more efficient encoder architecture of an (n, k) block code. This is the idea lying behind the so-called linear block codes, which we will encounter in Section 2.2.

2.1.1 Code Parameters

Channel codes are characterised by so-called *code parameters*. The most important code parameters of a general (n, k) block code that are introduced in the following are the code rate and the minimum Hamming distance (Bossert, 1999; Lin and Costello, 2004; Ling and Xing, 2004). With the help of these code parameters, the efficiency of the encoding process and the error detection and error correction capabilities can be evaluated for a given (n, k) block code.

ALGEBRAIC CODING THEORY

Code Rate

Under the assumption that each information symbol u_i of the (n, k) block code can assume q values, the number of possible information words and code words is given by[2]

$$M = q^k.$$

Since the code word length n is larger than the information word length k, the rate at which information is transmitted across the channel is reduced by the so-called *code rate*

$$R = \frac{\log_q(M)}{n} = \frac{k}{n}.$$

For the simple binary triple repetition code with $k = 1$ and $n = 3$, the code rate is $R = \frac{k}{n} = \frac{1}{3} \approx 0{,}3333$.

Weight and Hamming Distance

Each code word $\mathbf{b} = (b_0, b_1, \ldots, b_{n-1})$ can be assigned the *weight* wt($\mathbf{b}$) which is defined as the number of non-zero components $b_i \neq 0$ (Bossert, 1999), i.e.[3]

$$\text{wt}(\mathbf{b}) = |\{i : b_i \neq 0, 0 \leq i < n\}|.$$

Accordingly, the distance between two code words $\mathbf{b} = (b_0, b_1, \ldots, b_{n-1})$ and $\mathbf{b}' = (b'_0, b'_1, \ldots, b'_{n-1})$ is given by the so-called *Hamming distance* (Bossert, 1999)

$$\text{dist}(\mathbf{b}, \mathbf{b}') = |\{i : b_i \neq b'_i, 0 \leq i < n\}|.$$

The Hamming distance dist($\mathbf{b}, \mathbf{b}'$) provides the number of different components of $\mathbf{b}$ and $\mathbf{b}'$ and thus measures how close the code words $\mathbf{b}$ and $\mathbf{b}'$ are to each other. For a code $\mathbb{B}$ consisting of M code words $\mathbf{b}_1, \mathbf{b}_2, \ldots, \mathbf{b}_M$, the *minimum Hamming distance* is given by

$$d = \min_{\forall \mathbf{b} \neq \mathbf{b}'} \text{dist}(\mathbf{b}, \mathbf{b}').$$

We will denote the (n, k) block code $\mathbb{B} = \{\mathbf{b}_1, \mathbf{b}_2, \ldots, \mathbf{b}_M\}$ with $M = q^k$ q-nary code words of length n and minimum Hamming distance d by $\mathbb{B}(n, k, d)$. The *minimum weight* of the block code $\mathbb{B}$ is defined as $\min_{\forall \mathbf{b} \neq \mathbf{0}} \text{wt}(\mathbf{b})$. The code parameters of $\mathbb{B}(n, k, d)$ are summarised in Figure 2.4.

Weight Distribution

The so-called *weight distribution* $W(x)$ of an (n, k) block code $\mathbb{B} = \{\mathbf{b}_1, \mathbf{b}_2, \ldots, \mathbf{b}_M\}$ describes how many code words exist with a specific weight. Denoting the number of code words with weight i by

$$w_i = |\{\mathbf{b} \in \mathbb{B} : \text{wt}(\mathbf{b}) = i\}|$$

---

[2] In view of the linear block codes introduced in the following, we assume here that all possible information words $\mathbf{u} = (u_0, u_1, \ldots, u_{k-1})$ are encoded.

[3] $|\mathbb{B}|$ denotes the cardinality of the set $\mathbb{B}$.

Code parameters of (n, k) block codes $\mathbb{B}(n, k, d)$

- Code rate
$$R = \frac{k}{n} \qquad (2.1)$$

- Minimum weight
$$\min_{\forall \mathbf{b} \neq 0} \text{wt}(\mathbf{b}) \qquad (2.2)$$

- Minimum Hamming distance
$$d = \min_{\forall \mathbf{b} \neq \mathbf{b}'} \text{dist}(\mathbf{b}, \mathbf{b}') \qquad (2.3)$$

Figure 2.4: Code parameters of (n, k) block codes $\mathbb{B}(n, k, d)$

with $0 \leq i \leq n$ and $0 \leq w_i \leq M$, the weight distribution is defined by the polynomial (Bossert, 1999)

$$W(x) = \sum_{i=0}^{n} w_i\, x^i.$$

The minimum weight of the block code $\mathbb{B}$ can then be obtained from the weight distribution according to

$$\min_{\forall \mathbf{b} \neq 0} \text{wt}(\mathbf{b}) = \min_{i > 0} w_i.$$

Code Space

A q-nary block code $\mathbb{B}(n, k, d)$ with code word length n can be illustrated as a subset of the so-called *code space* $\mathbb{F}_q^n$. Such a code space is a graphical illustration of all possible q-nary words or vectors.[4] The total number of vectors of length n with weight w and q-nary components is given by

$$\binom{n}{w} (q-1)^w = \frac{n!}{w!\,(n-w)!}\, (q-1)^w$$

with the binomial coefficients

$$\binom{n}{w} = \frac{n!}{w!\,(n-w)!}.$$

The total number of vectors within $\mathbb{F}_q^n$ is then obtained from

$$|\mathbb{F}_q^n| = \sum_{w=0}^{n} \binom{n}{w} (q-1)^w = q^n.$$

---

[4]In Section 2.2 we will identify the code space with the finite vector space $\mathbb{F}_q^n$. For a brief overview of algebraic structures such as finite fields and vector spaces the reader is referred to Appendix A.

ALGEBRAIC CODING THEORY

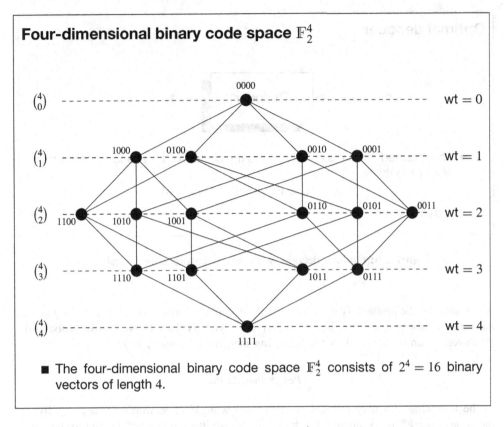

Figure 2.5: Four-dimensional binary code space $\mathbb{F}_2^4$. Reproduced by permission of J. Schlembach Fachverlag

The four-dimensional binary code space $\mathbb{F}_2^4$ with $q = 2$ and $n = 4$ is illustrated in Figure 2.5 (Neubauer, 2006b).

2.1.2 Maximum Likelihood Decoding

Channel codes are used in order to decrease the probability of incorrectly received code words or symbols. In this section we will derive a widely used decoding strategy. To this end, we will consider a decoding strategy to be optimal if the corresponding *word error probability*

$$p_{\text{err}} = \Pr\{\hat{\mathbf{u}} \neq \mathbf{u}\} = \Pr\{\hat{\mathbf{b}} \neq \mathbf{b}\}$$

is minimal (Bossert, 1999). The word error probability has to be distinguished from the *symbol error probability*

$$p_{\text{sym}} = \frac{1}{k} \sum_{i=0}^{k-1} \Pr\{\hat{u}_i \neq u_i\}$$

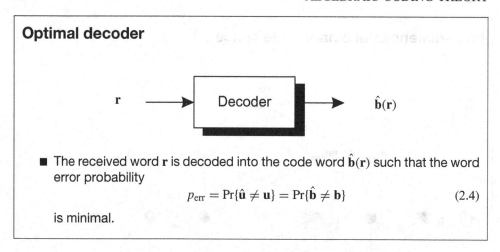

Figure 2.6: Optimal decoder with minimal word error probability

which denotes the probability of an incorrectly decoded information symbol u_i. In general, the symbol error probability is harder to derive analytically than the word error probability. However, it can be bounded by the following inequality (Bossert, 1999)

$$\frac{1}{k} p_{\text{err}} \leq p_{\text{sym}} \leq p_{\text{err}}.$$

In the following, a q-nary channel code $\mathbb{B} \in \mathbb{F}_q^n$ with M code words $\mathbf{b}_1, \mathbf{b}_2, \ldots, \mathbf{b}_M$ in the code space $\mathbb{F}_q^n$ is considered. Let $\mathbf{b}_j$ be the transmitted code word. Owing to the noisy channel, the received word $\mathbf{r}$ may differ from the transmitted code word $\mathbf{b}_j$. The task of the decoder in Figure 2.6 is to decode the transmitted code word based on the sole knowledge of $\mathbf{r}$ with minimal word error probability p_{err}.

This decoding step can be written according to the decoding rule $\mathbf{r} \mapsto \hat{\mathbf{b}} = \hat{\mathbf{b}}(\mathbf{r})$. For hard-decision decoding the received word $\mathbf{r}$ is an element of the discrete code space $\mathbb{F}_q^n$. To each code word $\mathbf{b}_j$ we assign a corresponding subspace $\mathbb{D}_j$ of the code space $\mathbb{F}_q^n$, the so-called *decision region*. These non-overlapping decision regions create the whole code space $\mathbb{F}_q^n$, i.e. $\bigcup_{j=1}^{M} \mathbb{D}_j = \mathbb{F}_q^n$ and $\mathbb{D}_i \cap \mathbb{D}_j = \emptyset$ for $i \neq j$ as illustrated in Figure 2.7. If the received word $\mathbf{r}$ lies within the decision region $\mathbb{D}_i$, the decoder decides in favour of the code word $\mathbf{b}_i$. That is, the decoding of the code word $\mathbf{b}_i$ according to the decision rule $\hat{\mathbf{b}}(\mathbf{r}) = \mathbf{b}_i$ is equivalent to the event $\mathbf{r} \in \mathbb{D}_i$. By properly choosing the decision regions $\mathbb{D}_i$, the decoder can be designed. For an optimal decoder the decision regions are chosen such that the word error probability p_{err} is minimal.

The probability of the event that the code word $\mathbf{b} = \mathbf{b}_j$ is transmitted and the code word $\hat{\mathbf{b}}(\mathbf{r}) = \mathbf{b}_i$ is decoded is given by

$$\Pr\{(\hat{\mathbf{b}}(\mathbf{r}) = \mathbf{b}_i) \wedge (\mathbf{b} = \mathbf{b}_j)\} = \Pr\{(\mathbf{r} \in \mathbb{D}_i) \wedge (\mathbf{b} = \mathbf{b}_j)\}.$$

We obtain the word error probability p_{err} by averaging over all possible events for which the transmitted code word $\mathbf{b} = \mathbf{b}_j$ is decoded into a different code word $\hat{\mathbf{b}}(\mathbf{r}) = \mathbf{b}_i$ with

ALGEBRAIC CODING THEORY

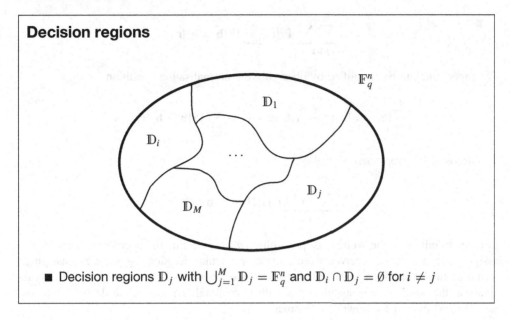

Figure 2.7: Non-overlapping decision regions $\mathbb{D}_j$ in the code space $\mathbb{F}_q^n$

$i \neq j$. This leads to (Neubauer, 2006b)

$$p_{\text{err}} = \Pr\{\hat{\mathbf{b}}(\mathbf{r}) \neq \mathbf{b}\}$$

$$= \sum_{i=1}^{M} \sum_{j \neq i} \Pr\{(\hat{\mathbf{b}}(\mathbf{r}) = \mathbf{b}_i) \wedge (\mathbf{b} = \mathbf{b}_j)\}$$

$$= \sum_{i=1}^{M} \sum_{j \neq i} \Pr\{(\mathbf{r} \in \mathbb{D}_i) \wedge (\mathbf{b} = \mathbf{b}_j)\}$$

$$= \sum_{i=1}^{M} \sum_{j \neq i} \sum_{\mathbf{r} \in \mathbb{D}_i} \Pr\{\mathbf{r} \wedge (\mathbf{b} = \mathbf{b}_j)\}.$$

With the help of Bayes' rule $\Pr\{\mathbf{r} \wedge (\mathbf{b} = \mathbf{b}_j)\} = \Pr\{\mathbf{b} = \mathbf{b}_j | \mathbf{r}\} \Pr\{\mathbf{r}\}$ and by changing the order of summation, we obtain

$$p_{\text{err}} = \sum_{i=1}^{M} \sum_{\mathbf{r} \in \mathbb{D}_i} \sum_{j \neq i} \Pr\{\mathbf{r} \wedge (\mathbf{b} = \mathbf{b}_j)\}$$

$$= \sum_{i=1}^{M} \sum_{\mathbf{r} \in \mathbb{D}_i} \sum_{j \neq i} \Pr\{\mathbf{b} = \mathbf{b}_j | \mathbf{r}\} \Pr\{\mathbf{r}\}$$

$$= \sum_{i=1}^{M} \sum_{\mathbf{r} \in \mathbb{D}_i} \Pr\{\mathbf{r}\} \sum_{j \neq i} \Pr\{\mathbf{b} = \mathbf{b}_j | \mathbf{r}\}.$$

The inner sum can be simplified by observing the normalisation condition

$$\sum_{j=1}^{M} \Pr\{\mathbf{b} = \mathbf{b}_j | \mathbf{r}\} = \Pr\{\mathbf{b} = \mathbf{b}_i | \mathbf{r}\} + \sum_{j \neq i} \Pr\{\mathbf{b} = \mathbf{b}_j | \mathbf{r}\} = 1.$$

This leads to the word error probability

$$p_{\text{err}} = \sum_{i=1}^{M} \sum_{\mathbf{r} \in \mathbb{D}_i} \Pr\{\mathbf{r}\} \left(1 - \Pr\{\mathbf{b} = \mathbf{b}_i | \mathbf{r}\}\right).$$

In order to minimise the word error probability p_{err}, we define the decision regions $\mathbb{D}_i$ by assigning each possible received word $\mathbf{r}$ to one particular decision region. If $\mathbf{r}$ is assigned to the particular decision region $\mathbb{D}_i$ for which the inner term $\Pr\{\mathbf{r}\} (1 - \Pr\{\mathbf{b} = \mathbf{b}_i | \mathbf{r}\})$ is smallest, the word error probability p_{err} will be minimal. Therefore, the decision regions are obtained from the following assignment

$$\mathbf{r} \in \mathbb{D}_j \quad \Leftrightarrow \quad \Pr\{\mathbf{r}\} \left(1 - \Pr\{\mathbf{b} = \mathbf{b}_j | \mathbf{r}\}\right) = \min_{1 \leq i \leq M} \Pr\{\mathbf{r}\} \left(1 - \Pr\{\mathbf{b} = \mathbf{b}_i | \mathbf{r}\}\right).$$

Since the probability $\Pr\{\mathbf{r}\}$ does not change with index i, this is equivalent to

$$\mathbf{r} \in \mathbb{D}_j \quad \Leftrightarrow \quad \Pr\{\mathbf{b} = \mathbf{b}_j | \mathbf{r}\} = \max_{1 \leq i \leq M} \Pr\{\mathbf{b} = \mathbf{b}_i | \mathbf{r}\}.$$

Finally, we obtain the optimal decoding rule according to

$$\hat{\mathbf{b}}(\mathbf{r}) = \mathbf{b}_j \quad \Leftrightarrow \quad \Pr\{\mathbf{b} = \mathbf{b}_j | \mathbf{r}\} = \max_{1 \leq i \leq M} \Pr\{\mathbf{b} = \mathbf{b}_i | \mathbf{r}\}.$$

The *optimal decoder* with minimal word error probability p_{err} emits the code word $\hat{\mathbf{b}} = \hat{\mathbf{b}}(\mathbf{r})$ for which the a-posteriori probability $\Pr\{\mathbf{b} = \mathbf{b}_i | \mathbf{r}\} = \Pr\{\mathbf{b}_i | \mathbf{r}\}$ is maximal. This decoding strategy

$$\hat{\mathbf{b}}(\mathbf{r}) = \underset{\mathbf{b} \in \mathbb{B}}{\operatorname{argmax}} \Pr\{\mathbf{b} | \mathbf{r}\}$$

is called MED (minimum error probability decoding) or MAP (maximum a-posteriori) decoding (Bossert, 1999).

For this MAP decoding strategy the a-posteriori probabilities $\Pr\{\mathbf{b} | \mathbf{r}\}$ have to be determined for all code words $\mathbf{b} \in \mathbb{B}$ and received words $\mathbf{r}$. With the help of Bayes' rule

$$\Pr\{\mathbf{b} | \mathbf{r}\} = \frac{\Pr\{\mathbf{r} | \mathbf{b}\} \Pr\{\mathbf{b}\}}{\Pr\{\mathbf{r}\}}$$

and by omitting the term $\Pr\{\mathbf{r}\}$ which does not depend on the specific code word $\mathbf{b}$, the decoding rule

$$\hat{\mathbf{b}}(\mathbf{r}) = \underset{\mathbf{b} \in \mathbb{B}}{\operatorname{argmax}} \Pr\{\mathbf{r} | \mathbf{b}\} \Pr\{\mathbf{b}\}$$

ALGEBRAIC CODING THEORY

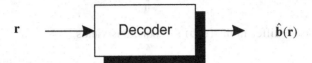

- For Minimum Error Probability Decoding (MED) or Maximum A-Posteriori (MAP) decoding the decoder rule is

$$\hat{\mathbf{b}}(\mathbf{r}) = \underset{\mathbf{b} \in \mathbb{B}}{\arg\max} \, \Pr\{\mathbf{b}|\mathbf{r}\} \qquad (2.5)$$

- For Maximum Likelihood Decoding (MLD) the decoder rule is

$$\hat{\mathbf{b}}(\mathbf{r}) = \underset{\mathbf{b} \in \mathbb{B}}{\arg\max} \, \Pr\{\mathbf{r}|\mathbf{b}\} \qquad (2.6)$$

- MLD is identical to MED if all code words are equally likely, i.e. $\Pr\{\mathbf{b}\} = \frac{1}{M}$.

Figure 2.8: Optimal decoding strategies

follows. For MAP decoding, the conditional probabilities $\Pr\{\mathbf{r}|\mathbf{b}\}$ as well as the a-priori probabilities $\Pr\{\mathbf{b}\}$ have to be known. If all M code words $\mathbf{b}$ appear with equal probability $\Pr\{\mathbf{b}\} = 1/M$, we obtain the so-called MLD (maximum likelihood decoding) strategy (Bossert, 1999)

$$\hat{\mathbf{b}}(\mathbf{r}) = \underset{\mathbf{b} \in \mathbb{B}}{\arg\max} \, \Pr\{\mathbf{r}|\mathbf{b}\}.$$

These decoding strategies are summarised in Figure 2.8. In the following, we will assume that all code words are equally likely, so that maximum likelihood decoding can be used as the optimal decoding rule. In order to apply the maximum likelihood decoding rule, the conditional probabilities $\Pr\{\mathbf{r}|\mathbf{b}\}$ must be available. We illustrate how this decoding rule can be further simplified by considering the binary symmetric channel.

2.1.3 Binary Symmetric Channel

In Section 1.2.3 we defined the binary symmetric channel as a memoryless channel with the conditional probabilities

$$\Pr\{r_i|b_i\} = \begin{cases} 1 - \varepsilon, & r_i = b_i \\ \varepsilon, & r_i \neq b_i \end{cases}$$

with channel bit error probability ε. Since the binary symmetric channel is assumed to be memoryless, the conditional probability $\Pr\{\mathbf{r}|\mathbf{b}\}$ can be calculated for code word $\mathbf{b} =$

$(b_0, b_1, \ldots, b_{n-1})$ and received word $\mathbf{r} = (r_0, r_1, \ldots, r_{n-1})$ according to

$$\Pr\{\mathbf{r}|\mathbf{b}\} = \prod_{i=0}^{n-1} \Pr\{r_i|b_i\}.$$

If the words $\mathbf{r}$ and $\mathbf{b}$ differ in $\mathrm{dist}(\mathbf{r}, \mathbf{b})$ symbols, this yields

$$\Pr\{\mathbf{r}|\mathbf{b}\} = (1-\varepsilon)^{n-\mathrm{dist}(\mathbf{r},\mathbf{b})} \varepsilon^{\mathrm{dist}(\mathbf{r},\mathbf{b})} = (1-\varepsilon)^n \left(\frac{\varepsilon}{1-\varepsilon}\right)^{\mathrm{dist}(\mathbf{r},\mathbf{b})}.$$

Taking into account $0 \leq \varepsilon < \frac{1}{2}$ and therefore $\frac{\varepsilon}{1-\varepsilon} < 1$, the MLD rule is given by

$$\hat{\mathbf{b}}(\mathbf{r}) = \underset{\mathbf{b} \in \mathbb{B}}{\mathrm{argmax}}\, \Pr\{\mathbf{r}|\mathbf{b}\} = \underset{\mathbf{b} \in \mathbb{B}}{\mathrm{argmax}}\, (1-\varepsilon)^n \left(\frac{\varepsilon}{1-\varepsilon}\right)^{\mathrm{dist}(\mathbf{r},\mathbf{b})} = \underset{\mathbf{b} \in \mathbb{B}}{\mathrm{argmin}}\, \mathrm{dist}(\mathbf{r}, \mathbf{b}),$$

i.e. for the binary symmetric channel the optimal maximum likelihood decoder (Bossert, 1999)

$$\hat{\mathbf{b}}(\mathbf{r}) = \underset{\mathbf{b} \in \mathbb{B}}{\mathrm{argmin}}\, \mathrm{dist}(\mathbf{r}, \mathbf{b})$$

emits that particular code word which differs in the smallest number of components from the received word $\mathbf{r}$, i.e. which has the smallest Hamming distance to the received word $\mathbf{r}$ (see Figure 2.9). This decoding rule is called *minimum distance decoding*. This minimum distance decoding rule is also optimal for a q-nary symmetric channel (Neubauer, 2006b). We now turn to the error probabilities for the binary symmetric channel during transmission before decoding. The probability of w errors at w given positions within the n-dimensional binary received word $\mathbf{r}$ is given by $\varepsilon^w (1-\varepsilon)^{n-w}$. Since there are $\binom{n}{w}$ different possibilities

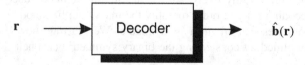

Minimum distance decoding for the binary symmetric channel

- The optimal maximum likelihood decoding rule for the binary symmetric channel is given by the minimum distance decoding rule

$$\hat{\mathbf{b}}(\mathbf{r}) = \underset{\mathbf{b} \in \mathbb{B}}{\mathrm{argmin}}\, \mathrm{dist}(\mathbf{r}, \mathbf{b}) \qquad (2.7)$$

Figure 2.9: Minimum distance decoding for the binary symmetric channel

ALGEBRAIC CODING THEORY

of choosing w out of n positions, the probability of w errors at arbitrary positions within an n-dimensional binary received word follows the binomial distribution

$$\Pr\{w \text{ errors}\} = \binom{n}{w} \varepsilon^w (1-\varepsilon)^{n-w}$$

with mean $n \varepsilon$. Because of the condition $\varepsilon < \frac{1}{2}$, the probability $\Pr\{w \text{ errors}\}$ decreases with increasing number of errors w, i.e. few errors are more likely than many errors.

The probability of error-free transmission is $\Pr\{0 \text{ errors}\} = (1-\varepsilon)^n$, whereas the probability of a disturbed transmission with $\mathbf{r} \neq \mathbf{b}$ is given by

$$\Pr\{\mathbf{r} \neq \mathbf{b}\} = \sum_{w=1}^{n} \binom{n}{w} \varepsilon^w (1-\varepsilon)^{n-w} = 1 - (1-\varepsilon)^n.$$

2.1.4 Error Detection and Error Correction

Based on the minimum distance decoding rule and the code space concept, we can assess the error detection and error correction capabilities of a given channel code. To this end, let $\mathbf{b}$ and $\mathbf{b}'$ be two code words of an (n, k) block code $\mathbb{B}(n, k, d)$. The distance of these code words shall be equal to the minimum Hamming distance, i.e. $\text{dist}(\mathbf{b}, \mathbf{b}') = d$. We are able to detect errors as long as the erroneously received word $\mathbf{r}$ is not equal to a code word different from the transmitted code word. This error detection capability is guaranteed as long as the number of errors is smaller than the minimum Hamming distance d, because another code word (e.g. $\mathbf{b}'$) can be reached from a given code word (e.g. $\mathbf{b}$) merely by changing at least d components. For an (n, k) block code $\mathbb{B}(n, k, d)$ with minimum Hamming distance d, the number of detectable errors is therefore given by (Bossert, 1999; Lin and Costello, 2004; Ling and Xing, 2004; van Lint, 1999)

$$e_{\text{det}} = d - 1.$$

For the analysis of the error correction capabilities of the (n, k) block code $\mathbb{B}(n, k, d)$ we define for each code word $\mathbf{b}$ the corresponding correction ball of radius ϱ as the subset of all words that are closer to the code word $\mathbf{b}$ than to any other code word $\mathbf{b}'$ of the block code $\mathbb{B}(n, k, d)$ (see Figure 2.10). As we have seen in the last section, for minimum distance decoding, all received words within a particular correction ball are decoded into the respective code word $\mathbf{b}$. According to the radius ϱ of the correction balls, besides the code word $\mathbf{b}$, all words that differ in $1, 2, \ldots, \varrho$ components from $\mathbf{b}$ are elements of the corresponding correction ball. We can uniquely decode all elements of a correction ball into the corresponding code word $\mathbf{b}$ as long as the correction balls do not intersect. This condition is true if $\varrho < \frac{d}{2}$ holds. Therefore, the number of correctable errors of a block code $\mathbb{B}(n, k, d)$ with minimum Hamming distance d is given by (Bossert, 1999; Lin and Costello, 2004; Ling and Xing, 2004; van Lint, 1999)[5]

$$e_{\text{cor}} = \left\lfloor \frac{d-1}{2} \right\rfloor.$$

---

[5]The term $\lfloor z \rfloor$ denotes the largest integer number that is not larger than z.

> **Error detection and error correction**
>
>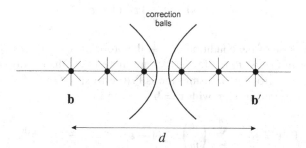
>
> - If the minimum Hamming distance between two arbitrary code words is d the code is able to detect up to
>
> $$e_{\text{det}} = d - 1 \qquad (2.8)$$
>
> errors.
>
> - If the minimum Hamming distance between two arbitrary code words is d the code is able to correct up to
>
> $$e_{\text{cor}} = \left\lfloor \frac{d-1}{2} \right\rfloor \qquad (2.9)$$
>
> errors.

Figure 2.10: Error detection and error correction

For the binary symmetric channel the number of errors w within the n-dimensional transmitted code word is binomially distributed according to $\Pr\{w \text{ errors}\} = \binom{n}{w} \varepsilon^w (1 - \varepsilon)^{n-w}$. Since an e_{det}-error detecting code is able to detect $w \leq e_{\text{det}} = d - 1$ errors, the remaining detection error probability is bounded by

$$p_{\text{det}} \leq \sum_{w=e_{\text{det}}+1}^{n} \binom{n}{w} \varepsilon^w (1-\varepsilon)^{n-w} = 1 - \sum_{w=0}^{e_{\text{det}}} \binom{n}{w} \varepsilon^w (1-\varepsilon)^{n-w}.$$

If an e_{cor}-error correcting code is used with $e_{\text{cor}} = \lfloor (d-1)/2 \rfloor$, the word error probability for a binary symmetric channel can be similarly bounded by

$$p_{\text{err}} \leq \sum_{w=e_{\text{cor}}+1}^{n} \binom{n}{w} \varepsilon^w (1-\varepsilon)^{n-w} = 1 - \sum_{w=0}^{e_{\text{cor}}} \binom{n}{w} \varepsilon^w (1-\varepsilon)^{n-w}.$$

2.2 Linear Block Codes

In the foregoing discussion of the fundamentals of general block codes we exclusively focused on the distance properties between code words and received words in the code space $\mathbb{F}_q^n$. If we consider the code words to be vectors in the finite vector space $\mathbb{F}_q^n$, taking into account the respective algebraic properties of vector spaces, we gain efficiency especially with regard to the encoding scheme.[6]

2.2.1 Definition of Linear Block Codes

The (n, k) block code $\mathbb{B}(n, k, d)$ with minimum Hamming distance d over the finite field $\mathbb{F}_q$ is called *linear*, if $\mathbb{B}(n, k, d)$ is a subspace of the vector space $\mathbb{F}_q^n$ of dimension k (Lin and Costello, 2004; Ling and Xing, 2004). The number of code words is then given by

$$M = q^k$$

according to the code rate

$$R = \frac{k}{n}.$$

Because of the linearity property, an arbitrary superposition of code words again leads to a valid code word of the linear block code $\mathbb{B}(n, k, d)$, i.e.

$$\alpha_2 \mathbf{b}_1 + \alpha_2 \mathbf{b}_2 + \cdots + \alpha_l \mathbf{b}_l \in \mathbb{B}(n, k, d)$$

with $\alpha_1, \alpha_2, \ldots, \alpha_l \in \mathbb{F}_q$ and $\mathbf{b}_1, \mathbf{b}_2, \ldots, \mathbf{b}_l \in \mathbb{B}(n, k, d)$. Owing to the linearity, the n-dimensional zero row vector $\mathbf{0} = (0, 0, \ldots, 0)$ consisting of n zeros is always a valid code word. It can be shown that the minimum Hamming distance of a linear block code $\mathbb{B}(n, k, d)$ is equal to the minimum weight of all non-zero code words, i.e.

$$d = \min_{\forall \mathbf{b} \neq \mathbf{b}'} \text{dist}(\mathbf{b}, \mathbf{b}') = \min_{\forall \mathbf{b} \neq \mathbf{0}} \text{wt}(\mathbf{b}).$$

These properties are summarised in Figure 2.11. As a simple example of a linear block code, the binary parity-check code is described in Figure 2.12 (Bossert, 1999).

For each linear block code an *equivalent code* can be found by rearranging the code word symbols.[7] This equivalent code is characterised by the same code parameters as the original code, i.e. the equivalent code has the same dimension k and the same minimum Hamming distance d.

2.2.2 Generator Matrix

The linearity property of a linear block code $\mathbb{B}(n, k, d)$ can be exploited for efficiently encoding a given information word $\mathbf{u} = (u_0, u_1, \ldots, u_{k-1})$. To this end, a basis $\{\mathbf{g}_0, \mathbf{g}_1, \ldots, \mathbf{g}_{k-1}\}$ of the subspace spanned by the linear block code is chosen, consisting of k linearly independent n-dimensional vectors

$$\mathbf{g}_i = (g_{i,0}, g_{i,1}, \cdots, g_{i,n-1})$$

---

[6] Finite fields and vector spaces are briefly reviewed in Sections A.1 and A.2 in Appendix A.

[7] In general, an equivalent code is obtained by suitable operations on the rows and columns of the generator matrix **G** which is defined in Section 2.2.2.

Linear block codes

- A linear q-nary (n, k) block code $\mathbb{B}(n, k, d)$ is defined as a subspace of the vector space $\mathbb{F}_q^n$.
- The number of code words is given by $M = q^k$.
- The minimum Hamming distance d is given by the minimum weight $\min_{\forall \mathbf{b} \neq 0} \text{wt}(\mathbf{b}) = d$.

Figure 2.11: Linear block codes $\mathbb{B}(n, k, d)$

Binary parity-check code

- The binary parity-check code is a linear block code over the finite field $\mathbb{F}_2$.
- This code takes a k-dimensional information word $\mathbf{u} = (u_0, u_1, \ldots, u_{k-1})$ and generates the code word $\mathbf{b} = (b_0, b_1, \ldots, b_{k-1}, b_k)$ with $b_i = u_i$ for $0 \leq i \leq k - 1$ and

$$b_k = u_0 + u_1 + \cdots + u_{k-1} = \sum_{i=0}^{k-1} u_i \qquad (2.10)$$

- The bit b_k is called the parity bit; it is chosen such that the resulting code word is of even parity.

Figure 2.12: Binary parity-check code

with $0 \leq i \leq k - 1$. The corresponding code word $\mathbf{b} = (b_0, b_1, \ldots, b_{n-1})$ is then given by

$$\mathbf{b} = u_0 \mathbf{g}_0 + u_1 \mathbf{g}_1 + \cdots + u_{k-1} \mathbf{g}_{k-1}$$

with the q-nary information symbols $u_i \in \mathbb{F}_q$. If we define the $k \times n$ matrix

$$\mathbf{G} = \begin{pmatrix} \mathbf{g}_0 \\ \mathbf{g}_1 \\ \vdots \\ \mathbf{g}_{k-1} \end{pmatrix} = \begin{pmatrix} g_{0,0} & g_{0,1} & \cdots & g_{0,n-1} \\ g_{1,0} & g_{1,1} & \cdots & g_{1,n-1} \\ \vdots & \vdots & \ddots & \vdots \\ g_{k-1,0} & g_{k-1,1} & \cdots & g_{k-1,n-1} \end{pmatrix}$$

ALGEBRAIC CODING THEORY

Generator matrix

- The generator matrix **G** of a linear block code is constructed by a suitable set of k linearly independent basis vectors $\mathbf{g}_i$ according to

$$\mathbf{G} = \begin{pmatrix} \mathbf{g}_0 \\ \mathbf{g}_1 \\ \vdots \\ \mathbf{g}_{k-1} \end{pmatrix} = \begin{pmatrix} g_{0,0} & g_{0,1} & \cdots & g_{0,n-1} \\ g_{1,0} & g_{1,1} & \cdots & g_{1,n-1} \\ \vdots & \vdots & \ddots & \vdots \\ g_{k-1,0} & g_{k-1,1} & \cdots & g_{k-1,n-1} \end{pmatrix} \qquad (2.11)$$

- The k-dimensional information word **u** is encoded into the n-dimensional code word **b** by the encoding rule

$$\mathbf{b} = \mathbf{u}\,\mathbf{G} \qquad (2.12)$$

Figure 2.13: Generator matrix **G** of a linear block code $\mathbb{B}(n, k, d)$

the information word $\mathbf{u} = (u_0, u_1, \ldots, u_{k-1})$ is encoded according to the matrix–vector multiplication

$$\mathbf{b} = \mathbf{u}\,\mathbf{G}.$$

Since all $M = q^k$ code words $\mathbf{b} \in \mathbb{B}(n, k, d)$ can be generated by this rule, the matrix **G** is called the *generator matrix* of the linear block code $\mathbb{B}(n, k, d)$ (see Figure 2.13). Owing to this property, the linear block code $\mathbb{B}(n, k, d)$ is completely defined with the help of the generator matrix **G** (Bossert, 1999; Lin and Costello, 2004; Ling and Xing, 2004).

In Figure 2.14 the so-called binary $(7, 4)$ Hamming code is defined by the given generator matrix **G**. Such a binary Hamming code has been defined by Hamming (Hamming, 1950). These codes or variants of them are used, e.g. in memories such as dynamic random access memories (DRAMs), in order to correct deteriorated data in memory cells.

For each linear block code $\mathbb{B}(n, k, d)$ an equivalent linear block code can be found that is defined by the $k \times n$ generator matrix

$$\mathbf{G} = \left(\mathbf{I}_k \,\middle|\, \mathbf{A}_{k,n-k}\right).$$

Owing to the $k \times k$ identity matrix $\mathbf{I}_k$ and the encoding rule

$$\mathbf{b} = \mathbf{u}\,\mathbf{G} = \left(\mathbf{u} \,\middle|\, \mathbf{u}\,\mathbf{A}_{k,n-k}\right)$$

the first k code symbols b_i are identical to the k information symbols u_i. Such an encoding scheme is called *systematic*. The remaining $m = n - k$ symbols within the vector $\mathbf{u}\,\mathbf{A}_{k,n-k}$ correspond to m *parity-check symbols* which are attached to the information vector **u** for the purpose of error detection or error correction.

> **Generator matrix of a binary Hamming code**
>
> ■ The generator matrix **G** of a binary (7,4) Hamming code is given by
> $$\mathbf{G} = \begin{pmatrix} 1 & 0 & 0 & 0 & 0 & 1 & 1 \\ 0 & 1 & 0 & 0 & 1 & 0 & 1 \\ 0 & 0 & 1 & 0 & 1 & 1 & 0 \\ 0 & 0 & 0 & 1 & 1 & 1 & 1 \end{pmatrix}$$
>
> ■ The code parameters of this binary Hamming code are $n = 7$, $k = 4$ and $d = 3$, i.e. this code is a binary $\mathbb{B}(7, 4, 3)$ code.
>
> ■ The information word $\mathbf{u} = (0, 0, 1, 1)$ is encoded into the code word
> $$\mathbf{b} = (0, 0, 1, 1) \begin{pmatrix} 1 & 0 & 0 & 0 & 0 & 1 & 1 \\ 0 & 1 & 0 & 0 & 1 & 0 & 1 \\ 0 & 0 & 1 & 0 & 1 & 1 & 0 \\ 0 & 0 & 0 & 1 & 1 & 1 & 1 \end{pmatrix} = (0, 0, 1, 1, 0, 0, 1)$$

Figure 2.14: Generator matrix of a binary $(7, 4)$ Hamming code

2.2.3 Parity-Check Matrix

With the help of the generator matrix $\mathbf{G} = (\mathbf{I}_k \,|\, \mathbf{A}_{k,n-k})$, the following $(n - k) \times n$ matrix – the so-called *parity-check matrix* – can be defined (Bossert, 1999; Lin and Costello, 2004; Ling and Xing, 2004)
$$\mathbf{H} = \left(\mathbf{B}_{n-k,k} \,|\, \mathbf{I}_{n-k} \right)$$
with the $(n - k) \times (n - k)$ identity matrix $\mathbf{I}_{n-k}$. The $(n - k) \times k$ matrix $\mathbf{B}_{n-k,k}$ is given by
$$\mathbf{B}_{n-k,k} = -\mathbf{A}_{k,n-k}^{\mathrm{T}}.$$
For the matrices $\mathbf{G}$ and $\mathbf{H}$ the following property can be derived
$$\mathbf{H}\mathbf{G}^{\mathrm{T}} = \mathbf{B}_{n-k,k} + \mathbf{A}_{k,n-k}^{\mathrm{T}} = \mathbf{0}_{n-k,k}$$
with the $(n - k) \times k$ zero matrix $\mathbf{0}_{n-k,k}$. The generator matrix $\mathbf{G}$ and the parity-check matrix $\mathbf{H}$ are *orthogonal*, i.e. all row vectors of $\mathbf{G}$ are orthogonal to all row vectors of $\mathbf{H}$.

Using the n-dimensional basis vectors $\mathbf{g}_0, \mathbf{g}_1, \ldots, \mathbf{g}_{k-1}$ and the transpose of the generator matrix $\mathbf{G}^{\mathrm{T}} = \left(\mathbf{g}_0^{\mathrm{T}}, \mathbf{g}_1^{\mathrm{T}}, \ldots, \mathbf{g}_{k-1}^{\mathrm{T}} \right)$, we obtain
$$\mathbf{H}\mathbf{G}^{\mathrm{T}} = \mathbf{H}\left(\mathbf{g}_0^{\mathrm{T}}, \mathbf{g}_1^{\mathrm{T}}, \ldots, \mathbf{g}_{k-1}^{\mathrm{T}} \right) = \left(\mathbf{H}\mathbf{g}_0^{\mathrm{T}}, \mathbf{H}\mathbf{g}_1^{\mathrm{T}}, \ldots, \mathbf{H}\mathbf{g}_{k-1}^{\mathrm{T}} \right) = (\mathbf{0}, \mathbf{0}, \ldots, \mathbf{0})$$
with the $(n - k)$-dimensional all-zero column vector $\mathbf{0} = (0, 0, \ldots, 0)^{\mathrm{T}}$. This is equivalent to $\mathbf{H}\mathbf{g}_i^{\mathrm{T}} = \mathbf{0}$ for $0 \leq i \leq k - 1$. Since each code vector $\mathbf{b} \in \mathbb{B}(n, k, d)$ can be written as
$$\mathbf{b} = \mathbf{u}\mathbf{G} = u_0\,\mathbf{g}_0 + u_1\,\mathbf{g}_1 + \cdots + u_{k-1}\,\mathbf{g}_{k-1}$$

ALGEBRAIC CODING THEORY

> **Parity-check matrix**
>
> - The parity-check matrix $\mathbf{H}$ of a linear block code $\mathbb{B}(n, k, d)$ with generator matrix $\mathbf{G} = \left(\mathbf{I}_k \,|\, \mathbf{A}_{k,n-k} \right)$ is defined by
>
> $$\mathbf{H} = \left(-\mathbf{A}_{k,n-k}^T \,|\, \mathbf{I}_{n-k} \right) \tag{2.13}$$
>
> - Generator matrix $\mathbf{G}$ and parity-check matrix $\mathbf{H}$ are orthogonal
>
> $$\mathbf{H}\mathbf{G}^T = \mathbf{0}_{n-k,k} \tag{2.14}$$
>
> - The system of parity-check equations is given by
>
> $$\mathbf{H}\mathbf{r}^T = \mathbf{0} \quad \Leftrightarrow \quad \mathbf{r} \in \mathbb{B}(n, k, d) \tag{2.15}$$

Figure 2.15: Parity-check matrix $\mathbf{H}$ of a linear block code $\mathbb{B}(n, k, d)$

with the information vector $\mathbf{u} = (u_0, u_1, \ldots, u_{k-1})$, it follows that

$$\mathbf{H}\mathbf{b}^T = u_0\,\mathbf{H}\mathbf{g}_0^T + u_1\,\mathbf{H}\mathbf{g}_1^T + \cdots + u_{k-1}\,\mathbf{H}\mathbf{g}_{k-1}^T = \mathbf{0}.$$

Each code vector $\mathbf{b} \in \mathbb{B}(n, k, d)$ of a linear (n, k) block code $\mathbb{B}(n, k, d)$ fulfils the condition

$$\mathbf{H}\mathbf{b}^T = \mathbf{0}.$$

Equivalently, if $\mathbf{H}\mathbf{r}^T \neq \mathbf{0}$, the vector $\mathbf{r}$ does not belong to the linear block code $\mathbb{B}(n, k, d)$. We arrive at the following *parity-check condition*

$$\mathbf{H}\mathbf{r}^T = \mathbf{0} \quad \Leftrightarrow \quad \mathbf{r} \in \mathbb{B}(n, k, d)$$

which amounts to a total of $n - k$ parity-check equations. Therefore, the matrix $\mathbf{H}$ is called the parity-check matrix of the linear (n, k) block code $\mathbb{B}(n, k, d)$ (see Figure 2.15).

There exists an interesting relationship between the minimum Hamming distance d and the parity-check matrix $\mathbf{H}$ which is stated in Figure 2.16 (Lin and Costello, 2004). In Figure 2.17 the parity-check matrix of the binary Hamming code with the generator matrix given in Figure 2.14 is shown. The corresponding parity-check equations of this binary Hamming code are illustrated in Figure 2.18.

2.2.4 Syndrome and Cosets

As we have seen in the last section, a vector $\mathbf{r}$ corresponds to a valid code word of a given linear block code $\mathbb{B}(n, k, d)$ with parity-check matrix $\mathbf{H}$ if and only if the parity-check equation $\mathbf{H}\mathbf{r}^T = \mathbf{0}$ is true. Otherwise, $\mathbf{r}$ is not a valid code word of $\mathbb{B}(n, k, d)$. Based on

Parity-check matrix and minimum Hamming distance

- Let $\mathbf{H}$ be the parity-check matrix of a linear block code $\mathbb{B}(n, k, d)$ with minimum Hamming distance d.

- The minimum Hamming distance d is equal to the smallest number of linearly dependent columns of the parity-check matrix $\mathbf{H}$.

Figure 2.16: Parity-check matrix $\mathbf{H}$ and minimum Hamming distance d of a linear block code $\mathbb{B}(n, k, d)$

Parity-check matrix of a binary Hamming code

- The parity-check matrix $\mathbf{H}$ of a binary (7,4) Hamming code is given by
$$\mathbf{H} = \begin{pmatrix} 0 & 1 & 1 & 1 & 1 & 0 & 0 \\ 1 & 0 & 1 & 1 & 0 & 1 & 0 \\ 1 & 1 & 0 & 1 & 0 & 0 & 1 \end{pmatrix}$$
It consists of all non-zero binary column vectors of length 3.

- Two arbitrary columns are linearly independent. However, the first three columns are linearly dependent. The minimum Hamming distance of a binary Hamming code is $d = 3$.

- The code word $\mathbf{b} = (0, 0, 1, 1, 0, 0, 1)$ fulfils the parity-check equation
$$\mathbf{H}\mathbf{b}^T = \begin{pmatrix} 0 & 1 & 1 & 1 & 1 & 0 & 0 \\ 1 & 0 & 1 & 1 & 0 & 1 & 0 \\ 1 & 1 & 0 & 1 & 0 & 0 & 1 \end{pmatrix} \begin{pmatrix} 0 \\ 0 \\ 1 \\ 1 \\ 0 \\ 0 \\ 1 \end{pmatrix} = \begin{pmatrix} 0 \\ 0 \\ 0 \end{pmatrix}$$

Figure 2.17: Parity-check matrix $\mathbf{H}$ of a binary Hamming code

ALGEBRAIC CODING THEORY

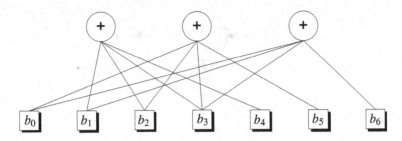

■ The parity-check equations of a binary (7,4) Hamming code with code word $\mathbf{b} = (b_0, b_1, b_2, b_3, b_4, b_5, b_6)$ are given by

$$b_1 + b_2 + b_3 + b_4 = 0$$
$$b_0 + b_2 + b_3 + b_5 = 0$$
$$b_0 + b_1 + b_3 + b_6 = 0$$

Figure 2.18: Graphical representation of the parity-check equations of a binary (7, 4) Hamming code $\mathbb{B}(7, 4, 3)$

the algebraic channel model shown in Figure 2.19, we interpret $\mathbf{r}$ as the received vector which is obtained from

$$\mathbf{r} = \mathbf{b} + \mathbf{e}$$

with the transmitted code vector $\mathbf{b}$ and the error vector $\mathbf{e}$. The jth component of the error vector $\mathbf{e}$ is $e_j = 0$ if no error has occurred at this particular position; otherwise the jth component is $e_j \neq 0$.

In view of the parity-check equation, we define the so-called *syndrome* (Lin and Costello, 2004; Ling and Xing, 2004)

$$\mathbf{s}^T = \mathbf{H}\mathbf{r}^T$$

which is used to check whether the received vector $\mathbf{r}$ belongs to the channel code $\mathbb{B}(n, k, d)$. Inserting the received vector $\mathbf{r}$ into this definition, we obtain

$$\mathbf{s}^T = \mathbf{H}\mathbf{r}^T = \mathbf{H}(\mathbf{b} + \mathbf{e})^T = \underbrace{\mathbf{H}\mathbf{b}^T}_{=\mathbf{0}} + \mathbf{H}\mathbf{e}^T = \mathbf{H}\mathbf{e}^T.$$

Here, we have taken into account that for each code vector $\mathbf{b} \in \mathbb{B}(n, k, d)$ the condition $\mathbf{H}\mathbf{b}^T = \mathbf{0}$ holds. Finally, we recognize that the syndrome does exclusively depend on the error vector $\mathbf{e}$, i.e.

$$\mathbf{s}^T = \mathbf{H}\mathbf{e}^T.$$

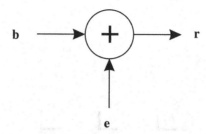

Algebraic channel model

- The transmitted n-dimensional code vector **b** is disturbed by the n-dimensional error vector **e**.
- The received vector **r** is given by
$$\mathbf{r} = \mathbf{b} + \mathbf{e} \tag{2.16}$$
- The syndrome
$$\mathbf{s}^T = \mathbf{H}\mathbf{r}^T \tag{2.17}$$
exclusively depends on the error vector **e** according to $\mathbf{s}^T = \mathbf{H}\mathbf{e}^T$.

Figure 2.19: Algebraic channel model

Thus, for the purpose of error detection the syndrome can be evaluated. If **s** is zero, the received vector **r** is equal to a valid code vector, i.e. $\mathbf{r} \in \mathbb{B}(n, k, d)$. In this case no error can be detected and it is assumed that the received vector corresponds to the transmitted code vector. If **e** is zero, the received vector $\mathbf{r} = \mathbf{b}$ delivers the transmitted code vector **b**. However, all non-zero error vectors **e** that fulfil the condition

$$\mathbf{H}\mathbf{e}^T = \mathbf{0}$$

also lead to a valid code word.[8] These errors cannot be detected.

In general, the $(n - k)$-dimensional syndrome $\mathbf{s}^T = \mathbf{H}\mathbf{e}^T$ of a linear (n, k) block code $\mathbb{B}(n, k, d)$ corresponds to $n - k$ scalar equations for the determination of the n-dimensional error vector **e**. The matrix equation

$$\mathbf{s}^T = \mathbf{H}\mathbf{e}^T$$

does not uniquely define the error vector **e**. All vectors **e** with $\mathbf{H}\mathbf{e}^T = \mathbf{s}^T$ form a set, the so-called *coset* of the k-dimensional subspace $\mathbb{B}(n, k, d)$ in the finite vector space $\mathbb{F}_q^n$. This coset has q^k elements. For an error-correcting q-nary block code $\mathbb{B}(n, k, d)$ and a given

---
[8]This property directly follows from the linearity of the block code $\mathbb{B}(n, k, d)$.

ALGEBRAIC CODING THEORY

syndrome **s**, the decoder has to choose one out of these q^k possible error vectors. As we have seen for the binary symmetric channel, it is often the case that few errors are more likely than many errors. An optimal decoder implementing the minimum distance decoding rule therefore chooses the error vector out of the coset that has the smallest number of non-zero components. This error vector is called the *coset leader* (Lin and Costello, 2004; Ling and Xing, 2004).

The decoding of the linear block code $\mathbb{B}(n, k, d)$ can be carried out by a table look-up procedure as illustrated in Figure 2.20. The $(n - k)$-dimensional syndrome **s** is used to address a table that for each particular syndrome contains the corresponding n-dimensional coset leader **e** at the respective address. Since there are q^{n-k} syndromes and each entry carries n q-nary symbols, the total size of the table is $n\,q^{n-k}$. The size of the table grows exponentially with $n - k$. For codes with large $n - k$, a decoding scheme based on a table look-up procedure is inefficient.[9] By introducing further algebraic properties, more efficient

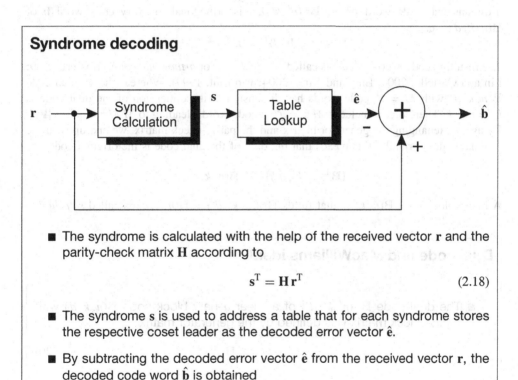

Syndrome decoding

- The syndrome is calculated with the help of the received vector **r** and the parity-check matrix **H** according to

$$\mathbf{s}^T = \mathbf{H}\,\mathbf{r}^T \qquad (2.18)$$

- The syndrome **s** is used to address a table that for each syndrome stores the respective coset leader as the decoded error vector **ê**.

- By subtracting the decoded error vector **ê** from the received vector **r**, the decoded code word **b̂** is obtained

$$\hat{\mathbf{b}} = \mathbf{r} - \hat{\mathbf{e}} \qquad (2.19)$$

Figure 2.20: Syndrome decoding of a linear block code $\mathbb{B}(n, k, d)$ with a table look-up procedure. Reproduced by permission of J. Schlembach Fachverlag

---

[9] It can be shown that in general the decoding of a linear block code is NP-complete.

decoder architectures can be defined. This will be apparent in the context of cyclic codes in Section 2.3.

2.2.5 Dual Code

So far we have described two equivalent ways to define a linear block code $\mathbb{B}(n, k, d)$ based on the $k \times n$ generator matrix $\mathbf{G}$ and the $(n - k) \times n$ parity-check matrix $\mathbf{H}$. By simply exchanging these matrices, we can define a new linear block code $\mathbb{B}^\perp(n', k', d')$ with generator matrix

$$\mathbf{G}^\perp = \mathbf{H}$$

and parity-check matrix

$$\mathbf{H}^\perp = \mathbf{G}$$

as shown in Figure 2.21. Because of the orthogonality condition $\mathbf{H}\mathbf{G}^T = \mathbf{0}$, each $n' = n$-dimensional code word $\mathbf{b}^\perp$ of $\mathbb{B}^\perp(n', k', d')$ is orthogonal to every code word $\mathbf{b}$ of $\mathbb{B}(n, k, d)$, i.e.

$$\mathbf{b}^\perp \mathbf{b}^T = 0.$$

The resulting code $\mathbb{B}^\perp(n', k', d')$ is called the *dual code* or *orthogonal code*, (Bossert, 1999; Lin and Costello, 2004; Ling and Xing, 2004; van Lint, 1999). Whereas the original code $\mathbb{B}(n, k, d)$ with $M = q^k$ code words has dimension k, the dimension of the dual code is $k' = n - k$. Therefore, it includes $M^\perp = q^{n-k}$ code words, leading to $M M^\perp = q^n = |\mathbb{F}_q^n|$. By again exchanging the generator matrix and the parity-check matrix, we end up with the original code. Formally, this means that the dual of the dual code is the original code

$$\left(\mathbb{B}^\perp(n', k', d')\right)^\perp = \mathbb{B}(n, k, d).$$

A linear block code $\mathbb{B}(n, k, d)$ that fulfils $\mathbb{B}^\perp(n', k', d') = \mathbb{B}(n, k, d)$ is called *self-dual*.

Dual code and MacWilliams identity

- The dual code $\mathbb{B}^\perp(n', k', d')$ of a linear q-nary block code $\mathbb{B}(n, k, d)$ with parity-check matrix $\mathbf{H}$ is defined by the generator matrix

$$\mathbf{G}^\perp = \mathbf{H} \qquad (2.20)$$

- The weight distribution $W^\perp(x)$ of the q-nary dual code $\mathbb{B}^\perp(n', k', d')$ follows from the MacWilliams identity

$$W^\perp(x) = q^{-k} \left(1 + (q - 1)x\right)^n W\left(\frac{1 - x}{1 + (q - 1)x}\right) \qquad (2.21)$$

Figure 2.21: Dual code $\mathbb{B}^\perp(n', k', d')$ and MacWilliams identity

ALGEBRAIC CODING THEORY

Because of the close relationship between $\mathbb{B}(n, k, d)$ and its dual $\mathbb{B}^{\perp}(n', k', d')$, we can derive the properties of the dual code $\mathbb{B}^{\perp}(n', k', d')$ from the properties of the original code $\mathbb{B}(n, k, d)$. In particular, the *MacWilliams identity* holds for the corresponding weight distributions $W^{\perp}(x)$ and $W(x)$ (Bossert, 1999; Lin and Costello, 2004; Ling and Xing, 2004; MacWilliams and Sloane, 1998). The MacWilliams identity states that

$$W^{\perp}(x) = q^{-k} \left(1 + (q-1)x\right)^n W\left(\frac{1-x}{1+(q-1)x}\right).$$

For linear binary block codes with $q = 2$ we obtain

$$W^{\perp}(x) = 2^{-k} (1+x)^n W\left(\frac{1-x}{1+x}\right).$$

The MacWilliams identity can, for example, be used to determine the minimum Hamming distance d' of the dual code $\mathbb{B}^{\perp}(n', k', d')$ on the basis of the known weight distribution of the original code $\mathbb{B}(n, k, d)$.

2.2.6 Bounds for Linear Block Codes

The code rate $R = k/n$ and the minimum Hamming distance d are important parameters of a linear block code $\mathbb{B}(n, k, d)$. It is therefore useful to know whether a linear block code theoretically exists for a given combination of R and d. In particular, for a given minimum Hamming distance d we will consider a linear block code to be better than another linear block code with the same minimum Hamming distance d if it has a higher code rate R. There exist several theoretical bounds which we will briefly discuss in the following (Berlekamp, 1984; Bossert, 1999; Ling and Xing, 2004; van Lint, 1999) (see also Figure 2.22). Block codes that fulfil a theoretical bound with equality are called *optimal codes*.

Singleton Bound

The simplest bound is the so-called *Singleton bound*. For a linear block code $\mathbb{B}(n, k, d)$ it is given by

$$k \leq n - d + 1.$$

A linear block code that fulfils the Singleton bound with equality according to $k = n - d + 1$ is called MDS (maximum distance separable). Important representatives of MDS codes are Reed–Solomon codes which are used, for example, in the channel coding scheme within the audio compact disc (see also Section 2.3.7).

Sphere Packing Bound

The so-called *sphere packing bound* or *Hamming bound* can be derived for a linear $e_{\text{cor}} = \lfloor (d-1)/2 \rfloor$-error correcting q-nary block code $\mathbb{B}(n, k, d)$ by considering the correction balls within the code space $\mathbb{F}_q^n$. Each correction ball encompasses a total of

$$\sum_{i=0}^{e_{\text{cor}}} \binom{n}{i} (q-1)^i$$

Bounds for linear block codes $\mathbb{B}(n, k, d)$

- **Singleton bound**
$$k \leq n - d + 1 \qquad (2.22)$$

- **Hamming bound**
$$\sum_{i=0}^{e_{\text{cor}}} \binom{n}{i} (q-1)^i \leq q^{n-k} \qquad (2.23)$$

with $e_{\text{cor}} = \lfloor (d-1)/2 \rfloor$

- **Plotkin bound**
$$q^k \leq \frac{d}{d - \theta n} \qquad (2.24)$$

for $d > \theta n$ with $\theta = 1 - \frac{1}{q}$

- **Gilbert–Varshamov bound**
$$\sum_{i=0}^{d-2} \binom{n-1}{i} (q-1)^i < q^{n-k} \qquad (2.25)$$

- **Griesmer bound**
$$n \geq \sum_{i=0}^{k-1} \left\lceil \frac{d}{q^i} \right\rceil \qquad (2.26)$$

Figure 2.22: Bounds for linear q-nary block codes $\mathbb{B}(n, k, d)$

vectors in the code space $\mathbb{F}_q^n$. Since there is exactly one correction ball for each code word, we have $M = q^k$ correction balls. The total number of vectors within any correction ball of radius e_{cor} is then given by

$$q^k \sum_{i=0}^{e_{\text{cor}}} \binom{n}{i} (q-1)^i.$$

Because this number must be smaller than or equal to the maximum number $|\mathbb{F}_q^n| = q^n$ of vectors in the finite vector space $\mathbb{F}_q^n$, the sphere packing bound

$$q^k \sum_{i=0}^{e_{\text{cor}}} \binom{n}{i} (q-1)^i \leq q^n$$

ALGEBRAIC CODING THEORY

follows. For binary block codes with $q = 2$ we obtain

$$\sum_{i=0}^{e_{\text{cor}}} \binom{n}{i} \leq 2^{n-k}.$$

A so-called *perfect code* fulfils the sphere packing bound with equality, i.e.

$$\sum_{i=0}^{e_{\text{cor}}} \binom{n}{i} (q-1)^i = q^{n-k}.$$

Perfect codes exploit the complete code space; every vector in the code space $\mathbb{F}_q^n$ is part of one and only one correction ball and can therefore uniquely be assigned to a particular code word. For perfect binary codes with $q = 2$ which are capable of correcting $e_{\text{cor}} = 1$ error we obtain

$$1 + n = 2^{n-k}.$$

Figure 2.23 shows the parameters of perfect single-error correcting binary block codes up to length 1023.

Plotkin Bound

Under the assumption that $d > \theta n$ with

$$\theta = \frac{q-1}{q} = 1 - \frac{1}{q}$$

Perfect binary block codes

- The following table shows the parameters of perfect single-error correcting binary block codes $\mathbb{B}(n, k, d)$ with minimum Hamming distance $d = 3$.

n	k	m
3	1	2
7	4	3
15	11	4
31	26	5
63	57	6
127	120	7
255	247	8
511	502	9
1023	1013	10

Figure 2.23: Code word length n, information word length k and number of parity bits $m = n - k$ of a perfect single-error correcting code

the *Plotkin bound* states that
$$q^k \leq \frac{d}{d - \theta n}.$$
For binary block codes with $q = 2$ and $\theta = \frac{2-1}{2} = \frac{1}{2}$ we obtain
$$2^k \leq \frac{2d}{2d - n}$$
for $2d > n$. The code words of a code that fulfils the Plotkin bound with equality all have the same distance d. Such codes are called *equidistant*.

Gilbert–Varshamov Bound

By making use of the fact that the minimal number of linearly dependent columns in the parity-check matrix is equal to the minimum Hamming distance, the *Gilbert–Varshamov bound* can be derived for a linear block code $\mathbb{B}(n, k, d)$. If
$$\sum_{i=0}^{d-2} \binom{n-1}{i} (q-1)^i < q^{n-k}$$
is fulfilled, it is possible to construct a linear q-nary block code $\mathbb{B}(n, k, d)$ with code rate $R = k/n$ and minimum Hamming distance d.

Griesmer Bound

The *Griesmer bound* yields a lower bound for the code word length n of a linear q-nary block code $\mathbb{B}(n, k, d)$ according to[10]
$$n \geq \sum_{i=0}^{k-1} \left\lceil \frac{d}{q^i} \right\rceil = d + \left\lceil \frac{d}{q} \right\rceil + \cdots + \left\lceil \frac{d}{q^{k-1}} \right\rceil.$$

Asymptotic Bounds

For codes with very large code word lengths n, asymptotic bounds are useful which are obtained for $n \to \infty$. These bounds relate the code rate
$$R = \frac{k}{n}$$
and the relative minimum Hamming distance
$$\delta = \frac{d}{n}.$$
In Figure 2.24 some asymptotic bounds are given for linear binary block codes with $q = 2$ (Bossert, 1999).

---

[10]The term $\lceil z \rceil$ denotes the smallest integer number that is not smaller than z.

Asymptotic bounds for linear binary block codes

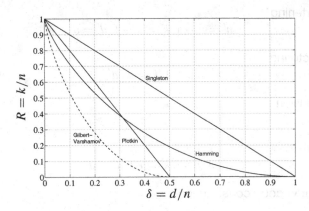

- Asymptotic Singleton bound

$$R \lesssim 1 - \delta \qquad (2.27)$$

- Asymptotic Hamming bound

$$R \lesssim 1 + \frac{\delta}{2} \log_2\left(\frac{\delta}{2}\right) + \left(1 - \frac{\delta}{2}\right) \log_2\left(1 - \frac{\delta}{2}\right) \qquad (2.28)$$

- Asymptotic Plotkin bound

$$R \lesssim \begin{cases} 1 - 2\delta, & 0 \leq \delta \leq \frac{1}{2} \\ 0, & \frac{1}{2} < \delta \leq 1 \end{cases} \qquad (2.29)$$

- Gilbert–Varshamov bound (with $0 \leq \delta \leq \frac{1}{2}$)

$$R \gtrsim 1 + \delta \log_2(\delta) + (1 - \delta) \log_2(1 - \delta) \qquad (2.30)$$

Figure 2.24: Asymptotic bounds for linear binary block codes $\mathbb{B}(n, k, d)$

2.2.7 Code Constructions

In this section we turn to the question of how new block codes can be generated on the basis of known block codes. We will consider the code constructions shown in Figure 2.25 (Berlekamp, 1984; Bossert, 1999; Lin and Costello, 2004; Ling and Xing, 2004). Some of these code constructions will reappear in later chapters (e.g. code puncturing for convolutional codes in Chapter 3 and code interleaving for Turbo codes in Chapter 4).

Code constructions for linear block codes

- **Code shortening**
$$n' = n - 1, k' = k - 1, d' = d \quad (2.31)$$

- **Code puncturing**
$$n' = n - 1, k' = k, d' \leq d \quad (2.32)$$

- **Code extension**
$$n' = n + 1, k' = k, d' \geq d \quad (2.33)$$

- **Code interleaving**
$$n' = nt, k' = kt, d' = d \quad (2.34)$$

- **Plotkin's (u|v) code construction**
$$n' = 2 \max\{n_1, n_2\}, k' = k_1 + k_2, d' = \min\{2 d_1, d_2\} \quad (2.35)$$

Figure 2.25: Code constructions for linear block codes and their respective code parameters

Code Shortening

Based on a linear block code $\mathbb{B}(n, k, d)$ with code word length n, information word length k and minimum Hamming distance d, we can create a new code by considering all code words $\mathbf{b} = (b_0, b_1, \ldots, b_{n-1})$ with $b_j = 0$. The so-called *shortened code* is then given by

$$\mathbb{B}'(n', k', d') = \Big\{ (b_0, \ldots, b_{j-1}, b_{j+1}, \ldots, b_{n-1}) : \\ (b_0, \ldots, b_{j-1}, b_j, b_{j+1}, \ldots, b_{n-1}) \in \mathbb{B}(n, k, d) \wedge b_j = 0 \Big\}.$$

This shortened code $\mathbb{B}'(n', k', d')$ is characterised by the following code parameters[11]

$$n' = n - 1,$$
$$k' = k - 1,$$
$$d' = d.$$

If $\mathbb{B}(n, k, d)$ is an MDS code, then the code $\mathbb{B}'(n', k', d')$ is also maximum distance separable because the Singleton bound is still fulfilled with equality: $d = n - k + 1 = (n - 1) - (k - 1) + 1 = n' - k' + 1 = d'$.

---

[11] We exclude the trivial case where all code words $\mathbf{b} \in \mathbb{B}(n, k, d)$ have a zero at the jth position.

ALGEBRAIC CODING THEORY

Code Puncturing

Starting from the systematic linear block code $\mathbb{B}(n, k, d)$ with code word length n, information word length k and minimum Hamming distance d, the so-called *punctured code* is obtained by eliminating the jth code position b_j of a code word $\mathbf{b} = (b_0, b_1, \ldots, b_{n-1}) \in \mathbb{B}(n, k, d)$ irrespective of its value. Here, we assume that b_j is a parity-check symbol. The resulting code consists of the code words $\mathbf{b}' = (b_0, \ldots, b_{j-1}, b_{j+1}, \ldots, b_{n-1})$. The punctured code

$$\mathbb{B}'(n', k', d') = \Big\{(b_0, \ldots, b_{j-1}, b_{j+1}, \ldots, b_{n-1}) :$$
$$(b_0, \ldots, b_{j-1}, b_j, b_{j+1}, \ldots, b_{n-1}) \in \mathbb{B}(n, k, d) \land b_j \text{ is parity-check symbol}\Big\}$$

is characterised by the following code parameters

$$n' = n - 1,$$
$$k' = k,$$
$$d' \leq d.$$

Code Extension

The so-called *extension* of a linear block code $\mathbb{B}(n, k, d)$ with code word length n, information word length k and minimum Hamming distance d is generated by attaching an additional parity-check symbol b_n to each code word $\mathbf{b} = (b_0, b_1, \ldots, b_{n-1}) \in \mathbb{B}(n, k, d)$. The additional parity-check symbol is calculated according to

$$b_n = -(b_0 + b_1 + \cdots + b_{n-1}) = -\sum_{i=0}^{n-1} b_i.$$

For a linear binary block code $\mathbb{B}(n, k, d)$ this yields

$$b_n = b_0 + b_1 + \cdots + b_{n-1} = \sum_{i=0}^{n-1} b_i.$$

We obtain the extended code

$$\mathbb{B}'(n', k', d') = \Big\{(b_0, b_1, \cdots, b_{n-1}, b_n) :$$
$$(b_0, b_1, \ldots, b_{n-1}) \in \mathbb{B}(n, k, d) \land b_0 + b_1 + \cdots + b_{n-1} + b_n = 0\Big\}$$

with code parameters

$$n' = n + 1,$$
$$k' = k,$$
$$d' \geq d.$$

In the case of a linear binary block code $\mathbb{B}(n,k,d)$ with $q=2$ and an odd minimum Hamming distance d, the extended code's minimum Hamming distance is increased by 1, i.e. $d' = d+1$.

The additional parity-check symbol b_n corresponds to an additional parity-check equation that augments the existing parity-check equations

$$\mathbf{H}\mathbf{r}^{\mathrm{T}} = \mathbf{0} \quad \Leftrightarrow \quad \mathbf{r} \in \mathbb{B}(n,k,d)$$

of the original linear block code $\mathbb{B}(n,k,d)$ with the $(n-k) \times n$ parity-check matrix $\mathbf{H}$. The $(n-k+1) \times n$ parity-check matrix $\mathbf{H}'$ of the extended block code $\mathbb{B}'(n',k',d')$ is then given by

$$\mathbf{H}' = \left(\begin{array}{c|c} & 0 \\ & 0 \\ \mathbf{H} & \vdots \\ & 0 \\ \hline 1\,1\,\cdots\,1 & 1 \end{array} \right).$$

Code Interleaving

Up to now we have considered the correction of independent errors within a given received word in accordance with a memoryless channel. In real channels, however, error bursts might occur, e.g. in fading channels in mobile communication systems. With the help of so-called *code interleaving*, an existing error-correcting channel code can be adapted such that error bursts up to a given length can also be corrected.

Starting with a linear block code $\mathbb{B}(n,k,d)$ with code word length n, information word length k and minimum Hamming distance d, the code interleaving to interleaving depth t is obtained as follows. We arrange t code words $\mathbf{b}_1$, $\mathbf{b}_2$, ..., $\mathbf{b}_t$ with $\mathbf{b}_j = (b_{j,0}, b_{j,1}, \ldots, b_{j,n-1})$ as rows in the $t \times n$ matrix

$$\begin{pmatrix} b_{1,0} & b_{1,1} & \cdots & b_{1,n-1} \\ b_{2,0} & b_{2,1} & \cdots & b_{2,n-1} \\ \vdots & \vdots & \ddots & \vdots \\ b_{t,0} & b_{t,1} & \cdots & b_{t,n-1} \end{pmatrix}.$$

The resulting code word

$$\mathbf{b}' = b_{1,0} b_{2,0} \cdots b_{t,0} b_{1,1} b_{2,1} \cdots b_{t,1} \cdots b_{1,n-1} b_{2,n-1} \cdots b_{t,n-1}$$

of length nt of the interleaved code $\mathbb{B}'(n',k',d')$ is generated by reading the symbols off the matrix column-wise. This encoding scheme is illustrated in Figure 2.26. The interleaved code is characterised by the following code parameters

$$n' = nt,$$
$$k' = kt,$$
$$d' = d.$$

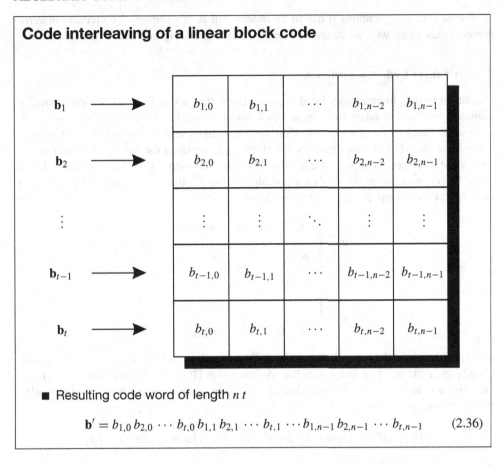

Figure 2.26: Code interleaving of a linear block code $\mathbb{B}(n, k, d)$

Neither the code rate

$$R' = \frac{k'}{n'} = \frac{kt}{nt} = \frac{k}{n} = R$$

nor the minimum Hamming distance $d' = d$ are changed. The latter means that the error detection and error correction capabilities of the linear block code $\mathbb{B}(n, k, d)$ with respect to independent errors are not improved by the interleaving step. However, error bursts consisting of ℓ_{burst} neighbouring errors within the received word can now be corrected. If the linear block code $\mathbb{B}(n, k, d)$ is capable of correcting up to e_{cor} errors, the interleaved code is capable of correcting error bursts up to length

$$\ell_{\text{burst}} = e_{\text{cor}} t$$

with interleaving depth t. This is achieved by spreading the error burst in the interleaved code word $\mathbf{b}'$ onto t different code words $\mathbf{b}_j$ such that in each code word a maximum of e_{cor} errors occurs. In this way, the code words $\mathbf{b}_j$ can be corrected by the original code $\mathbb{B}(n, k, d)$ without error if the length of the error burst does not exceed $e_{\text{cor}} t$.

For practical applications it has to be observed that this improved correction of error bursts comes along with an increased latency for the encoding and decoding steps.

Plotkin's (u|v) Code Construction

In contrast to the aforementioned code constructions, the **(u|v)** code construction originally proposed by Plotkin takes two linear block codes $\mathbb{B}(n_1, k_1, d_1)$ and $\mathbb{B}(n_2, k_2, d_2)$ with code word lengths n_1 and n_2, information word lengths k_1 and k_2 and minimum Hamming distances d_1 and d_2. For simplification, we fill the code words of the block code with smaller code word length by an appropriate number of zeros. This step is called *zero padding*. Let **u** and **v** be two arbitrary code words thus obtained from the linear block codes $\mathbb{B}(n_1, k_1, d_1)$ and $\mathbb{B}(n_2, k_2, d_2)$ respectively.[12] Then we have

$$\mathbf{u} = \begin{cases} (\mathbf{b}_1 \,|\, 0, \ldots, 0), & n_1 < n_2 \\ \mathbf{b}_1, & n_1 \geq n_2 \end{cases}$$

and

$$\mathbf{v} = \begin{cases} \mathbf{b}_2, & n_1 < n_2 \\ (\mathbf{b}_2 \,|\, 0, \ldots, 0), & n_1 \geq n_2 \end{cases}$$

with the arbitrarily chosen code words $\mathbf{b}_1 \in \mathbb{B}(n_1, k_1, d_1)$ and $\mathbf{b}_2 \in \mathbb{B}(n_2, k_2, d_2)$. We now identify the code words **u** and **v** with the original codes $\mathbb{B}(n_1, k_1, d_1)$ and $\mathbb{B}(n_2, k_2, d_2)$, i.e. we write $\mathbf{u} \in \mathbb{B}(n_1, k_1, d_1)$ and $\mathbf{v} \in \mathbb{B}(n_2, k_2, d_2)$. The code resulting from *Plotkin's* **(u|v)** *code construction* is then given by

$$\mathbb{B}'(n', k', d') = \{(\mathbf{u}|\mathbf{u} + \mathbf{v}) : \mathbf{u} \in \mathbb{B}(n_1, k_1, d_1) \wedge \mathbf{v} \in \mathbb{B}(n_2, k_2, d_2)\}.$$

This code construction creates all vectors of the form $(\mathbf{u}|\mathbf{u} + \mathbf{v})$ by concatenating all possible vectors **u** and $\mathbf{u} + \mathbf{v}$ with $\mathbf{u} \in \mathbb{B}(n_1, k_1, d_1)$ and $\mathbf{v} \in \mathbb{B}(n_2, k_2, d_2)$. The resulting code is characterised by the following code parameters

$$n' = 2 \max\{n_1, n_2\},$$
$$k' = k_1 + k_2,$$
$$d' = \min\{2 d_1, d_2\}.$$

2.2.8 Examples of Linear Block Codes

In this section we will present some important linear block codes. In particular, we will consider the already introduced repetition codes, parity-check codes and Hamming codes. Furthermore, simplex codes and Reed–Muller codes and their relationships are discussed (Berlekamp, 1984; Bossert, 1999; Lin and Costello, 2004; Ling and Xing, 2004).

---

[12]In the common notation of Plotkin's **(u|v)** code construction the vector **u** does not correspond to the information vector.

ALGEBRAIC CODING THEORY

Repetition code

- The repetition code $\mathbb{B}(n, 1, n)$ repeats the information symbol u_0 in the code vector $\mathbf{b} = (b_0, b_1, \ldots, b_{n-1})$, i.e.

$$b_0 = b_1 = \cdots = b_{n-1} = u_0 \qquad (2.37)$$

- Minimum Hamming distance

$$d = n \qquad (2.38)$$

- Code rate

$$R = \frac{1}{n} \qquad (2.39)$$

Figure 2.27: Repetition code

Repetition Codes

We have already introduced the binary triple repetition code in Section 1.3 as a simple introductory example of linear block codes. In this particular code the binary information symbol 0 or 1 is transmitted with the binary code word 000 or 111 respectively. In general, a *repetition code* over the alphabet $\mathbb{F}_q$ assigns to the q-nary information symbol u_0 the n-dimensional code word $\mathbf{b} = (u_0, u_0, \ldots, u_0)$. Trivially, this block code is linear. The minimum Hamming distance is

$$d = n.$$

Therefore, the repetition code in Figure 2.27 is a linear block code $\mathbb{B}(n, 1, n)$ that is able to detect $e_{\text{det}} = d - 1 = n - 1$ errors or to correct $e_{\text{cor}} = \lfloor (d-1)/2 \rfloor = \lfloor (n-1)/2 \rfloor$ errors. The code rate is

$$R = \frac{1}{n}.$$

For the purpose of error correction, the minimum distance decoding can be implemented by a majority decoding scheme. The decoder emits the information symbol $\hat{u}_0$ which appears most often in the received word.

The weight distribution of a repetition code is simply given by

$$W(x) = 1 + (q-1)x^n.$$

Although this repetition code is characterised by a low code rate R, it is used in some communication standards owing to its simplicity. For example, in the short-range wireless communication system Bluetooth^{TM} a triple repetition code is used as part of the coding scheme of the packet header of a transmitted baseband packet (Bluetooth, 2004).

Parity-check code

- The parity-check code $\mathbb{B}(n, n-1, 2)$ attaches a parity-check symbol so that the resulting code vector $\mathbf{b} = (b_0, b_1, \ldots, b_{n-1})$ fulfils the condition $b_0 + b_1 + \cdots + b_{n-1} = 0$, i.e.

$$b_{n-1} = -(u_0 + u_1 + \cdots + u_{n-2}) \qquad (2.40)$$

- Minimum Hamming distance
$$d = 2 \qquad (2.41)$$

- Code rate
$$R = \frac{n-1}{n} = 1 - \frac{1}{n} \qquad (2.42)$$

Figure 2.28: Parity-check code

Parity-Check Codes

Starting from the information word $\mathbf{u} = (u_0, u_1, \ldots, u_{k-1})$ with $u_i \in \mathbb{F}_q$, a *parity-check code* attaches a single parity-check symbol to the information word $\mathbf{u}$ as illustrated in Figure 2.28. The $n = (k+1)$-dimensional code word

$$\mathbf{b} = (b_0, b_1, \ldots, b_{n-2}, b_{n-1}) = (b_0, b_1, \ldots, b_{k-1}, b_k)$$

is given by $b_i = u_i$ for $0 \leq i \leq k-1$, and the parity-check symbol b_k which is chosen such that

$$\sum_{i=0}^{k} b_i = b_0 + b_1 + \cdots + b_k = 0$$

is fulfilled. Over the finite field $\mathbb{F}_q$ we have

$$b_k = -(u_0 + u_1 + \cdots + u_{k-1}) = -\sum_{i=0}^{k-1} u_i.$$

In the case of a binary parity-check code over the alphabet $\mathbb{F}_2$, the parity-check bit b_k is chosen such that the resulting code word $\mathbf{b}$ is of even parity, i.e. the number of binary symbols 1 is even. The parity-check code is a linear block code $\mathbb{B}(k+1, k, 2)$ with a minimum Hamming distance $d = 2$. This code is capable of detecting $e_{\text{det}} = d - 1 = 1$ error. The correction of errors is not possible. The code rate is

$$R = \frac{k}{k+1} = \frac{n-1}{n} = 1 - \frac{1}{n}.$$

ALGEBRAIC CODING THEORY

The error detection is carried out by checking whether the received vector $\mathbf{r} = (r_0, r_1, \ldots, r_{n-1})$ fulfils the parity-check condition

$$\sum_{i=0}^{n-1} r_i = r_0 + r_1 + \cdots + r_{n-1} = 0.$$

If this condition is not met, then at least one error has occurred.

The weight distribution of a binary parity-check code is equal to

$$W(x) = 1 + \binom{n}{2} x^2 + \binom{n}{4} x^4 + \cdots + x^n$$

for an even code word length n and

$$W(x) = 1 + \binom{n}{2} x^2 + \binom{n}{4} x^4 + \cdots + n x^{n-1}$$

for an odd code word length n. Binary parity-check codes are often applied in simple serial interfaces such as, for example, UART (Universal Asynchronous Receiver Transmitter).

Hamming Codes

Hamming codes can be defined as binary or q-nary block codes (Ling and Xing, 2004). In this section we will exclusively focus on binary Hamming codes, i.e. $q = 2$. Originally, these codes were developed by Hamming for error correction of faulty memory entries (Hamming, 1950). Binary Hamming codes are most easily defined by their corresponding parity-check matrix. Assume that we want to attach m parity-check symbols to an information word $\mathbf{u}$ of length k. The parity-check matrix is obtained by writing down all m-dimensional non-zero column vectors. Since there are $n = 2^m - 1$ binary column vectors of length m, the parity-check matrix

$$\begin{pmatrix} 1 & 0 & 1 & 0 & 1 & 0 & 1 & 0 & \cdots & 0 & 1 & 0 & 1 & 0 & 1 & 0 & 1 \\ 0 & 1 & 1 & 0 & 0 & 1 & 1 & 0 & \cdots & 0 & 0 & 1 & 1 & 0 & 0 & 1 & 1 \\ 0 & 0 & 0 & 1 & 1 & 1 & 1 & 0 & \cdots & 0 & 0 & 0 & 0 & 1 & 1 & 1 & 1 \\ 0 & 0 & 0 & 0 & 0 & 0 & 0 & 1 & \cdots & 1 & 1 & 1 & 1 & 1 & 1 & 1 & 1 \\ \vdots & \vdots & \vdots & \vdots & \vdots & \vdots & \vdots & \vdots & \ddots & \vdots & \vdots & \vdots & \vdots & \vdots & \vdots & \vdots & \vdots \\ 0 & 0 & 0 & 0 & 0 & 0 & 0 & 0 & \cdots & 1 & 1 & 1 & 1 & 1 & 1 & 1 & 1 \\ 0 & 0 & 0 & 0 & 0 & 0 & 0 & 0 & \cdots & 1 & 1 & 1 & 1 & 1 & 1 & 1 & 1 \end{pmatrix}$$

is of dimension $m \times (2^m - 1)$. By suitably rearranging the columns, we obtain the $(n-k) \times n$ or $m \times n$ parity-check matrix

$$\mathbf{H} = \left(\mathbf{B}_{n-k,k} \,\middle|\, \mathbf{I}_{n-k} \right)$$

of the equivalent binary Hamming code. The corresponding $k \times n$ generator matrix is then given by

$$\mathbf{G} = \left(\mathbf{I}_k \,\middle|\, -\mathbf{B}^{\mathrm{T}}_{n-k,k} \right).$$

Because of the number of columns within the parity-check matrix **H**, the code word length of the binary Hamming code is given by $n = 2^m - 1 = 2^{n-k} - 1$ with the number of binary information symbols $k = n - m = 2^m - m - 1$. Since the columns of the parity-check matrix are pairwise linearly independent and there exist three columns which sum up to the all-zero vector, the minimum Hamming distance of the binary Hamming code is $d = 3$.[13] The binary Hamming code can therefore be characterised as a linear block code $\mathbb{B}(2^m - 1, 2^m - m - 1, 3)$ that is able to detect $e_{\text{det}} = 2$ errors or to correct $e_{\text{cor}} = 1$ error.

Because of $n = 2^{n-k} - 1$ and $e_{\text{cor}} = 1$ we have

$$\sum_{i=0}^{e_{\text{cor}}} \binom{n}{i} = 1 + n = 2^{n-k},$$

i.e. the binary Hamming code is perfect because it fulfils the sphere packing bound.[14] Since the binary Hamming code $\mathbb{B}(2^m - 1, 2^m - m - 1, 3)$ only depends on the number of parity symbols m, we also call this code $\mathcal{H}(m)$ with the code parameters

$$n = 2^m - 1,$$
$$k = 2^m - m - 1,$$
$$d = 3.$$

For a binary Hamming code the decoding of an erroneously received vector $\mathbf{r} = \mathbf{b} + \mathbf{e}$ with error vector $\mathbf{e}$ of weight $\text{wt}(\mathbf{e}) = 1$ can be carried out by first calculating the syndrome

$$\mathbf{s}^{\text{T}} = \mathbf{H}\mathbf{r}^{\text{T}} = \mathbf{H}\mathbf{e}^{\text{T}}$$

which is then compared with all columns of the parity-check matrix **H**. If the non-zero syndrome $\mathbf{s}^{\text{T}}$ agrees with the jth column vector, the error vector $\mathbf{e}$ must have its single non-zero component at the jth position. The received vector

$$\mathbf{r} = (r_0, \ldots, r_{j-1}, r_j, r_{j+1}, \ldots, r_{n-1})$$

can therefore be decoded into the code word

$$\hat{\mathbf{b}} = (r_0, \ldots, r_{j-1}, r_j + 1, r_{j+1}, \ldots, r_{n-1})$$

by calculating $\hat{b}_j = r_j + 1$.

The weight distribution $W(x)$ of the binary Hamming code $\mathcal{H}(m)$ is given by

$$W(x) = \frac{1}{n+1}\left((1+x)^n + n(1-x)^{(n+1)/2}(1+x)^{(n-1)/2}\right).$$

The coefficients w_i can be recursively calculated according to

$$i\,w_i = \binom{n}{i-1} - w_{i-1} - (n - i + 2)\,w_{i-2}$$

with $w_0 = 1$ and $w_1 = 0$. Figure 2.29 summarises the properties of the binary Hamming code $\mathcal{H}(m)$.

---

[13] In the given matrix the first three columns sum up to the zero column vector of length m.

[14] There are only two other non-trivial perfect linear codes, the binary Golay code $\mathbb{B}(23, 12, 7)$ with $n = 23$, $k = 12$ and $d = 7$ and the ternary Golay code $\mathbb{B}(11, 6, 5)$ with $n = 11$, $k = 6$ and $d = 5$. The extended binary Golay code has been used, for example, for the Voyager 1 and 2 spacecrafts in deep-space communications.

ALGEBRAIC CODING THEORY

Binary Hamming code

- The binary Hamming code $\mathcal{H}(m)$ is a perfect single-error correcting code with parity-check-matrix **H** consisting of all $2^m - 1$ non-zero binary column vectors of length m.

- Code word length
$$n = 2^m - 1 \qquad (2.43)$$

- Minimum Hamming distance
$$d = 3 \qquad (2.44)$$

- Code rate
$$R = \frac{2^m - m - 1}{2^m - 1} = 1 - \frac{m}{2^m - 1} \qquad (2.45)$$

Figure 2.29: Binary Hamming code $\mathcal{H}(m) = \mathbb{B}(2^m - 1, 2^m - m - 1, 3)$

The extended binary Hamming code $\mathcal{H}'(m)$ is obtained from the binary Hamming code $\mathcal{H}(m)$ by attaching to each code word an additional parity-check symbol such that the resulting code word is of even parity. The corresponding non-systematic parity-check matrix is equal to

$$\begin{pmatrix} 1 & 0 & 1 & 0 & 1 & 0 & 1 & 0 & \cdots & 0 & 1 & 0 & 1 & 0 & 1 & 0 & 1 & 0 \\ 0 & 1 & 1 & 0 & 0 & 1 & 1 & 0 & \cdots & 0 & 0 & 1 & 1 & 0 & 0 & 1 & 1 & 0 \\ 0 & 0 & 0 & 1 & 1 & 1 & 1 & 0 & \cdots & 0 & 0 & 0 & 0 & 1 & 1 & 1 & 1 & 0 \\ 0 & 0 & 0 & 0 & 0 & 0 & 0 & 1 & \cdots & 1 & 1 & 1 & 1 & 1 & 1 & 1 & 1 & 0 \\ \vdots & \vdots & \vdots & \vdots & \vdots & \vdots & \vdots & \vdots & \ddots & \vdots & \vdots & \vdots & \vdots & \vdots & \vdots & \vdots & \vdots & \vdots \\ 0 & 0 & 0 & 0 & 0 & 0 & 0 & 0 & \cdots & 1 & 1 & 1 & 1 & 1 & 1 & 1 & 1 & 0 \\ 0 & 0 & 0 & 0 & 0 & 0 & 0 & 0 & \cdots & 1 & 1 & 1 & 1 & 1 & 1 & 1 & 1 & 0 \\ 1 & 1 & 1 & 1 & 1 & 1 & 1 & 1 & \cdots & 1 & 1 & 1 & 1 & 1 & 1 & 1 & 1 & 1 \end{pmatrix}$$

The minimum Hamming distance of the extended Hamming code is $d' = 4$.

Hamming codes are used, for example, in semiconductor memories such as DRAMs for error correction of single bit errors or in the short-range wireless communication system Bluetooth^{TM} as part of the coding scheme of the packet header of a baseband packet (Bluetooth, 2004).

Simplex Codes

The dual code of the binary Hamming code $\mathcal{H}(m)$ is the so-called *simplex code* $\mathcal{S}(m)$, i.e. $\mathcal{H}^\perp(m) = \mathcal{S}(m)$. This binary code over the finite field $\mathbb{F}_2$ consists of the zero code word **0** and $2^m - 1$ code words of code word length n and weight 2^{m-1}, i.e. wt(**b**) $= 2^{m-1}$ for all

Simplex code

- The simplex code $\mathcal{S}(m)$ is the dual of the binary Hamming code $\mathcal{H}(m)$, i.e. $\mathcal{S}(m) = \mathcal{H}^\perp(m)$.
- Code word length
$$n = 2^m - 1 \qquad (2.46)$$
- Minimum Hamming distance
$$d = 2^{m-1} \qquad (2.47)$$
- Code rate
$$R = \frac{m}{2^m - 1} \qquad (2.48)$$

Figure 2.30: Simplex code $\mathcal{S}(m) = \mathbb{B}(2^m - 1, m, 2^{m-1})$

code words $\mathbf{b} \in \mathcal{S}(m)$ with $\mathbf{b} \neq \mathbf{0}$. The total number of code words is $M = 2^m$ and the code word length is $n = 2^m - 1$. The minimum weight of the simplex code is $\min_{\forall \mathbf{b} \neq \mathbf{0}} \text{wt}(\mathbf{b}) = 2^{m-1}$, which is equal to the minimum Hamming distance $d = 2^{m-1}$ owing to the linearity of the simplex code. From these considerations, the weight distribution $W(x)$ of the simplex code

$$W(x) = 1 + (2^m - 1) x^{2^{m-1}} = 1 + n x^{(n+1)/2}$$

easily follows.

According to Figure 2.30 the binary simplex code is characterised by the code parameters

$$n = 2^m - 1,$$
$$k = m,$$
$$d = 2^{m-1}.$$

Since the simplex code $\mathcal{S}(m) = \mathcal{H}^\perp(m)$ is the dual of the binary Hamming code $\mathcal{H}(m)$, the $m \times n$ parity-check matrix

$$\begin{pmatrix}
1 & 0 & 1 & 0 & 1 & 0 & 1 & 0 & \cdots & 0 & 1 & 0 & 1 & 0 & 1 & 0 & 1 \\
0 & 1 & 1 & 0 & 0 & 1 & 1 & 0 & \cdots & 0 & 0 & 1 & 1 & 0 & 0 & 1 & 1 \\
0 & 0 & 0 & 1 & 1 & 1 & 1 & 0 & \cdots & 0 & 0 & 0 & 0 & 1 & 1 & 1 & 1 \\
0 & 0 & 0 & 0 & 0 & 0 & 0 & 1 & \cdots & 1 & 1 & 1 & 1 & 1 & 1 & 1 & 1 \\
\vdots & \vdots & \vdots & \vdots & \vdots & \vdots & \vdots & \vdots & \ddots & \vdots & \vdots & \vdots & \vdots & \vdots & \vdots & \vdots & \vdots \\
0 & 0 & 0 & 0 & 0 & 0 & 0 & 0 & \cdots & 1 & 1 & 1 & 1 & 1 & 1 & 1 & 1 \\
0 & 0 & 0 & 0 & 0 & 0 & 0 & 0 & \cdots & 1 & 1 & 1 & 1 & 1 & 1 & 1 & 1
\end{pmatrix}$$

ALGEBRAIC CODING THEORY

> **Hadamard matrix H_m**
>
>
>
> ■ Recursive definition of $2^m \times 2^m$ Hadamard matrix
>
> $$H_m = \begin{pmatrix} H_{m-1} & H_{m-1} \\ H_{m-1} & -H_{m-1} \end{pmatrix} \qquad (2.49)$$
>
> ■ Initialisation
>
> $$H_0 = (1) \qquad (2.50)$$

Figure 2.31: Hadamard matrix H_m. Reproduced by permission of J. Schlembach Fachverlag

of the Hamming code $\mathcal{H}(m)$ yields the generator matrix of the simplex code $\mathcal{S}(m)$. If required, the generator matrix can be brought into the systematic form

$$G = \left(I_m \,\middle|\, A_{m,n-m} \right).$$

There exists an interesting relationship between the binary simplex code $\mathcal{S}(m)$ and the so-called *Hadamard matrix* H_m. This $2^m \times 2^m$ matrix is recursively defined according to

$$H_m = \begin{pmatrix} H_{m-1} & H_{m-1} \\ H_{m-1} & -H_{m-1} \end{pmatrix}$$

with the initialisation $H_0 = (1)$. In Figure 2.31 the Hadamard matrix H_m is illustrated for $2 \leq m \leq 5$; a black rectangle corresponds to the entry 1 whereas a white rectangle corresponds to the entry -1.

As an example we consider $m = 2$ for which we obtain the 4×4 matrix

$$H_2 = \begin{pmatrix} 1 & 1 & 1 & 1 \\ 1 & -1 & 1 & -1 \\ 1 & 1 & -1 & -1 \\ 1 & -1 & -1 & 1 \end{pmatrix}.$$

Replacing all 1s with 0s and all -1s with 1s yields the matrix

$$\begin{pmatrix} 0 & 0 & 0 & 0 \\ 0 & 1 & 0 & 1 \\ 0 & 0 & 1 & 1 \\ 0 & 1 & 1 & 0 \end{pmatrix}.$$

If the first column is deleted, the resulting matrix

$$\begin{pmatrix} 0 & 0 & 0 \\ 1 & 0 & 1 \\ 0 & 1 & 1 \\ 1 & 1 & 0 \end{pmatrix}$$

includes as its rows all code vectors of the simplex code $\mathcal{S}(2)$.

In general, the simplex code $\mathcal{S}(m)$ can be obtained from the $2^m \times 2^m$ Hadamard matrix $\mathbf{H}_m$ by first applying the mapping

$$1 \mapsto 0 \quad \text{and} \quad -1 \mapsto 1$$

and then deleting the first column. The rows of the resulting matrix deliver all 2^m code words of the binary simplex code $\mathcal{S}(m)$.

As a further example we consider the $2^3 \times 2^3 = 8 \times 8$ Hadamard matrix

$$\mathbf{H}_3 = \begin{pmatrix} 1 & 1 & 1 & 1 & 1 & 1 & 1 & 1 \\ 1 & -1 & 1 & -1 & 1 & -1 & 1 & -1 \\ 1 & 1 & -1 & -1 & 1 & 1 & -1 & -1 \\ 1 & -1 & -1 & 1 & 1 & -1 & -1 & 1 \\ 1 & 1 & 1 & 1 & -1 & -1 & -1 & -1 \\ 1 & -1 & 1 & -1 & -1 & 1 & -1 & 1 \\ 1 & 1 & -1 & -1 & -1 & -1 & 1 & 1 \\ 1 & -1 & -1 & 1 & -1 & 1 & 1 & -1 \end{pmatrix}.$$

With the mapping $1 \mapsto 0$ and $-1 \mapsto 1$, the matrix

$$\begin{pmatrix} 0 & 0 & 0 & 0 & 0 & 0 & 0 & 0 \\ 0 & 1 & 0 & 1 & 0 & 1 & 0 & 1 \\ 0 & 0 & 1 & 1 & 0 & 0 & 1 & 1 \\ 0 & 1 & 1 & 0 & 0 & 1 & 1 & 0 \\ 0 & 0 & 0 & 0 & 1 & 1 & 1 & 1 \\ 0 & 1 & 0 & 1 & 1 & 0 & 1 & 0 \\ 0 & 0 & 1 & 1 & 1 & 1 & 0 & 0 \\ 0 & 1 & 1 & 0 & 1 & 0 & 0 & 1 \end{pmatrix}$$

follows. If we delete the first column, the rows of the resulting matrix

$$\begin{pmatrix} 0 & 0 & 0 & 0 & 0 & 0 & 0 \\ 1 & 0 & 1 & 0 & 1 & 0 & 1 \\ 0 & 1 & 1 & 0 & 0 & 1 & 1 \\ 1 & 1 & 0 & 0 & 1 & 1 & 0 \\ 0 & 0 & 0 & 1 & 1 & 1 & 1 \\ 1 & 0 & 1 & 1 & 0 & 1 & 0 \\ 0 & 1 & 1 & 1 & 1 & 0 & 0 \\ 1 & 1 & 0 & 1 & 0 & 0 & 1 \end{pmatrix}$$

yield the binary simplex code $\mathcal{S}(3)$ with code word length $n = 2^3 - 1 = 7$ and minimum Hamming distance $d = 2^{3-1} = 4$.

ALGEBRAIC CODING THEORY

Reed–Muller Codes

The Hadamard matrix can further be used for the construction of other codes. The so-called *Reed–Muller codes* form a class of error-correcting codes that are capable of correcting more than one error. Although these codes do not have the best code parameters, they are characterised by a simple and efficiently implementable decoding strategy both for hard-decision as well as soft-decision decoding. There exist first-order and higher-order Reed–Muller codes, as we will explain in the following.

First-Order Reed–Muller Codes $\mathcal{R}(1, m)$ The binary first-order Reed–Muller code $\mathcal{R}(1, m)$ can be defined with the help of the following $m \times 2^m$ matrix which contains all possible 2^m binary column vectors of length m

$$\begin{pmatrix} 0 & 0 & 0 & 0 & 0 & 0 & 0 & 0 & \cdots & 1 & 1 & 1 & 1 & 1 & 1 & 1 & 1 \\ 0 & 0 & 0 & 0 & 0 & 0 & 0 & 0 & \cdots & 1 & 1 & 1 & 1 & 1 & 1 & 1 & 1 \\ \vdots & \vdots & \vdots & \vdots & \vdots & \vdots & \vdots & \vdots & \ddots & \vdots & \vdots & \vdots & \vdots & \vdots & \vdots & \vdots & \vdots \\ 0 & 0 & 0 & 0 & 0 & 0 & 0 & 1 & \cdots & 1 & 1 & 1 & 1 & 1 & 1 & 1 & 1 \\ 0 & 0 & 0 & 0 & 1 & 1 & 1 & 1 & 0 & \cdots & 0 & 0 & 0 & 0 & 1 & 1 & 1 & 1 \\ 0 & 0 & 1 & 1 & 0 & 0 & 1 & 1 & 0 & \cdots & 0 & 0 & 1 & 1 & 0 & 0 & 1 & 1 \\ 0 & 1 & 0 & 1 & 0 & 1 & 0 & 1 & 0 & \cdots & 0 & 1 & 0 & 1 & 0 & 1 & 0 & 1 \end{pmatrix}.$$

By attaching the 2^m-dimensional all-one vector

$$\begin{pmatrix} 1 & 1 & 1 & 1 & 1 & 1 & 1 & 1 & \cdots & 1 & 1 & 1 & 1 & 1 & 1 & 1 & 1 \end{pmatrix}$$

as the first row, we obtain the $(m + 1) \times 2^m$ generator matrix

$$\mathbf{G} = \begin{pmatrix} 1 & 1 & 1 & 1 & 1 & 1 & 1 & 1 & \cdots & 1 & 1 & 1 & 1 & 1 & 1 & 1 & 1 \\ 0 & 0 & 0 & 0 & 0 & 0 & 0 & 0 & \cdots & 1 & 1 & 1 & 1 & 1 & 1 & 1 & 1 \\ 0 & 0 & 0 & 0 & 0 & 0 & 0 & 0 & \cdots & 1 & 1 & 1 & 1 & 1 & 1 & 1 & 1 \\ \vdots & \vdots & \vdots & \vdots & \vdots & \vdots & \vdots & \vdots & \ddots & \vdots & \vdots & \vdots & \vdots & \vdots & \vdots & \vdots & \vdots \\ 0 & 0 & 0 & 0 & 0 & 0 & 0 & 1 & \cdots & 1 & 1 & 1 & 1 & 1 & 1 & 1 & 1 \\ 0 & 0 & 0 & 0 & 1 & 1 & 1 & 1 & 0 & \cdots & 0 & 0 & 0 & 0 & 1 & 1 & 1 & 1 \\ 0 & 0 & 1 & 1 & 0 & 0 & 1 & 1 & 0 & \cdots & 0 & 0 & 1 & 1 & 0 & 0 & 1 & 1 \\ 0 & 1 & 0 & 1 & 0 & 1 & 0 & 1 & 0 & \cdots & 0 & 1 & 0 & 1 & 0 & 1 & 0 & 1 \end{pmatrix}.$$

The binary first-order Reed–Muller code $\mathcal{R}(1, m)$ in Figure 2.32 is characterised by the following code parameters

$$n = 2^m,$$
$$k = 1 + m,$$
$$d = 2^{m-1}.$$

The corresponding weight distribution $W(x)$ is given by

$$W(x) = 1 + \left(2^{m+1} - 2\right) x^{2^{m-1}} + x^{2^m},$$

> **First-order Reed–Muller code $\mathcal{R}(1, m)$**
>
> - The first-order Reed–Muller code $\mathcal{R}(1, m)$ is a binary code that is able to correct more than one error.
> - Code word length
> $$n = 2^m \tag{2.51}$$
> - Minimum Hamming distance
> $$d = 2^{m-1} \tag{2.52}$$
> - Code rate
> $$R = \frac{1 + \binom{m}{1}}{2^m} \tag{2.53}$$

Figure 2.32: First-order Reed–Muller code $\mathcal{R}(1, m)$

i.e. the binary first-order Reed–Muller code $\mathcal{R}(1, m)$ consists of code vectors with weights 0, 2^{m-1} and 2^m. The hard-decision decoding of a Reed–Muller code can be carried out by a majority logic decoding scheme.

There is a close relationship between the binary first-order Reed–Muller code $\mathcal{R}(1, m)$ and the simplex code $\mathcal{S}(m)$ with code word length $n = 2^m - 1$ and $k = m$ information symbols as well as the Hamming code $\mathcal{H}(m)$. The $m \times 2^m$ matrix used in the construction of the generator matrix of the first-order Reed–Muller code $\mathcal{R}(1, m)$ corresponds to the parity-check matrix of the binary Hamming code $\mathcal{H}(m)$ with the all-zero column attached. This parity-check matrix in turn yields the generator matrix of the simplex code $\mathcal{S}(m)$. Therefore, this $m \times 2^m$ matrix generates all code words of the simplex code $\mathcal{S}(m)$ with a zero attached in the first position. The attachment of the all-one row vector to the $m \times 2^m$ matrix is equivalent to adding all inverted code words to the code. In summary, the simplex code $\mathcal{S}(m)$ yields the Reed–Muller code $\mathcal{R}(1, m)$ by attaching a zero to all code words in the first position and adding all inverted vectors to the code (Bossert, 1999).

If we compare this code construction scheme with the relationship between simplex codes $\mathcal{S}(m)$ and Hadamard matrices $\mathbf{H}_m$, we recognise that the rows of the $2^{m+1} \times 2^m$ matrix

$$\begin{pmatrix} \mathbf{H}_m \\ -\mathbf{H}_m \end{pmatrix}$$

subjected to the mapping $1 \mapsto 0$ and $-1 \mapsto 1$ yield the code vectors of the first-order Reed–Muller code $\mathcal{R}(1, m)$.

This observation in combination with the orthogonality property

$$\mathbf{H}_m \mathbf{H}_m = 2^m \mathbf{I}_{2^m}$$

of the Hadamard matrix $\mathbf{H}_m$ with the $2^m \times 2^m$ identity matrix $\mathbf{I}_{2^m}$ can be used to implement a *soft-decision decoding* algorithm. To this end, we assume that the received vector

$$\mathbf{r} = \mathbf{x} + \mathbf{e}$$

corresponds to the real-valued output signal of an AWGN channel (see also Figure 1.5) with n-dimensional normally distributed noise vector $\mathbf{e}$.[15] The bipolar signal vector $\mathbf{x} = \mathbf{x}_j$ with components ± 1 is given by the jth row of the matrix

$$\begin{pmatrix} \mathbf{H}_m \\ -\mathbf{H}_m \end{pmatrix}.$$

This bipolar signal vector $\mathbf{x}$ is obtained from the binary code vector $\mathbf{b}$ by the mapping $1 \mapsto -1$ and $0 \mapsto 1$. With the help of the 2^{m+1}-dimensional row vector

$$\mathbf{v} = (0, \ldots, 0, 1, 0, \ldots, 0),$$

the components of which are 0 except for the jth component which is equal to 1, we can express the bipolar signal vector according to

$$\mathbf{x} = \mathbf{v} \begin{pmatrix} \mathbf{H}_m \\ -\mathbf{H}_m \end{pmatrix}.$$

Within the soft-decision decoder architecture the real-valued received vector $\mathbf{r}$ is transformed with the Hadamard matrix $\mathbf{H}_m$, leading to

$$\begin{aligned}
\mathbf{r}\,\mathbf{H}_m &= \left[\mathbf{v} \begin{pmatrix} \mathbf{H}_m \\ -\mathbf{H}_m \end{pmatrix} + \mathbf{e} \right] \mathbf{H}_m \\
&= \mathbf{v} \begin{pmatrix} \mathbf{H}_m \\ -\mathbf{H}_m \end{pmatrix} \mathbf{H}_m + \mathbf{e}\,\mathbf{H}_m \\
&= \mathbf{v} \begin{pmatrix} \mathbf{H}_m \mathbf{H}_m \\ -\mathbf{H}_m \mathbf{H}_m \end{pmatrix} + \mathbf{e}\,\mathbf{H}_m \\
&= \mathbf{v} \begin{pmatrix} 2^m \mathbf{I}_{2^m} \\ -2^m \mathbf{I}_{2^m} \end{pmatrix} + \mathbf{e}\,\mathbf{H}_m \\
&= 2^m\,\mathbf{v} \begin{pmatrix} \mathbf{I}_{2^m} \\ -\mathbf{I}_{2^m} \end{pmatrix} + \mathbf{e}\,\mathbf{H}_m \\
&= \pm 2^m\,\mathbf{v} + \mathbf{e}\,\mathbf{H}_m.
\end{aligned}$$

Because of $\mathbf{r}\,\mathbf{H}_m = \pm 2^m\,\mathbf{v} + \mathbf{e}\,\mathbf{H}_m$, the soft-decision decoder searches for the largest modulus of all components in the transformed received vector $\mathbf{r}\,\mathbf{H}_m$. This component in conjunction with the respective sign delivers the decoded signal vector $\hat{\mathbf{x}}$ or code vector $\hat{\mathbf{b}}$. The transform $\mathbf{r}\,\mathbf{H}_m$ can be efficiently implemented with the help of the fast Hadamard transform (FHT). In Figure 2.33 this soft-decision decoding is compared with the optimal hard-decision minimum distance decoding of a first-order Reed–Muller code $\mathcal{R}(1, 4)$.

---

[15]Here, the arithmetics are carried out in the vector space $\mathbb{R}^n$.

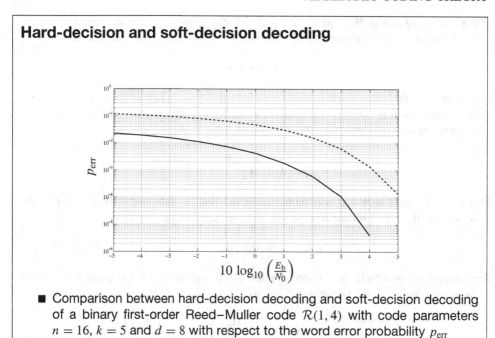

Figure 2.33: Hard-decision and soft-decision decoding of a binary first-order Reed–Muller code $\mathcal{R}(1,4)$

The FHT is used, for example, in UMTS (Holma and Toskala, 2004) receivers for the decoding of TFCI (transport format combination indicator) symbols which are encoded with the help of a subset of a second-order Reed–Muller code $\mathcal{R}(2,5)$ with code word length $n = 2^5 = 32$ (see, for example, (3GPP, 1999)). Reed–Muller codes have also been used in deep-space explorations, e.g. the first-order Reed–Muller code $\mathcal{R}(1,5)$ in the Mariner spacecraft (Costello *et al.*, 1998).

Higher-order Reed–Muller Codes $\mathcal{R}(r, m)$ The construction principle of binary first-order Reed–Muller codes $\mathcal{R}(1, m)$ can be extended to higher-order Reed–Muller codes $\mathcal{R}(r, m)$. For this purpose we consider the 5×16 generator matrix

$$\mathbf{G} = \begin{pmatrix} 1 & 1 & 1 & 1 & 1 & 1 & 1 & 1 & 1 & 1 & 1 & 1 & 1 & 1 & 1 & 1 \\ 0 & 0 & 0 & 0 & 0 & 0 & 0 & 0 & 1 & 1 & 1 & 1 & 1 & 1 & 1 & 1 \\ 0 & 0 & 0 & 0 & 1 & 1 & 1 & 1 & 0 & 0 & 0 & 0 & 1 & 1 & 1 & 1 \\ 0 & 0 & 1 & 1 & 0 & 0 & 1 & 1 & 0 & 0 & 1 & 1 & 0 & 0 & 1 & 1 \\ 0 & 1 & 0 & 1 & 0 & 1 & 0 & 1 & 0 & 1 & 0 & 1 & 0 & 1 & 0 & 1 \end{pmatrix}$$

of a first-order Reed–Muller code $\mathcal{R}(1, 4)$. The rows of this matrix correspond to the Boolean functions f_0, f_1, f_2, f_3 and f_4 shown in Figure 2.34. If we also consider the

ALGEBRAIC CODING THEORY

Boolean functions for the definition of a first-order Reed–Muller code $\mathcal{R}(1,4)$

x_1	0	0	0	0	0	0	0	0	1	1	1	1	1	1	1	1
x_2	0	0	0	0	1	1	1	1	0	0	0	0	1	1	1	1
x_3	0	0	1	1	0	0	1	1	0	0	1	1	0	0	1	1
x_4	0	1	0	1	0	1	0	1	0	1	0	1	0	1	0	1
f_0	1	1	1	1	1	1	1	1	1	1	1	1	1	1	1	1
f_1	0	0	0	0	0	0	0	0	1	1	1	1	1	1	1	1
f_2	0	0	0	0	1	1	1	1	0	0	0	0	1	1	1	1
f_3	0	0	1	1	0	0	1	1	0	0	1	1	0	0	1	1
f_4	0	1	0	1	0	1	0	1	0	1	0	1	0	1	0	1

■ Boolean functions with $x_1, x_2, x_3, x_4 \in \{0, 1\}$

$$f_0(x_1, x_2, x_3, x_4) = 1$$
$$f_1(x_1, x_2, x_3, x_4) = x_1$$
$$f_2(x_1, x_2, x_3, x_4) = x_2$$
$$f_3(x_1, x_2, x_3, x_4) = x_3$$
$$f_4(x_1, x_2, x_3, x_4) = x_4$$

Figure 2.34: Boolean functions for the definition of a first-order Reed–Muller code $\mathcal{R}(1,4)$

additional Boolean functions $f_{1,2}$, $f_{1,3}$, $f_{1,4}$, $f_{2,3}$, $f_{2,4}$ and $f_{3,4}$ in Figure 2.35, we obtain the 11×16 generator matrix

$$\mathbf{G} = \begin{pmatrix} 1 & 1 & 1 & 1 & 1 & 1 & 1 & 1 & 1 & 1 & 1 & 1 & 1 & 1 & 1 & 1 \\ 0 & 0 & 0 & 0 & 0 & 0 & 0 & 0 & 1 & 1 & 1 & 1 & 1 & 1 & 1 & 1 \\ 0 & 0 & 0 & 0 & 1 & 1 & 1 & 1 & 0 & 0 & 0 & 0 & 1 & 1 & 1 & 1 \\ 0 & 0 & 1 & 1 & 0 & 0 & 1 & 1 & 0 & 0 & 1 & 1 & 0 & 0 & 1 & 1 \\ 0 & 1 & 0 & 1 & 0 & 1 & 0 & 1 & 0 & 1 & 0 & 1 & 0 & 1 & 0 & 1 \\ 0 & 0 & 0 & 0 & 0 & 0 & 0 & 0 & 0 & 0 & 0 & 0 & 1 & 1 & 1 & 1 \\ 0 & 0 & 0 & 0 & 0 & 0 & 0 & 0 & 0 & 1 & 1 & 0 & 0 & 1 & 1 \\ 0 & 0 & 0 & 0 & 0 & 0 & 0 & 0 & 1 & 0 & 1 & 0 & 1 & 0 & 1 \\ 0 & 0 & 0 & 0 & 0 & 1 & 1 & 0 & 0 & 0 & 0 & 0 & 0 & 1 & 1 \\ 0 & 0 & 0 & 0 & 1 & 0 & 1 & 0 & 0 & 0 & 0 & 0 & 1 & 0 & 1 \\ 0 & 0 & 0 & 1 & 0 & 0 & 0 & 1 & 0 & 0 & 0 & 1 & 0 & 0 & 0 & 1 \end{pmatrix}$$

of the second-order Reed–Muller code $\mathcal{R}(2,4)$ from the table in Figure 2.35.

Boolean functions for the definition of a second-order Reed–Muller code $\mathcal{R}(2,4)$

x_1	0	0	0	0	0	0	0	0	1	1	1	1	1	1	1	1
x_2	0	0	0	0	1	1	1	1	0	0	0	0	1	1	1	1
x_3	0	0	1	1	0	0	1	1	0	0	1	1	0	0	1	1
x_4	0	1	0	1	0	1	0	1	0	1	0	1	0	1	0	1
f_0	1	1	1	1	1	1	1	1	1	1	1	1	1	1	1	1
f_1	0	0	0	0	0	0	0	0	1	1	1	1	1	1	1	1
f_2	0	0	0	0	1	1	1	1	0	0	0	0	1	1	1	1
f_3	0	0	1	1	0	0	1	1	0	0	1	1	0	0	1	1
f_4	0	1	0	1	0	1	0	1	0	1	0	1	0	1	0	1
$f_{1,2}$	0	0	0	0	0	0	0	0	0	0	0	0	1	1	1	1
$f_{1,3}$	0	0	0	0	0	0	0	0	0	0	1	1	0	0	1	1
$f_{1,4}$	0	0	0	0	0	0	0	0	0	1	0	1	0	1	0	1
$f_{2,3}$	0	0	0	0	0	0	1	1	0	0	0	0	0	0	1	1
$f_{2,4}$	0	0	0	0	0	1	0	1	0	0	0	0	0	1	0	1
$f_{3,4}$	0	0	0	1	0	0	0	1	0	0	0	1	0	0	0	1

■ Boolean functions with $x_1, x_2, x_3, x_4 \in \{0,1\}$

$$f_{1,2}(x_1, x_2, x_3, x_4) = x_1 x_2$$

$$f_{1,3}(x_1, x_2, x_3, x_4) = x_1 x_3$$

$$f_{1,4}(x_1, x_2, x_3, x_4) = x_1 x_4$$

$$f_{2,3}(x_1, x_2, x_3, x_4) = x_2 x_3$$

$$f_{2,4}(x_1, x_2, x_3, x_4) = x_2 x_4$$

$$f_{3,4}(x_1, x_2, x_3, x_4) = x_3 x_4$$

Figure 2.35: Boolean functions for the definition of a second-order Reed–Muller code $\mathcal{R}(2,4)$

This code construction methodology can be extended to the definition of higher-order Reed–Muller codes $\mathcal{R}(r,m)$ based on the Boolean functions

$$x_1^{i_1} x_2^{i_2} \cdots x_m^{i_m}$$

with

$$i_1 + i_2 + \cdots + i_m \leq r$$

> **Reed–Muller code $\mathcal{R}(r, m)$**
>
> - The Reed–Muller code $\mathcal{R}(r, m)$ is a binary code that is able to correct more than one error.
> - Code word length
> $$n = 2^m \qquad (2.54)$$
> - Minimum Hamming distance
> $$d = 2^{m-r} \qquad (2.55)$$
> - Code rate
> $$R = \frac{1 + \binom{m}{1} + \binom{m}{2} + \cdots + \binom{m}{r}}{2^m} \qquad (2.56)$$

Figure 2.36: Reed–Muller code $\mathcal{R}(r, m)$

and $i_j \in \{0, 1\}$. These Boolean functions are used to define the rows of the generator matrix **G**. The resulting Reed–Muller code $\mathcal{R}(r, m)$ of order r in Figure 2.36 is characterised by the following code parameters

$$n = 2^m,$$
$$k = 1 + \binom{m}{1} + \binom{m}{2} + \cdots + \binom{m}{r},$$
$$d = 2^{m-r}.$$

There also exists a recursive definition of Reed–Muller codes based on Plotkin's $(\mathbf{u}|\mathbf{v})$ code construction (Bossert, 1999)

$$\mathcal{R}(r+1, m+1) = \{(\mathbf{u}|\mathbf{u}+\mathbf{v}) : \mathbf{u} \in \mathcal{R}(r+1, m) \wedge \mathbf{v} \in \mathcal{R}(r, m)\}.$$

The code vectors $\mathbf{b} = (\mathbf{u}|\mathbf{u}+\mathbf{v})$ of the Reed–Muller code $\mathcal{R}(r+1, m+1)$ are obtained from all possible combinations of the code vectors $\mathbf{u} \in \mathcal{R}(r+1, m)$ of the Reed–Muller code $\mathcal{R}(r+1, m)$ and $\mathbf{v} \in \mathcal{R}(r, m)$ of the Reed–Muller code $\mathcal{R}(r, m)$. Furthermore, the dual code of the Reed–Muller code $\mathcal{R}(r, m)$ yields the Reed–Muller code

$$\mathcal{R}^\perp(r, m) = \mathcal{R}(m - r - 1, m).$$

Finally, Figure 2.37 summarises the relationship between first-order Reed–Muller codes $\mathcal{R}(1, m)$, binary Hamming codes $\mathcal{H}(m)$ and simplex codes $\mathcal{S}(m)$ (Bossert, 1999).

> **First-order Reed–Muller codes $\mathcal{R}(1, m)$, binary Hamming codes $\mathcal{H}(m)$ and simplex codes $\mathcal{S}(m)$**
>
> - Reed–Muller code $\mathcal{R}(1, m)$ and extended Hamming code $\mathcal{H}'(m)$
>   - (i) $\mathcal{R}(1, m) = \mathcal{H}'^{\perp}(m)$
>   - (ii) $\mathcal{H}'(m) = \mathcal{R}^{\perp}(1, m) = \mathcal{R}(m - 2, m)$
> - Hamming code $\mathcal{H}(m)$ and simplex code $\mathcal{S}(m)$
>   - (i) $\mathcal{S}(m) = \mathcal{H}^{\perp}(m)$
>   - (ii) $\mathcal{H}(m) = \mathcal{S}^{\perp}(m)$
> - Simplex code $\mathcal{S}(m)$ and Reed–Muller code $\mathcal{R}(1, m)$
>   - (i) $\mathcal{S}(m) \mapsto \mathcal{R}(1, m)$
>   1. All code words from $\mathcal{S}(m)$ are extended by the first position 0.
>   2. $\mathcal{R}(1, m)$ consists of all corresponding code words including the inverted code words.
>   - (ii) $\mathcal{R}(1, m) \mapsto \mathcal{S}(m)$
>   1. All code words with the first component 0 are removed from $\mathcal{R}(1, m)$.
>   2. The first component is deleted.

Figure 2.37: Relationship between first-order Reed–Muller codes $\mathcal{R}(1, m)$, binary Hamming codes $\mathcal{H}(m)$ and simplex codes $\mathcal{S}(m)$

2.3 Cyclic Codes

Linear block codes make it possible efficiently to implement the encoding of information words with the help of the generator matrix **G**. In general, however, the problem of decoding an arbitrary linear block code is difficult. For that reason we now turn to cyclic codes as special linear block codes. These codes introduce further algebraic properties in order to be able to define more efficient algebraic decoding algorithms (Berlekamp, 1984; Lin and Costello, 2004; Ling and Xing, 2004).

2.3.1 Definition of Cyclic Codes

A *cyclic code* is characterised as a linear block code $\mathbb{B}(n, k, d)$ with the additional property that for each code word
$$\mathbf{b} = (b_0, b_1, \ldots, b_{n-2}, b_{n-1})$$

ALGEBRAIC CODING THEORY

all cyclically shifted words

$$(b_{n-1}, b_0, \ldots, b_{n-3}, b_{n-2}),$$
$$(b_{n-2}, b_{n-1}, \ldots, b_{n-4}, b_{n-3}),$$
$$\vdots$$
$$(b_2, b_3, \ldots, b_0, b_1),$$
$$(b_1, b_2, \ldots, b_{n-1}, b_0)$$

are also valid code words of $\mathbb{B}(n, k, d)$ (Lin and Costello, 2004; Ling and Xing, 2004). This property can be formulated concisely if a code word $\mathbf{b} \in \mathbb{F}_q^n$ is represented as a polynomial

$$b(z) = b_0 + b_1 z + \cdots + b_{n-2} z^{n-2} + b_{n-1} z^{n-1}$$

over the finite field $\mathbb{F}_q$.[16] A cyclic shift

$$(b_0, b_1, \ldots, b_{n-2}, b_{n-1}) \mapsto (b_{n-1}, b_0, b_1, \ldots, b_{n-2})$$

of the code polynomial $b(z) \in \mathbb{F}_q[z]$ can then be expressed as

$$b_0 + b_1 z + \cdots + b_{n-2} z^{n-2} + b_{n-1} z^{n-1} \mapsto b_{n-1} + b_0 z + b_1 z^2 + \cdots + b_{n-2} z^{n-1}.$$

Because of

$$b_{n-1} + b_0 z + b_1 z^2 + \cdots + b_{n-2} z^{n-1} = z\,b(z) - b_{n-1}\left(z^n - 1\right)$$

and by observing that a code polynomial $b(z)$ is of maximal degree $n-1$, we represent the cyclically shifted code polynomial modulo $z^n - 1$, i.e.

$$b_{n-1} + b_0 z + b_1 z^2 + \cdots + b_{n-2} z^{n-1} \equiv z\,b(z) \mod z^n - 1.$$

Cyclic codes $\mathbb{B}(n, k, d)$ therefore fulfil the following algebraic property

$$b(z) \in \mathbb{B}(n, k, d) \quad \Leftrightarrow \quad z\,b(z) \mod z^n - 1 \in \mathbb{B}(n, k, d).$$

For that reason – if not otherwise stated – we consider polynomials in the factorial ring $\mathbb{F}_q[z]/(z^n - 1)$. Figure 2.38 summarises the definition of cyclic codes.

Similarly to general linear block codes, which can be defined by the generator matrix **G** or the corresponding parity-check matrix **H**, cyclic codes can be characterised by the generator polynomial $g(z)$ and the parity-check polynomial $h(z)$, as we will show in the following (Berlekamp, 1984; Bossert, 1999; Lin and Costello, 2004; Ling and Xing, 2004).

2.3.2 Generator Polynomial

A linear block code $\mathbb{B}(n, k, d)$ is defined by the $k \times n$ generator matrix

$$\mathbf{G} = \begin{pmatrix} \mathbf{g}_0 \\ \mathbf{g}_1 \\ \vdots \\ \mathbf{g}_{k-1} \end{pmatrix} = \begin{pmatrix} g_{0,0} & g_{0,1} & \cdots & g_{0,n-1} \\ g_{1,0} & g_{1,1} & \cdots & g_{1,n-1} \\ \vdots & \vdots & \ddots & \vdots \\ g_{k-1,0} & g_{k-1,1} & \cdots & g_{k-1,n-1} \end{pmatrix}$$

---

[16]Polynomials over finite fields are explained in Section A.3 in Appendix A.

> **Definition of cyclic codes**
>
> ■ Each code word $\mathbf{b} = (b_0, b_1, \ldots, b_{n-2}, b_{n-1})$ of a cyclic code $\mathbb{B}(n,k,d)$ is represented by the polynomial
>
> $$b(z) = b_0 + b_1 z + \cdots + b_{n-2} z^{n-2} + b_{n-1} z^{n-1} \qquad (2.57)$$
>
> ■ All cyclic shifts of a code word $\mathbf{b}$ are also valid code words in the cyclic code $\mathbb{B}(n,k,d)$, i.e.
>
> $$b(z) \in \mathbb{B}(n,k,d) \quad \Leftrightarrow \quad z\,b(z) \mod z^n - 1 \in \mathbb{B}(n,k,d) \qquad (2.58)$$

Figure 2.38: Definition of cyclic codes

with k linearly independent basis vectors $\mathbf{g}_0, \mathbf{g}_1, \ldots, \mathbf{g}_{k-1}$ which themselves are valid code vectors of the linear block code $\mathbb{B}(n,k,d)$. Owing to the algebraic properties of a cyclic code there exists a unique polynomial

$$g(z) = g_0 + g_1 z + \cdots + g_{n-k-1} z^{n-k-1} + g_{n-k} z^{n-k}$$

of minimal degree $\deg(g(z)) = n - k$ with $g_{n-k} = 1$ such that the corresponding generator matrix can be written as

$$\mathbf{G} = \begin{pmatrix} g_0 & g_1 & \cdots & g_{n-k} & 0 & \cdots & 0 & 0 & \cdots & 0 & 0 \\ 0 & g_0 & \cdots & g_{n-k-1} & g_{n-k} & \cdots & 0 & 0 & \cdots & 0 & 0 \\ \vdots & \vdots & \ddots & \vdots & \vdots & \ddots & \vdots & \vdots & \ddots & \vdots & \vdots \\ 0 & 0 & \cdots & 0 & 0 & \cdots & g_0 & g_1 & \cdots & g_{n-k} & 0 \\ 0 & 0 & \cdots & 0 & 0 & \cdots & 0 & g_0 & \cdots & g_{n-k-1} & g_{n-k} \end{pmatrix}.$$

This polynomial $g(z)$ is called the *generator polynomial* of the cyclic code $\mathbb{B}(n,k,d)$ (Berlekamp, 1984; Bossert, 1999; Lin and Costello, 2004; Ling and Xing, 2004). The rows of the generator matrix $\mathbf{G}$ are obtained from the generator polynomial $g(z)$ and all cyclic shifts $z\,g(z), z^2\,g(z), \ldots, z^{k-1}\,g(z)$ which correspond to valid code words of the cyclic code. Formally, we can write the generator matrix as

$$\mathbf{G} = \begin{pmatrix} g(z) \\ z\,g(z) \\ \vdots \\ z^{k-2}\,g(z) \\ z^{k-1}\,g(z) \end{pmatrix}.$$

ALGEBRAIC CODING THEORY

In view of the encoding rule for linear block codes $\mathbf{b} = \mathbf{u}\,\mathbf{G}$, we can write

$$(u_0, u_1, \ldots, u_{k-1}) \begin{pmatrix} g(z) \\ z\,g(z) \\ \vdots \\ z^{k-2}\,g(z) \\ z^{k-1}\,g(z) \end{pmatrix} = u_0\,g(z) + u_1\,z\,g(z) + \cdots + u_{k-1}\,z^{k-1}\,g(z).$$

For the information word $\mathbf{u} = (u_0, u_1, \cdots, u_{k-1})$ we define the corresponding information polynomial

$$u(z) = u_0 + u_1\,z + u_2\,z^2 + \cdots + u_{k-1}\,z^{k-1}.$$

This information polynomial $u(z)$ can thus be encoded according to the polynomial multiplication

$$b(z) = u(z)\,g(z).$$

Because of $b(z) = u(z)\,g(z)$, the generator polynomial $g(z)$ divides every code polynomial $b(z)$. If $g(z)$ does not divide a given polynomial, this polynomial is not a valid code polynomial, i.e.

$$g(z) \mid b(z) \quad \Leftrightarrow \quad b(z) \in \mathbb{B}(n, k, d)$$

or equivalently

$$b(z) \equiv 0 \mod g(z) \quad \Leftrightarrow \quad b(z) \in \mathbb{B}(n, k, d).$$

The simple multiplicative encoding rule $b(z) = u(z)\,g(z)$, however, does not lead to a systematic encoding scheme where all information symbols are found at specified positions.

By making use of the relation $b(z) \equiv 0$ modulo $g(z)$, we can derive a systematic encoding scheme. To this end, we place the k information symbols u_i in the k upper positions in the code word

$$\mathbf{b} = (b_0, b_1, \ldots, b_{n-k-1}, \underbrace{u_0, u_1, \ldots, u_{k-1}}_{=\mathbf{u}}).$$

The remaining code symbols $b_0, b_1, \ldots, b_{n-k-1}$ correspond to the $n - k$ parity-check symbols which have to be determined. By applying the condition $b(z) \equiv 0$ modulo $g(z)$ to the code polynomial

$$b(z) = b_0 + b_1\,z + \cdots + b_{n-k-1}\,z^{n-k-1} + u_0\,z^{n-k} + u_1\,z^{n-k+1} + \cdots + u_{k-1}\,z^{n-1}$$
$$= b_0 + b_1\,z + \cdots + b_{n-k-1}\,z^{n-k-1} + z^{n-k}\,u(z)$$

we obtain

$$b_0 + b_1\,z + \cdots + b_{n-k-1}\,z^{n-k-1} \equiv -z^{n-k}\,u(z) \mod g(z).$$

The parity-check symbols $b_0, b_1, \ldots, b_{n-k-1}$ are determined from the remainder of the division of the shifted information polynomial $z^{n-k}\,u(z)$ by the generator polynomial $g(z)$. Figure 2.39 summarises the non-systematic and systematic encoding schemes for cyclic codes.

It can be shown that the binary Hamming code in Figure 2.29 is equivalent to a cyclic code. The cyclic binary $(7, 4)$ Hamming code, for example, is defined by the generator polynomial

$$g(z) = 1 + z + z^3 \in \mathbb{F}_2[z].$$

> **Generator polynomial**
>
> - The cyclic code $\mathbb{B}(n, k, d)$ is defined by the unique generator polynomial
> $$g(z) = g_0 + g_1 z + \cdots + g_{n-k-1} z^{n-k-1} + g_{n-k} z^{n-k} \qquad (2.59)$$
> of minimal degree $\deg(g(z)) = n - k$ with $g_{n-k} = 1$.
> - Non-systematic encoding
> $$b(z) = u(z)\, g(z) \qquad (2.60)$$
> - Systematic encoding
> $$b_0 + b_1 z + \cdots + b_{n-k-1} z^{n-k-1} \equiv -z^{n-k}\, u(z) \mod g(z) \qquad (2.61)$$

Figure 2.39: Encoding of a cyclic code with the help of the generator polynomial $g(z)$

The non-systematic and systematic encoding schemes for this cyclic binary Hamming code are illustrated in Figure 2.40.

Cyclic Redundancy Check

With the help of the generator polynomial $g(z)$ of a cyclic code $\mathbb{B}(n, k, d)$, the so-called *cyclic redundancy check* (CRC) can be defined for the detection of errors (Lin and Costello, 2004). Besides the detection of $e_{\text{det}} = d - 1$ errors by a cyclic code $\mathbb{B}(n, k, d)$ with minimum Hamming distance d, cyclic error bursts can also be detected. With a generator polynomial $g(z)$ of degree $\deg(g(z)) = n - k$, all cyclic error bursts of length

$$\ell_{\text{burst}} \leq n - k$$

can be detected (Jungnickel, 1995). This can be seen by considering the error model $r(z) = b(z) + e(z)$ with the received polynomial $r(z)$, the code polynomial $b(z)$ and the error polynomial $e(z)$ (see also Figure 2.46). Errors can be detected as long as the parity-check equation

$$g(z) \mid r(z) \quad \Leftrightarrow \quad r(z) \in \mathbb{B}(n, k, d)$$

of the cyclic code $\mathbb{B}(n, k, d)$ is fulfilled. Since $g(z)|b(z)$, all errors for which the error polynomial $e(z)$ is not divisible by the generator polynomial $g(z)$ can be detected. As long as the degree $\deg(e(z))$ is smaller than $\deg(g(z)) = n - k$, the error polynomial $e(z)$ cannot be divided by the generator polynomial. This is also true if cyclically shifted variants $z^i\, e(z)$ of such an error polynomial are considered. Since for an error burst of length ℓ_{burst} the degree of the error polynomial is equal to $\ell_{\text{burst}} - 1$, the error detection is possible if

$$\deg(e(z)) = \ell_{\text{burst}} - 1 < n - k = \deg(g(z)).$$

ALGEBRAIC CODING THEORY

Cyclic binary Hamming code

- The generator polynomial of a cyclic binary (7,4) Hamming code is given by
$$g(z) = 1 + z + z^3 \in \mathbb{F}_2[z]$$

- Non-systematic encoding of the information polynomial $u(z) = 1 + z^3$ yields
$$b(z) = u(z)\,g(z) = \left(1 + z^3\right)\left(1 + z + z^3\right) = 1 + z + z^4 + z^6$$

- Systematic encoding of the information polynomial $u(z) = 1 + z^3$ yields
$$z^3\left(z^3 + 1\right) \equiv z^2 + z \quad \text{mod } 1 + z + z^3$$
leading to the code polynomial
$$b(z) = z + z^2 + z^3 + z^6$$

Figure 2.40: Cyclic binary $(7, 4)$ Hamming code

Figure 2.41 gives some commonly used CRC generator polynomials (Lin and Costello, 2004).

2.3.3 Parity-Check Polynomial

It can be shown that the generator polynomial $g(z)$ of a cyclic code divides the polynomial $z^n - 1$ in the polynomial ring $\mathbb{F}_q[z]$, i.e.
$$g(z)\,h(z) = z^n - 1.$$
In the factorial ring $\mathbb{F}_q[z]/(z^n - 1)$ this amounts to $g(z)\,h(z) = 0$ or equivalently
$$g(z)\,h(z) \equiv 0 \quad \text{mod } z^n - 1.$$
The polynomial
$$h(z) = \frac{z^n - 1}{g(z)}$$
is the so-called *parity-check polynomial* (Berlekamp, 1984; Bossert, 1999; Lin and Costello, 2004; Ling and Xing, 2004). Since every code polynomial $b(z)$ is a multiple of the generator polynomial $g(z)$, the parity-check equation can also be written as (see Figure 2.42)
$$b(z)\,h(z) \equiv 0 \quad \text{mod } z^n - 1 \quad \Leftrightarrow \quad b(z) \in \mathbb{B}(n,k,d).$$
This parity-check equation is in correspondence with the matrix equation
$$\mathbf{H}\,\mathbf{b}^\mathrm{T} = \mathbf{0} \quad \Leftrightarrow \quad \mathbf{b} \in \mathbb{B}(n,k,d).$$

Cyclic redundancy check

- The generator polynomial $g(z)$ can be used for the detection of errors by making use of the cyclic redundancy check (CRC)

$$r(z) \equiv 0 \mod g(z) \quad \Leftrightarrow \quad r(z) \in \mathbb{B}(n, k, d) \qquad (2.62)$$

- The received polynomial

$$r(z) = r_0 + r_1 z + \cdots + r_{n-2} z^{n-2} + r_{n-1} z^{n-1} \qquad (2.63)$$

of the received word $\mathbf{r} = (r_0, r_1, \ldots, r_{n-2}, r_{n-1})$ is divided by the generator polynomial $g(z)$. If the remainder is zero, the received word is a valid code word, otherwise transmission errors have been detected. In the so-called ARQ (automatic repeat request) scheme, the receiver can prompt the transmitter to retransmit the code word.

- The following table shows some generator polynomials used for CRC.

CRC-12	$g(z) = 1 + z + z^2 + z^3 + z^4 + z^{12}$
CRC-16	$g(z) = 1 + z^2 + z^{15} + z^{16}$
CRC-CCITT	$g(z) = 1 + z^5 + z^{12} + z^{16}$

Figure 2.41: Cyclic redundancy check

Parity-check polynomial

- The parity-check polynomial $h(z)$ of a cyclic code $\mathbb{B}(n, k, d)$ with generator polynomial $g(z)$ is given by

$$h(z) = \frac{z^n - 1}{g(z)} \qquad (2.64)$$

- Parity-check equation

$$b(z) h(z) \equiv 0 \mod z^n - 1 \quad \Leftrightarrow \quad b(z) \in \mathbb{B}(n, k, d) \qquad (2.65)$$

Figure 2.42: Parity-check polynomial $h(z)$

Parity-check polynomial of the cyclic binary Hamming code

- The generator polynomial of the cyclic binary (7,4) Hamming code is given by
$$g(z) = 1 + z + z^3 \in \mathbb{F}_2[z]$$

- The parity-check polynomial of this cyclic binary Hamming code is equal to
$$h(z) = 1 + z + z^2 + z^4$$

Figure 2.43: Parity-check polynomial of the cyclic binary $(7, 4)$ Hamming code

for general linear block codes. Figure 2.43 gives the parity-check polynomial of the cyclic $(7, 4)$ binary Hamming code.

Because of $\deg(g(z)) = n - k$ and $\deg(z^n - 1) = n = \deg(g(z)) + \deg(h(z))$, the degree of the parity-check polynomial is given by $\deg(h(z)) = k$. Taking into account the normalisation $g_{n-k} = 1$, we see that $h_k = 1$, i.e.

$$h(z) = h_0 + h_1 z + \cdots + h_{k-1} z^{k-1} + z^k.$$

Based on the parity-check polynomial $h(z)$, yet another systematic encoding algorithm can be derived. To this end, we make use of $g(z) h(z) = z^n - 1$ and $b(z) = u(z) g(z)$. This yields

$$b(z) h(z) = u(z) g(z) h(z) = u(z) \left(z^n - 1 \right) = -u(z) + z^n u(z).$$

The degree of the information polynomial $u(z)$ is bounded by $\deg(u(z)) \leq k - 1$, whereas the minimal exponent of the polynomial $z^n u(z)$ is n. Therefore, the polynomial $b(z) h(z)$ does not contain the exponentials $z^k, z^{k+1}, \ldots, z^{n-1}$. This yields the $n - k$ parity-check equations

$$b_0 h_k + b_1 h_{k-1} + \cdots + b_k h_0 = 0,$$
$$b_1 h_k + b_2 h_{k-1} + \cdots + b_{k+1} h_0 = 0,$$
$$\vdots$$
$$b_{n-k-2} h_k + b_{n-k-1} h_{k-1} + \cdots + b_{n-2} h_0 = 0,$$
$$b_{n-k-1} h_k + b_{n-k} h_{k-1} + \cdots + b_{n-1} h_0 = 0$$

which can be written as the discrete convolution

$$\sum_{j=0}^{k} h_j b_{i-j} = 0$$

for $k \le i \le n-1$. This corresponds to the matrix equation $\mathbf{H}\mathbf{b}^T = \mathbf{0}$ for general linear block codes with the $(n-k) \times n$ parity-check matrix

$$\mathbf{H} = \begin{pmatrix} h_k & h_{k-1} & \cdots & h_0 & 0 & \cdots & 0 & 0 & \cdots & 0 & 0 \\ 0 & h_k & \cdots & h_1 & h_0 & \cdots & 0 & 0 & \cdots & 0 & 0 \\ \vdots & \vdots & \ddots & \vdots & \vdots & \ddots & \vdots & \vdots & \ddots & \vdots & \vdots \\ 0 & 0 & \cdots & 0 & 0 & \cdots & h_k & h_{k-1} & \cdots & h_0 & 0 \\ 0 & 0 & \cdots & 0 & 0 & \cdots & 0 & h_k & \cdots & h_1 & h_0 \end{pmatrix}.$$

For the systematic encoding scheme the k code symbols $b_{n-k}, b_{n-k+1}, \ldots, b_{n-1}$ are set equal to the respective information symbols $u_0, u_1, \ldots, u_{k-1}$. This yields the following system of equations taking into account the normalisation $h_k = 1$

$$b_0 + b_1 h_{k-1} + \cdots + b_k h_0 = 0,$$
$$b_1 + b_2 h_{k-1} + \cdots + b_{k+1} h_0 = 0,$$
$$\vdots$$
$$b_{n-k-2} + b_{n-k-1} h_{k-1} + \cdots + b_{n-2} h_0 = 0,$$
$$b_{n-k-1} + b_{n-k} h_{k-1} + \cdots + b_{n-1} h_0 = 0$$

which is recursively solved for the parity-check symbols $b_{n-k-1}, b_{n-k-2}, \ldots, b_1, b_0$. This leads to the systematic encoding scheme

$$b_{n-k-1} = -(b_{n-k} h_{k-1} + \cdots + b_{n-1} h_0),$$
$$b_{n-k-2} = -(b_{n-k-1} h_{k-1} + \cdots + b_{n-2} h_0),$$
$$\vdots$$
$$b_1 = -(b_2 h_{k-1} + \cdots + b_{k+1} h_0),$$
$$b_0 = -(b_1 h_{k-1} + \cdots + b_k h_0).$$

2.3.4 Dual Codes

Similarly to general linear block codes, the dual code $\mathbb{B}^\perp(n', k', d')$ of the cyclic code $\mathbb{B}(n, k, d)$ with generator polynomial $g(z)$ and parity-check polynomial $h(z)$ can be defined by changing the role of these polynomials (Jungnickel, 1995). Here, we have to take into account that the generator polynomial must be normalised such that the highest exponent has coefficient 1. To this end, we make use of the already derived $(n-k) \times n$ parity-check matrix

$$\mathbf{H} = \begin{pmatrix} h_k & h_{k-1} & \cdots & h_0 & 0 & \cdots & 0 & 0 & \cdots & 0 & 0 \\ 0 & h_k & \cdots & h_1 & h_0 & \cdots & 0 & 0 & \cdots & 0 & 0 \\ \vdots & \vdots & \ddots & \vdots & \vdots & \ddots & \vdots & \vdots & \ddots & \vdots & \vdots \\ 0 & 0 & \cdots & 0 & 0 & \cdots & h_k & h_{k-1} & \cdots & h_0 & 0 \\ 0 & 0 & \cdots & 0 & 0 & \cdots & 0 & h_k & \cdots & h_1 & h_0 \end{pmatrix}$$

ALGEBRAIC CODING THEORY

of a cyclic code. By comparing this matrix with the generator matrix

$$G = \begin{pmatrix} g_0 & g_1 & \cdots & g_{n-k} & 0 & \cdots & 0 & 0 & \cdots & 0 & 0 \\ 0 & g_0 & \cdots & g_{n-k-1} & g_{n-k} & \cdots & 0 & 0 & \cdots & 0 & 0 \\ \vdots & \vdots & \ddots & \vdots & \vdots & \ddots & \vdots & \vdots & \ddots & \vdots & \vdots \\ 0 & 0 & \cdots & 0 & 0 & \cdots & g_0 & g_1 & \cdots & g_{n-k} & 0 \\ 0 & 0 & \cdots & 0 & 0 & \cdots & 0 & g_0 & \cdots & g_{n-k-1} & g_{n-k} \end{pmatrix}$$

of a cyclic code, we observe that, besides different dimensions and a different order of the matrix elements, the structure of these matrices is similar. With this observation, the generator polynomial $g^\perp(z)$ of the dual cyclic code $\mathbb{B}^\perp(n', k', d')$ can be obtained from the reversed and normalised parity-check polynomial according to

$$g^\perp(z) = \frac{h_k + h_{k-1} z + \cdots + h_1 z^{k-1} + h_0 z^k}{h_0}$$

$$= z^k \frac{h_0 + h_1 z^{-1} + \cdots + h_{k-1} z^{-k+1} + h_k z^{-k}}{h_0}$$

$$= z^k \frac{h(z^{-1})}{h_0}.$$

Figure 2.44 summarises the definition of the dual cyclic code $\mathbb{B}^\perp(n', k', d')$ of the cyclic code $\mathbb{B}(n, k, d)$.

2.3.5 Linear Feedback Shift Registers

As we have seen, the systematic encoding of a cyclic code can be carried out by dividing the shifted information polynomial $z^{n-k} u(z)$ by the generator polynomial $g(z)$. The respective

---

Dual cyclic code

- Let $h(z)$ be the parity-check polynomial of the cyclic code $\mathbb{B}(n, k, d)$.
- The dual code $\mathbb{B}^\perp(n', k', d')$ is the cyclic code defined by the generator polynomial

$$g^\perp(z) = z^k \frac{h(z^{-1})}{h_0} \tag{2.66}$$

- For cyclic binary codes $h_0 = 1$, and therefore

$$g^\perp(z) = z^k h(z^{-1}) \tag{2.67}$$

Figure 2.44: Dual cyclic code $\mathbb{B}^\perp(n', k', d')$ of the cyclic code $\mathbb{B}(n, k, d)$

remainder yields the sought parity-check symbols. As is well known, polynomial division can be carried out with the help of *linear feedback shift registers*. Based on these linear feedback shift registers, efficient encoding and decoding architectures for cyclic codes can be derived (Berlekamp, 1984; Lin and Costello, 2004).

Encoding Architecture

For the information polynomial

$$u(z) = u_0 + u_1 z + \cdots + u_{k-2} z^{k-2} + u_{k-1} z^{k-1}$$

and the generator polynomial

$$g(z) = g_0 + g_1 z + \cdots + g_{n-k-1} z^{n-k-1} + z^{n-k}$$

the encoder architecture of a cyclic code over the finite field $\mathbb{F}_q$ can be derived by making use of a linear feedback shift register with $n - k$ registers as shown in Figure 2.45 (Neubauer, 2006b).

After the q-nary information symbols $u_{k-1}, u_{k-2}, \cdots, u_1, u_0$ have been shifted into the linear feedback shift register, the registers contain the components of the remainder

$$s(z) = s_0 + s_1 z + \cdots + s_{n-k-2} z^{n-k-2} + s_{n-k-1} z^{n-k-1}$$
$$\equiv z^{n-k} u(z) \mod g(z).$$

Besides an additional sign, this term yields the remainder that is needed for the systematic encoding of the information polynomial $u(z)$. Therefore, after k clock cycles, the linear feedback shift register contains the negative parity-check symbols within the registers that can subsequently be emitted by the shift register. In summary, this encoding architecture yields the code polynomial

$$b(z) = z^{n-k} u(z) - s(z).$$

Decoding Architecture

The linear feedback shift register can also be used for the decoding of a cyclic code. Starting from the polynomial channel model illustrated in Figure 2.46

$$r(z) = b(z) + e(z)$$

with the error polynomial

$$e(z) = e_0 + e_1 z + e_2 z^2 + \cdots + e_{n-1} z^{n-1}$$

the *syndrome polynomial* is defined according to

$$s(z) \equiv r(z) \mod g(z).$$

The degree of the syndrome polynomial $s(z)$ is bounded by $\deg(s(z)) \leq n - k - 1$.

ALGEBRAIC CODING THEORY

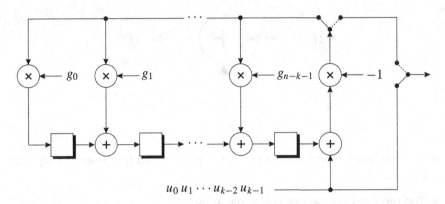

Encoder architecture for a cyclic code

- The sequence of information symbols $u_{k-1}, u_{k-2}, \cdots, u_1, u_0$ is shifted into the linear feedback shift register within the first k clock cycles (switch positions according to the solid lines).

- At the end of the kth clock cycle the registers contain the negative parity-check symbols.

- The parity-check symbols are emitted during the next $n-k$ clock cycles (switch positions according to the dashed lines).

Figure 2.45: Encoding of a cyclic code with a linear feedback shift register. Reproduced by permission of J. Schlembach Fachverlag

If the received polynomial $r(z)$ is error free and therefore corresponds to the transmitted code polynomial $b(z)$, the syndrome polynomial $s(z)$ is zero. If, however, the received polynomial $r(z)$ is disturbed, the syndrome polynomial

$$s(z) \equiv r(z) \equiv b(z) + e(z) \equiv e(z) \mod g(z)$$

exclusively depends on the error polynomial $e(z)$. Because of $s(z) \equiv r(z)$ modulo $g(z)$, the syndrome polynomial is obtained as the polynomial remainder of the division of the received polynomial $r(z)$ by the generator polynomial $g(z)$. This division operation can again be carried out by a linear feedback shift register as illustrated in Figure 2.47 (Neubauer, 2006b). Similarly to the syndrome decoding of linear block codes, we obtain the decoder architecture shown in Figure 2.48 which is now based on a linear feedback shift register and a table look-up procedure (Neubauer, 2006b). Further information about encoder and decoder architectures based on linear feedback shift registers can be found elsewhere (Berlekamp, 1984; Lin and Costello, 2004).

Polynomial channel model

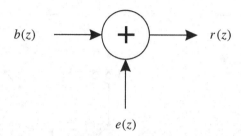

- The transmitted code polynomial $b(z)$ is disturbed by the error polynomial $e(z)$.
- The received polynomial $r(z)$ is given by

$$r(z) = b(z) + e(z) \tag{2.68}$$

- The syndrome polynomial

$$s(z) \equiv r(z) \mod g(z) \tag{2.69}$$

exclusively depends on the error polynomial $e(z)$.

Figure 2.46: Polynomial channel model

2.3.6 BCH Codes

We now turn to the so-called *BCH codes* as an important class of cyclic codes (Berlekamp, 1984; Bossert, 1999; Lin and Costello, 2004; Ling and Xing, 2004). These codes make it possible to derive an efficient algebraic decoding algorithm. In order to be able to define BCH codes, we first have to introduce zeros of a cyclic code.[17]

Zeros of Cyclic Codes

A cyclic code $\mathbb{B}(n, k, d)$ over the finite field $\mathbb{F}_q$ is defined by a unique generator polynomial $g(z)$ which divides each code polynomial $b(z) \in \mathbb{B}(n, k, d)$ according to $g(z) \mid b(z)$ or equivalently

$$b(z) \equiv 0 \mod g(z).$$

The generator polynomial itself is a divisor of the polynomial $z^n - 1$, i.e.

$$g(z) h(z) = z^n - 1.$$

---

[17]For further details about the arithmetics in finite fields the reader is referred to Section A.3 in Appendix A.

ALGEBRAIC CODING THEORY

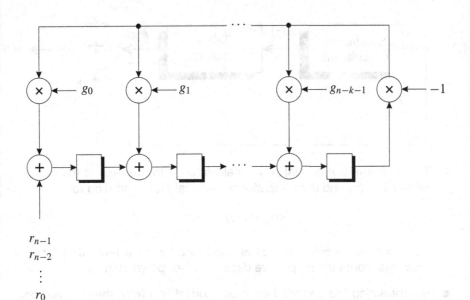

Decoder architecture for a cyclic code

- The sequence of received symbols $r_{n-1}, r_{n-2}, \ldots, r_1, r_0$ is shifted into the linear feedback shift register.
- At the end of the nth clock cycle the registers contain the syndrome polynomial coefficients $s_0, s_1, \ldots, s_{n-k-2}, s_{n-k-1}$ from left to right.

Figure 2.47: Decoding of a cyclic code with a linear feedback shift register. Reproduced by permission of J. Schlembach Fachverlag

In the following, we assume that the polynomial $z^n - 1$ has only single zeros which is equivalent to the condition (McEliece, 1987)

$$\gcd(q, n) = 1.$$

This means that the cardinality q of the finite field $\mathbb{F}_q$ and the code word length n have to be relatively prime. In this case, there exists a primitive nth root of unity $\alpha \in \mathbb{F}_{q^l}$ in a suitable extension field $\mathbb{F}_{q^l}$ with $\alpha^n = 1$ and $\alpha^\nu \neq 1$ for $\nu < n$. For the extension field $\mathbb{F}_{q^l}$ it follows that $n \mid q^l - 1$. With the help of such a primitive nth root of unity α, the n zeros of the polynomial $z^n - 1$ are given by $1, \alpha, \alpha^2, \ldots, \alpha^{n-1}$.

The polynomial $z^n - 1$ can thus be factored into

$$z^n - 1 = (z - 1)(z - \alpha)(z - \alpha^2) \cdots (z - \alpha^{n-1}).$$

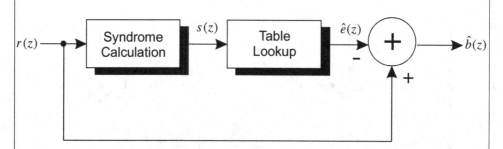

- The syndrome polynomial $s(z)$ is calculated with the help of the received polynomial $r(z)$ and the generator polynomial $g(z)$ according to

$$s(z) \equiv r(z) \mod g(z) \tag{2.70}$$

- The syndrome polynomial $s(z)$ is used to address a table that for each syndrome stores the respective decoded error polynomial $\hat{e}(z)$.

- By subtracting the decoded error polynomial $\hat{e}(z)$ from the received polynomial $r(z)$, the decoded code polynomial is obtained by

$$\hat{b}(z) = r(z) - \hat{e}(z) \tag{2.71}$$

Figure 2.48: Syndrome decoding of a cyclic code with a linear feedback shift register and a table look-up procedure. Reproduced by permission of J. Schlembach Fachverlag

Since $g(z)$ divides the polynomial $z^n - 1$, the generator polynomial of degree $\deg(g(z)) = n - k$ can be defined by the corresponding set of zeros $\alpha^{i_1}, \alpha^{i_2}, \ldots, \alpha^{i_{n-k}}$. This yields

$$g(z) = (z - \alpha^{i_1})(z - \alpha^{i_2}) \cdots (z - \alpha^{i_{n-k}}).$$

However, not all possible choices for the zeros are allowed because the generator polynomial $g(z)$ of a cyclic code $\mathbb{B}(n, k, d)$ over the finite field $\mathbb{F}_q$ must be an element of the polynomial ring $\mathbb{F}_q[z]$, i.e. the polynomial coefficients must be elements of the finite field $\mathbb{F}_q$. This is guaranteed if for each root α^i its corresponding conjugate roots $\alpha^{iq}, \alpha^{iq^2}, \ldots$ are also zeros of the generator polynomial $g(z)$. The product over all respective linear factors yields the minimal polynomial

$$m_i(z) = (z - \alpha^i)(z - \alpha^{iq})(z - \alpha^{iq^2}) \cdots$$

ALGEBRAIC CODING THEORY

Cyclotomic cosets and minimal polynomials

- Factorisation of $z^7 - 1 = z^7 + 1$ over the finite field $\mathbb{F}_2$ into minimal polynomials
$$z^7 + 1 = (z+1)(z^3 + z + 1)(z^3 + z^2 + 1)$$

- Cyclotomic cosets $C_i = \{i\,2^j \mod 7 : 0 \le j \le 2\}$

$$C_0 = \{0\}$$
$$C_1 = \{1, 2, 4\}$$
$$C_3 = \{3, 6, 5\}$$

Figure 2.49: Cyclotomic cosets and minimal polynomials over the finite field $\mathbb{F}_2$

with coefficients in $\mathbb{F}_q$. The set of exponents $i, iq, iq^2, \ldots$ of the primitive nth root of unity $\alpha \in \mathbb{F}_{q^l}$ corresponds to the so-called *cyclotomic coset* (Berlekamp, 1984; Bossert, 1999; Ling and Xing, 2004; McEliece, 1987)

$$C_i = \{i\,q^j \mod q^l - 1 : 0 \le j \le l - 1\}$$

which can be used in the definition of the minimal polynomial

$$m_i(z) = \prod_{\kappa \in C_i} (z - \alpha^\kappa).$$

Figure 2.49 illustrates the cyclotomic cosets and minimal polynomials over the finite field $\mathbb{F}_2$. The generator polynomial $g(z)$ can thus be written as the product of the corresponding minimal polynomials. Since each minimal polynomial occurs only once, the generator polynomial is given by the least common multiple

$$g(z) = \mathrm{lcm}\left(m_{i_1}(z), m_{i_2}(z), \ldots, m_{i_{n-k}}(z)\right).$$

The characteristics of the generator polynomial $g(z)$ and the respective cyclic code $\mathbb{B}(n, k, d)$ are determined by the minimal polynomials and the cyclotomic cosets respectively.

A cyclic code $\mathbb{B}(n, k, d)$ with generator polynomial $g(z)$ can now be defined by the set of minimal polynomials or the corresponding roots $\alpha_1, \alpha_2, \ldots, \alpha_{n-k}$. Therefore, we will denote the cyclic code by its zeros according to

$$\mathbb{B}(n, k, d) = \mathcal{C}(\alpha_1, \alpha_2, \ldots, \alpha_{n-k}).$$

Because of $g(\alpha_1) = g(\alpha_2) = \cdots = g(\alpha_{n-k}) = 0$ and $g(z) \,|\, b(z)$, the zeros of the generator polynomial $g(z)$ are also zeros

$$b(\alpha_1) = b(\alpha_2) = \cdots = b(\alpha_{n-k}) = 0$$

of each code polynomial $b(z) \in \mathcal{C}(\alpha_1, \alpha_2, \ldots, \alpha_{n-k})$.

BCH Bound

Based on the zeros $\alpha_1, \alpha_2, \ldots, \alpha_{n-k}$ of the generator polynomial $g(z)$, a lower bound for the minimum Hamming distance d of a cyclic code $\mathcal{C}(\alpha_1, \alpha_2, \ldots, \alpha_{n-k})$ has been derived by Bose, Ray-Chaudhuri and Hocquenghem. This so-called *BCH bound*, which is given in Figure 2.50, states that the minimum Hamming distance d is at least equal to δ if there are $\delta - 1$ successive zeros α^β, $\alpha^{\beta+1}$, $\alpha^{\beta+2}$, ..., $\alpha^{\beta+\delta-2}$ (Berlekamp, 1984; Jungnickel, 1995; Lin and Costello, 2004).

Because of $g(z) \,|\, b(z)$ for every code polynomial $b(z)$, the condition in the BCH bound also amounts to $b(\alpha^\beta) = b(\alpha^{\beta+1}) = b(\alpha^{\beta+2}) = \cdots = b(\alpha^{\beta+\delta-2}) = 0$. With the help of the code polynomial $b(z) = b_0 + b_1 z + b_2 z^2 + \cdots + b_{n-1} z^{n-1}$, this yields

$$\begin{aligned}
b_0 + b_1 \alpha^\beta + b_2 \alpha^{\beta \cdot 2} + \cdots + b_{n-1} \alpha^{\beta(n-1)} &= 0, \\
b_0 + b_1 \alpha^{\beta+1} + b_2 \alpha^{(\beta+1) \cdot 2} + \cdots + b_{n-1} \alpha^{(\beta+1)(n-1)} &= 0, \\
b_0 + b_1 \alpha^{\beta+2} + b_2 \alpha^{(\beta+2) \cdot 2} + \cdots + b_{n-1} \alpha^{(\beta+2)(n-1)} &= 0, \\
&\vdots \\
b_0 + b_1 \alpha^{\beta+\delta-2} + b_2 \alpha^{(\beta+\delta-2) \cdot 2} + \cdots + b_{n-1} \alpha^{(\beta+\delta-2)(n-1)} &= 0
\end{aligned}$$

BCH bound

- Let $\mathcal{C}(\alpha_1, \alpha_2, \ldots, \alpha_{n-k})$ be a cyclic code of code word length n over the finite field $\mathbb{F}_q$ with generator polynomial $g(z)$, and let $\alpha \in \mathbb{F}_{q^l}$ be an nth root of unity in the extension field $\mathbb{F}_{q^l}$ with $\alpha^n = 1$.

- If the cyclic code incorporates $\delta - 1$ zeros

$$\alpha^\beta, \alpha^{\beta+1}, \alpha^{\beta+2}, \ldots, \alpha^{\beta+\delta-2}$$

according to

$$g(\alpha^\beta) = g(\alpha^{\beta+1}) = g(\alpha^{\beta+2}) = \cdots = g(\alpha^{\beta+\delta-2}) = 0$$

the minimum Hamming distance d of the cyclic code is bounded below by

$$d \geq \delta \tag{2.72}$$

Figure 2.50: BCH bound

which corresponds to the system of equations

$$\begin{pmatrix} 1 & \alpha^\beta & \alpha^{\beta 2} & \cdots & \alpha^{\beta(n-1)} \\ 1 & \alpha^{\beta+1} & \alpha^{(\beta+1)2} & \cdots & \alpha^{(\beta+1)(n-1)} \\ 1 & \alpha^{\beta+2} & \alpha^{(\beta+2)2} & \cdots & \alpha^{(\beta+2)(n-1)} \\ \vdots & \vdots & \vdots & \ddots & \vdots \\ 1 & \alpha^{\beta+\delta-2} & \alpha^{(\beta+\delta-2)2} & \cdots & \alpha^{(\beta+\delta-2)(n-1)} \end{pmatrix} \begin{pmatrix} b_0 \\ b_1 \\ b_2 \\ \vdots \\ b_{n-1} \end{pmatrix} = \begin{pmatrix} 0 \\ 0 \\ 0 \\ \vdots \\ 0 \end{pmatrix}.$$

By comparing this matrix equation with the parity-check equation $\mathbf{H}\mathbf{b}^T = \mathbf{0}$ of general linear block codes, we observe that the $(\delta - 1) \times n$ matrix in the above matrix equation corresponds to a part of the parity-check matrix $\mathbf{H}$. If this matrix has at least $\delta - 1$ linearly independent columns, then the parity-check matrix $\mathbf{H}$ also has at least $\delta - 1$ linearly independent columns. Therefore, the smallest number of linearly dependent columns of $\mathbf{H}$, and thus the minimum Hamming distance, is not smaller than δ, i.e. $d \geq \delta$. If we consider the determinant of the $(\delta - 1) \times (\delta - 1)$ matrix consisting of the first $\delta - 1$ columns, we obtain (Jungnickel, 1995)

$$\begin{vmatrix} 1 & \alpha^\beta & \alpha^{\beta 2} & \cdots & \alpha^{\beta(\delta-2)} \\ 1 & \alpha^{\beta+1} & \alpha^{(\beta+1)2} & \cdots & \alpha^{(\beta+1)(\delta-2)} \\ 1 & \alpha^{\beta+2} & \alpha^{(\beta+2)2} & \cdots & \alpha^{(\beta+2)(\delta-2)} \\ \vdots & \vdots & \vdots & \ddots & \vdots \\ 1 & \alpha^{\beta+\delta-2} & \alpha^{(\beta+\delta-2)2} & \cdots & \alpha^{(\beta+\delta-2)(\delta-2)} \end{vmatrix}$$

$$= \begin{vmatrix} 1 & 1 & 1 & \cdots & 1 \\ 1 & \alpha^1 & \alpha^2 & \cdots & \alpha^{\delta-2} \\ 1 & \alpha^2 & \alpha^4 & \cdots & \alpha^{2(\delta-2)} \\ \vdots & \vdots & \vdots & \ddots & \vdots \\ 1 & \alpha^{\delta-2} & \alpha^{(\delta-2)2} & \cdots & \alpha^{(\delta-2)(\delta-2)} \end{vmatrix} \alpha^{\beta(\delta-1)(\delta-2)/2}.$$

The resulting determinant on the right-hand side corresponds to a so-called *Vandermonde matrix*, the determinant of which is different from 0. Taking into account that $\alpha^{\beta(\delta-1)(\delta-2)/2} \neq 0$, the $(\delta - 1) \times (\delta - 1)$ matrix consisting of the first $\delta - 1$ columns is regular with $\delta - 1$ linearly independent columns. This directly leads to the BCH bound $d \geq \delta$.

According to the BCH bound, the minimum Hamming distance of a cyclic code is determined by the properties of a subset of the zeros of the respective generator polynomial. In order to define a cyclic code by prescribing a suitable set of zeros, we will therefore merely note this specific subset. A cyclic binary Hamming code, for example, is determined by a single zero α; the remaining conjugate roots α^2, α^4, ... follow from the condition that the coefficients of the generator polynomial are elements of the finite field $\mathbb{F}_2$. The respective cyclic code will therefore be denoted by $\mathcal{C}(\alpha)$.

Definition of BCH Codes

In view of the BCH bound in Figure 2.50, a cyclic code with a guaranteed minimum Hamming distance d can be defined by prescribing $\delta - 1$ successive powers

$$\alpha^\beta, \alpha^{\beta+1}, \alpha^{\beta+2}, \ldots, \alpha^{\beta+\delta-2}$$

BCH codes

- Let $\alpha \in \mathbb{F}_{q^l}$ be an nth root of unity in the extension field $\mathbb{F}_{q^l}$ with $\alpha^n = 1$.
- The cyclic code $\mathcal{C}(\alpha^\beta, \alpha^{\beta+1}, \alpha^{\beta+2}, \ldots, \alpha^{\beta+\delta-2})$ over the finite field $\mathbb{F}_q$ is called the BCH code to the design distance δ.
- The minimum Hamming distance is bounded below by

$$d \geq \delta \qquad (2.73)$$

- A narrow-sense BCH code is obtained for $\beta = 1$.
- If $n = q^l - 1$, the BCH code is called primitive.

Figure 2.51: Definition of BCH codes

of an appropriate nth root of unity α as zeros of the generator polynomial $g(z)$. Because of

$$d \geq \delta$$

the parameter δ is called the *design distance* of the cyclic code. The resulting cyclic code over the finite field $\mathbb{F}_q$ is the so-called *BCH code* $\mathcal{C}(\alpha^\beta, \alpha^{\beta+1}, \alpha^{\beta+2}, \ldots, \alpha^{\beta+\delta-2})$ to the design distance δ (see Figure 2.51) (Berlekamp, 1984; Lin and Costello, 2004; Ling and Xing, 2004). If we choose $\beta = 1$, we obtain the *narrow-sense BCH code* to the design distance δ. For the code word length

$$n = q^l - 1$$

the primitive nth root of unity α is a primitive element in the extension field $\mathbb{F}_{q^l}$ due to $\alpha^n = \alpha^{q^l-1} = 1$. The corresponding BCH code is called a *primitive BCH code*. BCH codes are often used in practical applications because they are easily designed for a wanted minimum Hamming distance d (Benedetto and Biglieri, 1999; Proakis, 2001). Furthermore, efficient algebraic decoding schemes exist, as we will see in Section 2.3.8.

As an example, we consider the *cyclic binary Hamming code* over the finite field $\mathbb{F}_2$ with $n = 2^m - 1$ and $k = 2^m - m - 1$. Let α be a primitive nth root of unity in the extension field $\mathbb{F}_{2^m}$. With the conjugate roots $\alpha, \alpha^2, \alpha^{2^2}, \ldots, \alpha^{2^{m-1}}$, the cyclic code

$$\mathcal{C}(\alpha) = \left\{ b(z) \in \mathbb{F}_2[z]/(z^n - 1) : b(\alpha) = 0 \right\}$$

is defined by the generator polynomial

$$g(z) = (z - \alpha)(z - \alpha^2)(z - \alpha^{2^2}) \cdots (z - \alpha^{2^{m-1}}).$$

Owing to the roots α and α^2 there exist $\delta - 1 = 2$ successive roots. According to the BCH bound, the minimum Hamming distance is bounded below by $d \geq \delta = 3$. In fact, as we already know, Hamming codes have a minimum Hamming distance $d = 3$.

In general, for the definition of a cyclic BCH code we prescribe $\delta - 1$ successive zeros $\alpha^\beta, \alpha^{\beta+1}, \alpha^{\beta+2}, \ldots, \alpha^{\beta+\delta-2}$. By adding the corresponding conjugate roots, we obtain the generator polynomial $g(z)$ which can be written as

$$g(z) = \operatorname{lcm}\left(m_\beta(z), m_{\beta+1}(z), \ldots, m_{\beta+\delta-2}(z)\right).$$

The generator polynomial $g(z)$ is equal to the least common multiple of the respective polynomials $m_i(z)$ which denote the minimal polynomials for α^i with $\beta \leq i \leq \beta + \delta - 2$.

2.3.7 Reed–Solomon Codes

As an important special case of primitive BCH codes we now consider BCH codes over the finite field $\mathbb{F}_q$ with code word length

$$n = q - 1.$$

These codes are called *Reed–Solomon codes* (Berlekamp, 1984; Bossert, 1999; Lin and Costello, 2004; Ling and Xing, 2004); they are used in a wide range of applications ranging from communication systems to the encoding of audio data in a compact disc (Costello *et al.*, 1998). Because of $\alpha^n = \alpha^{q-1} = 1$, the nth root of unity α is an element of the finite field $\mathbb{F}_q$. Since the corresponding minimal polynomial of α^i over the finite field $\mathbb{F}_q$ is simply given by

$$m_i(z) = z - \alpha^i$$

the generator polynomial $g(z)$ of such a primitive BCH code to the design distance δ is

$$g(z) = (z - \alpha^\beta)(z - \alpha^{\beta+1}) \cdots (z - \alpha^{\beta+\delta-2}).$$

The degree of the generator polynomial is equal to

$$\deg(g(z)) = n - k = \delta - 1.$$

Because of the BCH bound, the minimum Hamming distance is bounded below by $d \geq \delta = n - k + 1$ whereas the Singleton bound delivers the upper bound $d \leq n - k + 1$. Therefore, the minimum Hamming distance of a Reed–Solomon code is given by

$$d = n - k + 1 = q - k.$$

Since the Singleton bound is fulfilled with equality, a Reed–Solomon code is an MDS (maximum distance separable) code. In general, a Reed–Solomon code over the finite field $\mathbb{F}_q$ is characterised by the following code parameters

$$n = q - 1,$$
$$k = q - \delta,$$
$$d = \delta.$$

In Figure 2.52 the characteristics of a Reed–Solomon code are summarised. For practically relevant code word lengths n, the cardinality q of the finite field $\mathbb{F}_q$ is large. In practical applications $q = 2^l$ is usually chosen. The respective elements of the finite field $\mathbb{F}_{2^l}$ are then represented as l-dimensional binary vectors over $\mathbb{F}_2$.

Reed–Solomon codes

- Let α be a primitive element of the finite field $\mathbb{F}_q$ with $n = q - 1$.
- The Reed–Solomon code is defined as the primitive BCH code to the design distance δ over the finite field $\mathbb{F}_q$ with generator polynomial

$$g(z) = (z - \alpha^\beta)(z - \alpha^{\beta+1}) \cdots (z - \alpha^{\beta+\delta-2}) \qquad (2.74)$$

- Code parameters

$$n = q - 1 \qquad (2.75)$$
$$k = q - \delta \qquad (2.76)$$
$$d = \delta \qquad (2.77)$$

- Because a Reed–Solomon code fulfils the Singleton bound with equality, it is an MDS code.

Figure 2.52: Reed–Solomon codes over the finite field $\mathbb{F}_q$

Spectral Encoding

We now turn to an interesting relationship between Reed–Solomon codes and the discrete Fourier transform (DFT) over the finite field $\mathbb{F}_q$ (Bossert, 1999; Lin and Costello, 2004; Neubauer, 2006b) (see also Section A.4 in Appendix A). This relationship leads to an efficient encoding algorithm based on FFT (fast Fourier transform) algorithms. The respective encoding algorithm is called *spectral encoding*.

To this end, let $\alpha \in \mathbb{F}_q$ be a primitive nth root of unity in the finite field $\mathbb{F}_q$ with $n = q - 1$. Starting from the code polynomial

$$b(z) = b_0 + b_1 z + \cdots + b_{n-2} z^{n-2} + b_{n-1} z^{n-1}$$

the discrete Fourier transform of length n over the finite field $\mathbb{F}_q$ is defined by

$$B_j = b(\alpha^j) = \sum_{i=0}^{n-1} b_i \alpha^{ij} \quad \circ\!\!-\!\!\bullet \quad b_i = n^{-1} B(\alpha^{-i}) = n^{-1} \sum_{j=0}^{n-1} B_j \alpha^{-ij}$$

with $b_i \in \mathbb{F}_q$ and $B_j \in \mathbb{F}_q$. The spectral polynomial is given by

$$B(z) = B_0 + B_1 z + \cdots + B_{n-2} z^{n-2} + B_{n-1} z^{n-1}.$$

Since every code polynomial $b(z)$ is divided by the generator polynomial

$$g(z) = (z - \alpha^\beta)(z - \alpha^{\beta+1}) \cdots (z - \alpha^{\beta+\delta-2})$$

ALGEBRAIC CODING THEORY

which is characterised by the zeros $\alpha^\beta, \alpha^{\beta+1}, \ldots, \alpha^{\beta+\delta-2}$, every code polynomial $b(z)$ also has zeros at $\alpha^\beta, \alpha^{\beta+1}, \ldots, \alpha^{\beta+\delta-2}$, i.e.

$$b(\alpha^\beta) = b(\alpha^{\beta+1}) = b(\alpha^{\beta+2}) = \cdots = b(\alpha^{\beta+\delta-2}) = 0.$$

In view of the discrete Fourier transform and the spectral polynomial $B(z) \circ\!\!-\!\!\bullet\, b(z)$, this can be written as

$$B_j = b(\alpha^j) = 0$$

for $\beta \leq j \leq \beta + \delta - 2$. These spectral coefficients are called *parity frequencies*.

If we choose $\beta = q - \delta$, we obtain the Reed–Solomon code of length $n = q - 1$, dimension $k = q - \delta$ and minimum Hamming distance $d = \delta$. The information polynomial

$$u(z) = u_0 + u_1 z + u_2 z^2 + \cdots + u_{q-\delta-1} z^{q-\delta-1}$$

with $k = q - \delta$ information symbols u_j is now used to define the spectral polynomial $B(z)$ according to

$$B(z) = u(z),$$

i.e. $B_j = u_j$ for $0 \leq j \leq k - 1$ and $B_j = 0$ for $k \leq j \leq n - 1$. This setting yields $B_j = b(\alpha^j) = 0$ for $q - \delta \leq j \leq q - 2$. The corresponding code polynomial $b(z)$ is obtained from the inverse discrete Fourier transform according to

$$b_i = -B(\alpha^{-i}) = -\sum_{j=0}^{q-2} B_j \alpha^{-ij}.$$

Here, we have used the fact that $n = q - 1 \equiv -1$ modulo p, where p denotes the characteristic of the finite field $\mathbb{F}_q$, i.e. $q = p^l$ with the prime number p. Finally, the spectral encoding rule reads

$$b_i = -\sum_{j=0}^{q-\delta-1} u_j \alpha^{-ij}.$$

Because there are fast algorithms available for the calculation of the discrete Fourier transform, this encoding algorithm can be carried out efficiently. It has to be noted, however, that the resulting encoding scheme is not systematic. The spectral encoding algorithm of a Reed–Solomon code is summarised in Figure 2.53 (Neubauer, 2006b). A respective decoding algorithm is given elsewhere (Lin and Costello, 2004).

Reed–Solomon codes are used in a wide range of applications, e.g. in communication systems, deep-space applications, digital video broadcasting (DVB) or consumer systems (Costello *et al.*, 1998). In the DVB system, for example, a shortened Reed–Solomon code with $n = 204, k = 188$ and $t = 8$ is used which is derived from a Reed–Solomon code over the finite field $\mathbb{F}_{2^8} = \mathbb{F}_{256}$ (ETSI, 2006). A further important example is the encoding of the audio data in the compact disc with the help of the cross-interleaved Reed–Solomon code (CIRC) which is briefly summarised in Figure 2.54 (Costello *et al.*, 1998; Hoeve *et al.*, 1982).

Spectral encoding of a Reed–Solomon code

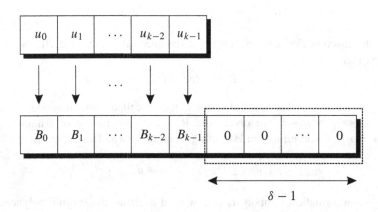

- The k information symbols are chosen as the first k spectral coefficients, i.e.
$$B_j = u_j \qquad (2.78)$$
for $0 \leq j \leq k-1$.

- The remaining $n - k$ spectral coefficients are set to 0 according to
$$B_j = 0 \qquad (2.79)$$
for $k \leq j \leq n-1$.

- The code symbols are calculated with the help of the inverse discrete Fourier transform according to
$$b_i = -B(\alpha^{-i}) = -\sum_{j=0}^{q-2} B_j \alpha^{-ij} = -\sum_{j=0}^{q-\delta-1} u_j \alpha^{-ij} \qquad (2.80)$$
for $0 \leq i \leq n-1$.

Figure 2.53: Spectral encoding of a Reed–Solomon code over the finite field $\mathbb{F}_q$. Reproduced by permission of J. Schlembach Fachverlag

2.3.8 Algebraic Decoding Algorithm

Having defined BCH codes and Reed–Solomon codes, we now discuss an algebraic decoding algorithm that can be used for decoding a received polynomial $r(z)$ (Berlekamp, 1984; Bossert, 1999; Jungnickel, 1995; Lin and Costello, 2004; Neubauer, 2006b). To this end, without loss of generality we restrict the derivation to narrow-sense BCH codes

ALGEBRAIC CODING THEORY

Reed–Solomon codes and the compact disc

- In the compact disc the encoding of the audio data is done with the help of two interleaved Reed–Solomon codes.

- The Reed–Solomon code with minimum distance $d = \delta = 5$ over the finite field $\mathbb{F}_{256} = \mathbb{F}_{2^8}$ with length $n = q - 1 = 255$ and $k = q - \delta = 251$ is shortened such that two linear codes $\mathbb{B}(28, 24, 5)$ and $\mathbb{B}(32, 28, 5)$ over the finite field $\mathbb{F}_{2^8}$ arise.

- The resulting interleaved coding scheme is called CIRC (cross-interleaved Reed–Solomon code).

- For each stereo channel, the audio signal is sampled with 16-bit resolution and a sampling frequency of 44.1 kHz, leading to a total of $2 \times 16 \times 44\,100 = 1\,411\,200$ bits per second. Each 16-bit stereo sample represents two 8-bit symbols in the field $\mathbb{F}_{2^8}$.

- The inner shortened Reed–Solomon code $\mathbb{B}(28, 24, 5)$ encodes 24 information symbols according to six stereo sample pairs.

- The outer shortened Reed–Solomon code $\mathbb{B}(32, 28, 5)$ encodes the resulting 28 symbols, leading to 32 code symbols.

- In total, the CIRC leads to $4\,231\,800$ channel bits on a compact disc which are further modulated and represented as so-called pits on the compact disc carrier.

Figure 2.54: Reed–Solomon codes and the compact disc

$\mathcal{C}(\alpha, \alpha^2, \ldots, \alpha^{\delta-1})$ with $\alpha \in \mathbb{F}_{q^l}$ of a given designed distance

$$\delta = 2t + 1.$$

It is important to note that the algebraic decoding algorithm we are going to derive is only capable of correcting up to t errors even if the true minimum Hamming distance d is larger than the designed distance δ. For the derivation of the algebraic decoding algorithm we make use of the fact that the generator polynomial $g(z)$ has as zeros $\delta - 1 = 2t$ successive powers of a primitive nth root of unity α, i.e.

$$g(\alpha) = g(\alpha^2) = \cdots = g(\alpha^{2t}) = 0.$$

Since each code polynomial

$$b(z) = b_0 + b_1 z + \cdots + b_{n-2} z^{n-2} + b_{n-1} z^{n-1}$$

is a multiple of the generator polynomial $g(z)$, this property also translates to the code polynomial $b(z)$ as well, i.e.

$$b(\alpha) = b(\alpha^2) = \cdots = b(\alpha^{2t}) = 0.$$

This condition will be used in the following for the derivation of an algebraic decoding algorithm for BCH and Reed–Solomon codes.

The polynomial channel model illustrated in Figure 2.46 according to

$$r(z) = b(z) + e(z)$$

with error polynomial

$$e(z) = \sum_{i=0}^{n-1} e_i z^i$$

will be presupposed. If we assume that $w \leq t$ errors have occurred at the error positions i_j, the error polynomial can be written as

$$e(z) = \sum_{j=1}^{w} e_{i_j} z^{i_j}.$$

With the help of these error positions i_j we define the so-called *error locators* according to

$$X_j = \alpha^{i_j}$$

as well as the *error values*

$$Y_j = e_{i_j}$$

for $1 \leq j \leq w$ (Berlekamp, 1984). As shown in Figure 2.55, the error polynomial $e(z)$ can then be written as

$$e(z) = \sum_{j=1}^{w} Y_j z^{i_j}.$$

The algebraic decoding algorithm has to determine the number of errors w as well as the error positions i_j or error locators X_j and the corresponding error values Y_j. This will be done in a two-step approach by first calculating the error locators X_j; based on the calculated error locators X_j, the error values Y_j are determined next.

By analogy with our treatment of the decoding schemes for linear block codes and cyclic codes, we will introduce the syndromes on which the derivation of the algebraic decoding algorithm rests (see Figure 2.56). Here, the *syndromes* are defined according to

$$S_j = r(\alpha^j) = \sum_{i=0}^{n-1} r_i \alpha^{ij}$$

for $1 \leq j \leq 2t$. For a valid code polynomial $b(z) \in \mathcal{C}(\alpha, \alpha^2, \ldots, \alpha^{2t})$ of the BCH code $\mathcal{C}(\alpha, \alpha^2, \ldots, \alpha^{2t})$, the respective syndromes $b(\alpha^j)$, which are obtained by evaluating the polynomial $b(z)$ at the given powers of the primitive nth root of unity α, are identically

ALGEBRAIC CODING THEORY

Error polynomial, error locators and error values

- Error polynomial
$$e(z) = e_{i_1} z^{i_1} + e_{i_2} z^{i_2} + \cdots + e_{i_w} z^{i_w} \qquad (2.81)$$
with error positions i_j for $1 \leq j \leq w$

- Error locators
$$X_j = \alpha^{i_j} \quad \text{for} \quad 1 \leq j \leq w \qquad (2.82)$$

- Error values
$$Y_j = e_{i_j} \quad \text{for} \quad 1 \leq j \leq w \qquad (2.83)$$

- Error polynomial
$$e(z) = Y_1 z^{i_1} + Y_2 z^{i_2} + \cdots + Y_w z^{i_w} \qquad (2.84)$$

Figure 2.55: Algebraic characterisation of transmission errors based on the error polynomial $e(z)$, error locators X_j and error values Y_j

---

Syndromes

- Received polynomial
$$r(z) = b(z) + e(z) \qquad (2.85)$$

- Syndromes of received polynomial $r(z)$
$$S_j = r(\alpha^j) = \sum_{i=0}^{n-1} r_i \alpha^{ij} \quad \text{for} \quad 1 \leq j \leq 2t \qquad (2.86)$$

- Since $b(\alpha^j) = 0$ for the transmitted code polynomial $b(z) \in \mathcal{C}(\alpha, \alpha^2, \ldots, \alpha^{2t})$ and $1 \leq j \leq 2t$, the syndromes exclusively depend on the error polynomial $e(z)$, i.e.
$$S_j = r(\alpha^j) = b(\alpha^j) + e(\alpha^j) = e(\alpha^j) \quad \text{for} \quad 1 \leq j \leq 2t \qquad (2.87)$$

Figure 2.56: Syndromes in the algebraic decoding algorithm

Syndromes and error locator polynomial

■ Syndromes
$$S_j = r(\alpha^j) = \sum_{i=1}^{w} Y_i X_i^j \qquad (2.88)$$
for $1 \leq j \leq 2t$

■ Error locator polynomial
$$\lambda(z) = \prod_{j=1}^{w}(1 - X_j z) = \sum_{i=0}^{w} \lambda_i z^i \qquad (2.89)$$
of degree $\deg(\lambda(z)) = w$

Figure 2.57: Syndromes S_j and error locator polynomial $\lambda(z)$

zero, i.e. $b(\alpha^j) = 0$ for $1 \leq j \leq 2t$. This is true because the powers α^j correspond to the zeros used to define the BCH code $\mathcal{C}(\alpha, \alpha^2, \ldots, \alpha^{2t})$. Therefore, the syndromes
$$S_j = r(\alpha^j) = b(\alpha^j) + e(\alpha^j) = e(\alpha^j)$$
for $1 \leq j \leq 2t$ merely depend on the error polynomial $e(z)$. The syndrome S_j is an element of the extension field $\mathbb{F}_{q^l}$ which also includes the primitive nth root of unity α. The consequence of this is that we need to carry out the arithmetics of the algebraic decoding algorithm in the extension field $\mathbb{F}_{q^l}$. For Reed–Solomon codes, however, the calculations can be done in the finite field $\mathbb{F}_q$ because $\alpha \in \mathbb{F}_q$ and therefore $S_j \in \mathbb{F}_q$.

Since the syndromes $S_j = e(\alpha^j)$ are completely determined by the error polynomial $e(z)$, they can be expressed using the error locators X_j and error values Y_j (see Figure 2.57). With
$$S_j = e(\alpha^j) = Y_1 \alpha^{i_1 j} + Y_2 \alpha^{i_2 j} + \cdots + Y_w \alpha^{i_w j}$$
and by making use of the error locators $X_j = \alpha^{i_j}$, we obtain
$$S_j = Y_1 X_1^j + Y_2 X_2^j + \cdots + Y_w X_w^j = \sum_{i=1}^{w} Y_i X_i^j.$$

In order to determine the error polynomial $e(z)$ and to decode the transmitted code polynomial $b(z)$, we have to calculate the error locators X_j and error values Y_j on the basis of the syndromes S_j. The latter is the only information available to the algebraic decoding algorithm. To this end, we define the so-called *error locator polynomial* (Berlekamp, 1984)
$$\lambda(z) = \prod_{j=1}^{w}(1 - X_j z)$$

ALGEBRAIC CODING THEORY

as a polynomial of degree w by prescribing the w zeros as the inverses of the error locators X_j. Expanding the respective product, we arrive at the polynomial representation

$$\lambda(z) = \sum_{i=0}^{w} \lambda_i z^i$$

with coefficients λ_i.[18] If these coefficients λ_i or the error locator polynomial $\lambda(z)$ are known, the error locators X_j can be determined by searching for the zeros of the error locator polynomial $\lambda(z)$.

According to the defining equation, the error locator polynomial $\lambda(z)$ has its zeros at the inverse error locators X_k^{-1}, i.e.

$$\lambda(X_k^{-1}) = \sum_{i=0}^{w} \lambda_i X_k^{-i} = 0$$

for $1 \leq k \leq w$. This expression is multiplied by $Y_k X_k^{j+w}$ and accumulated with respect to the summation index k, leading to

$$\sum_{k=1}^{w} Y_k X_k^{j+w} \sum_{i=0}^{w} \lambda_i X_k^{-i} = 0.$$

By exchanging the order of summation, we obtain

$$\sum_{i=0}^{w} \lambda_i \underbrace{\sum_{k=1}^{w} Y_k X_k^{j+w-i}}_{= S_{j+w-i}} = 0.$$

In this expression the syndromes S_{j+w-i} are identified, leading to the so-called *key equation*

$$\sum_{i=0}^{w} \lambda_i S_{j+w-i} = 0.$$

The key equation in Figure 2.58 relates the known syndromes S_j to the unknown coefficients λ_i of the error locator polynomial $\lambda(z)$ with $\lambda_0 = 1$ by a linear recurrence equation. The solution to this key equation can be formulated as finding the shortest linear recursive filter over the finite field $\mathbb{F}_{q^l}$ with filter coefficients λ_i. For this purpose, the so-called *Berlekamp–Massey algorithm* can be used (Berlekamp, 1984; Bossert, 1999; Lin and Costello, 2004; Massey, 1969). As soon as the error locator polynomial $\lambda(z)$ is known, we can determine its zeros as the inverses of the error locators X_j, leading to the error positions i_j.

As an alternative to the Berlekamp–Massey algorithm, *Euclid's algorithm* for calculating the greatest common divisor of two polynomials can be used for determining the error locator polynomial $\lambda(z)$ (Bossert, 1999; Dornstetter, 1987; Jungnickel, 1995; Lin and Costello, 2004; Neubauer, 2006b). In order to formulate the respective algorithm, we define the *syndrome polynomial* $S(z)$ as well as the so-called *error evaluator polynomial* $\omega(z)$ as shown in Figure 2.59 (Berlekamp, 1984).

---

[18]The coefficients λ_i relate to the so-called elementary symmetric polynomials.

Key equation and Berlekamp–Massey algorithm

- Key equation
$$S_{j+w} + \lambda_1 S_{j+w-1} + \lambda_2 S_{j+w-2} + \cdots + S_j = 0 \qquad (2.90)$$
for $j = 1, 2, \ldots$

- The key equation can be solved with the help of the Berlekamp–Massey algorithm which finds the shortest linear recursive filter with filter coefficients λ_i emitting the syndromes S_j.

Figure 2.58: Key equation and Berlekamp–Massey algorithm

Syndrome and error evaluator polynomial

- Syndrome polynomial
$$S(z) = \sum_{j=1}^{2t} S_j z^{j-1} = S_1 + S_2 z + \cdots + S_{2t} z^{2t-1} \qquad (2.91)$$

- Error evaluator polynomial
$$\omega(z) \equiv S(z)\lambda(z) \mod z^{2t} \qquad (2.92)$$

Figure 2.59: Syndrome polynomial $S(z)$ and error evaluator polynomial $\omega(z)$

With the help of the syndromes S_j, the error evaluator polynomial $\omega(z)$ can be expressed as follows (Jungnickel, 1995; Neubauer, 2006b)

$$\begin{aligned}
\omega(z) &\equiv S(z)\lambda(z) \mod z^{2t} \\
&\equiv \sum_{j=1}^{2t} S_j z^{j-1} \prod_{k=1}^{w}(1 - X_k z) \mod z^{2t} \\
&\equiv \sum_{j=1}^{2t} \sum_{i=1}^{w} Y_i X_i^j z^{j-1} \prod_{k=1}^{w}(1 - X_k z) \mod z^{2t}.
\end{aligned}$$

By exchanging the order of summation, we obtain

$$\omega(z) \equiv \sum_{i=1}^{w} Y_i X_i \sum_{j=1}^{2t} (X_i z)^{j-1} \prod_{k=1}^{w} (1 - X_k z) \mod z^{2t}.$$

With the help of the formula for the finite geometric series

$$\sum_{j=1}^{2t} (X_i z)^{j-1} = \frac{1 - (X_i z)^{2t}}{1 - X_i z}$$

which is also valid over finite fields, we have

$$\omega(z) \equiv \sum_{i=1}^{w} Y_i X_i \frac{1 - (X_i z)^{2t}}{1 - X_i z} \prod_{k=1}^{w} (1 - X_k z) \mod z^{2t}$$

$$\equiv \sum_{i=1}^{w} Y_i X_i \left(1 - (X_i z)^{2t}\right) \prod_{k=1, k \neq i}^{w} (1 - X_k z) \mod z^{2t}.$$

On account of the factor $(1 - (X_i z)^{2t})$ and the modulo operation, the second term $(X_i z)^{2t}$ can be neglected, i.e.

$$\omega(z) \equiv \sum_{i=1}^{w} Y_i X_i \prod_{k=1, k \neq i}^{w} (1 - X_k z) \mod z^{2t}$$

$$= \sum_{i=1}^{w} Y_i X_i \prod_{k=1, k \neq i}^{w} (1 - X_k z).$$

As can be seen from this expression, the degree of the error evaluator polynomial $\omega(z)$ is equal to $\deg(\omega(z)) = w - 1$, and therefore less than $2t$, so that the modulo operation can be removed from the above equation. It can be shown that the error evaluator polynomial $\omega(z)$ and the error locator polynomial $\lambda(z)$ are relatively prime.

According to Figure 2.60, the error locator polynomial $\lambda(z)$ of degree $\deg(\lambda(z)) = w \leq t$ and the error evaluator polynomial $\omega(z)$ of degree $\deg(\omega(z)) = w - 1 \leq t - 1$ can be determined with the help of Euclid's algorithm. To this end, the equation $\omega(z) \equiv S(z)\lambda(z)$ modulo z^{2t}, which is also called the *key equation* of the algebraic decoding problem, is expressed by

$$\omega(z) = S(z)\lambda(z) + T(z) z^{2t}$$

with a suitable polynomial $T(z)$. Apart from a common factor, this equation corresponds to the relationship between the polynomials $S(z)$ and z^{2t} and their greatest common divisor $\gcd(z^{2t}, S(z))$. By iterating Euclid's algorithm (see also Figure A.3) until the degree $\deg(r_i(z))$ of the remainder

$$r_i(z) = f_i(z) S(z) + g_i(z) z^{2t}$$

> **Euclid's algorithm for solving the key equation**
>
> ■ Key equation
> $$\omega(z) \equiv S(z)\,\lambda(z) \mod z^{2t} \qquad (2.93)$$
> with
> $$\deg(\lambda(z)) = w \leq t \quad \text{and} \quad \deg(\omega(z)) = w - 1 \leq t - 1$$
>
> ■ On account of
> $$\omega(z) = S(z)\,\lambda(z) + T(z)\,z^{2t}$$
> the error evaluator polynomial $\omega(z)$ and the error locator polynomial $\lambda(z)$ can be determined with the help of Euclid's algorithm for calculating the greatest common divisor $\gcd(S(z), z^{2t})$ of the polynomials z^{2t} and $S(z)$.

Figure 2.60: Euclid's algorithm for solving the key equation by calculating the error evaluator polynomial $\omega(z)$ and the error locator polynomial $\lambda(z)$

in iteration i is smaller than t, the error locator polynomial $\lambda(z)$ and the error evaluator polynomial $\omega(z)$ are obtained, apart from a constant factor c, from[19]

$$\lambda(z) = c\,\frac{g_i(z)}{\gcd(r_i(z), g_i(z))},$$
$$\omega(z) = c\,\frac{r_i(z)}{\gcd(r_i(z), g_i(z))}.$$

In order to derive the error values Y_j, we evaluate the error evaluator polynomial $\omega(z)$ at the inverses of the error locators X_j. This yields

$$\omega(X_j^{-1}) = Y_j\,X_j \prod_{k=1, k \neq j}^{w} (1 - X_k\,X_j^{-1}).$$

We also calculate the formal derivative

$$\lambda'(z) = \left(\prod_{k=1}^{w}(1 - X_k z)\right)' = -\sum_{i=1}^{w} X_i \prod_{k=1, k \neq i}^{w} (1 - X_k z)$$

of the error locator polynomial $\lambda(z)$ with respect to z. Evaluating this formal derivative at the inverses of the error locators X_j, we arrive at

$$\lambda'(X_j^{-1}) = -X_j \prod_{k=1, k \neq j}^{w} (1 - X_k\,X_j^{-1}).$$

---

[19] As we will see in the following, this constant factor c does not matter in the calculation of the error values Y_j when using Forney's formula derived below.

Besides the factor $-Y_j$, this expression is identical to the expression derived above for the error evaluator polynomial. By relating these expressions, *Forney's formula*

$$Y_j = -\frac{\omega(X_j^{-1})}{\lambda'(X_j^{-1})}$$

results for calculating the error values Y_j.

In summary, we obtain the algebraic decoding algorithm for narrow-sense BCH codes in Figure 2.61 (Neubauer, 2006b). This algebraic decoding algorithm can be illustrated by the block diagram in Figure 2.62 showing the individual steps of the algorithm (Lee, 2003). In this block diagram the main parameters and polynomials determined during the course of the algorithm are shown. Here, *Chien search* refers to the sequential search for the zeros of the error locator polynomial (Berlekamp, 1984). The arithmetics are carried out in the extension field $\mathbb{F}_{q^l}$. For a Reed–Solomon code as a special case of primitive BCH codes with code word length $n = q - 1$ the calculations are executed in the finite field $\mathbb{F}_q$. Further details of Reed–Solomon decoders, including data about the implementation complexity which is measured by the number of gates needed to realise the corresponding integrated circuit module, can be found elsewhere (Lee, 2003, 2005).

Erasure Correction

Erasures are defined as errors with known error positions i_j but unknown error values Y_j. For the correction of these erasures, the algebraic decoding algorithm can be simplified. Since the error positions i_j as well as the number of errors w are known, the error locators $X_j = \alpha^{i_j}$ and the error locator polynomial $\lambda(z)$ can be directly formulated. Figure 2.63 illustrates the algebraic decoding algorithm for the correction of erasures.

2.4 Summary

In this chapter we have introduced the basic concepts of algebraic coding theory. Linear block codes have been discussed which can be defined by their respective generator and parity-check matrices. Several code construction techniques have been presented. As important examples of linear block codes we have treated the repetition code, parity-check code, Hamming code, simplex code and Reed–Muller code. So-called low density parity-check or LDPC codes, which are currently under research, will be presented in Section 4.1. Further information about other important linear and non-linear codes can be found elsewhere (Berlekamp, 1984; Bossert, 1999; Lin and Costello, 2004; Ling and Xing, 2004; MacWilliams and Sloane, 1998; McEliece, 2002; van Lint, 1999).

By introducing further algebraic structures, cyclic codes were presented as an important subclass of linear block codes for which efficient algebraic decoding algorithms exist. Cyclic codes can be defined with the help of suitable generator or parity-check polynomials. Owing to their specific properties, efficient encoding and decoding architectures are available, based on linear feedback shift registers. With the help of the zeros of the respective generator polynomial, BCH codes and Reed–Solomon codes were defined. Further details about cyclic codes can be found elsewhere (Berlekamp, 1984; Bossert, 1999; Lin and Costello, 2004; Ling and Xing, 2004).

Algebraic decoding algorithm

Notation

received word $r(z) \in \mathbb{F}_q[z]/(z^n - 1)$;

decoded code word $\hat{b}(z) \in \mathbb{F}_q[z]/(z^n - 1)$;

Algebraic decoding algorithm

calculate syndromes $S_j = r(\alpha^j) = \sum_{i=0}^{n-1} r_i \alpha^{ij}$ for $1 \leq j \leq 2t$;

if $S_j = 0$ for $1 \leq j \leq 2t$

then

$\hat{e}(z) = 0$;

else

calculate syndrome polynomial $S(z) = \sum_{j=1}^{2t} S_j z^{j-1}$;

calculate relatively prime polynomials $\omega(z)$, $\lambda(z)$ with $\deg(\omega(z)) \leq t - 1$, $\deg(\lambda(z)) \leq t$ and $\omega(z) \equiv S(z) \lambda(z)$ modulo z^{2t} using Euclid's algorithm;

determine zeros $X_j^{-1} = \alpha^{-i_j}$ of $\lambda(z)$ for $1 \leq j \leq w$ with $w = \deg(\lambda(z))$;

calculate error values $Y_j = -\frac{\omega(X_j^{-1})}{\lambda'(X_j^{-1})}$ for $1 \leq j \leq w$;

calculate error positions i_j from $X_j = \alpha^{i_j}$ for $1 \leq j \leq w$;

calculate error polynomial $\hat{e}(z) = \sum_{j=1}^{w} Y_j z^{i_j}$;

end

calculate code polynomial $\hat{b}(z) = r(z) - \hat{e}(z)$;

Figure 2.61: Algebraic decoding algorithm for narrow-sense BCH codes

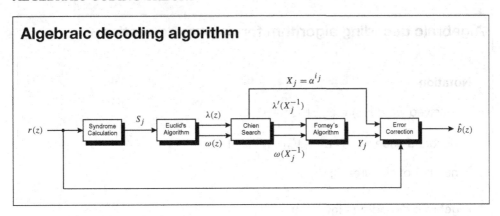

Figure 2.62: Block diagram of algebraic decoding algorithm

The maximum likelihood decoding strategy has been derived in this chapter as an optimal decoding algorithm that minimises the word error probability (Bossert, 1999). Instead of maximum likelihood decoding, a symbol-by-symbol MAP decoding algorithm can be implemented on the basis of the BCJR algorithm (Bahl *et al.*, 1974). This algorithm is derived by representing block codes with the help of trellis diagrams which will be discussed in detail in the context of convolutional codes in the next chapter. Furthermore, we have mainly discussed hard-decision decoding schemes because they prevail in today's applications of linear block codes and cyclic codes. However, as was indicated for the decoding of Reed–Muller codes, soft-decision decoding schemes usually lead to a smaller error probability. Further information about the decoding of linear block codes, including hard-decision and soft-decision decoding algorithms, is given elsewhere (Bossert, 1999; Lin and Costello, 2004).

The algebraic coding theory as treated in this chapter is nowadays often called the 'classical' coding theory. After Shannon's seminal work (Shannon, 1948), which laid the foundation of information theory, the first class of systematic single-error correcting channel codes was invented by Hamming (Hamming, 1950). Channel codes that are capable of correcting more than a single error were presented by Reed and Muller, leading to the Reed–Muller codes which have been applied, for example, in space communications within the Mariner and Viking Mars mission (Costello *et al.*, 1998; Muller, 1954). Several years later, BCH codes were developed by Bose, Ray-Chaudhuri and Hocquenghem (Bose *et al.*, 1960). In the same year, Reed and Solomon introduced Reed–Solomon codes which found a wide area of applications ranging from space communications to digital video broadcasting to the compact disc (Reed and Solomon, 1960). Algebraic decoding algorithms were found by Peterson for binary codes as well as by Gorenstein and Zierler for q-nary codes (Gorenstein and Zierler, 1961; Peterson, 1960). With the help of efficiently implementable iterative decoding algorithms proposed by Berlekamp and Massey, these

Algebraic decoding algorithm for erasure correction

Notation

received word $r(z) \in \mathbb{F}_q[z]/(z^n - 1)$;

decoded code word $\hat{b}(z) \in \mathbb{F}_q[z]/(z^n - 1)$;

number of erasures w;

Algebraic decoding algorithm

calculate syndromes $S_j = r(\alpha^j) = \sum_{i=0}^{n-1} r_i \alpha^{ij}$ for $1 \leq j \leq 2t$;

calculate syndrome polynomial $S(z) = \sum_{j=1}^{2t} S_j z^{j-1}$;

calculate error locators $X_j = \alpha^{i_j}$;

calculate error locator polynomial $\lambda(z) = \prod_{j=1}^{w} (1 - X_j z)$;

calculate error evaluator polynomial $\omega(z) \equiv S(z)\lambda(z)$ modulo z^{2t};

calculate error values $Y_j = -\frac{\omega(X_j^{-1})}{\lambda'(X_j^{-1})}$ for $1 \leq j \leq w$;

calculate error polynomial $\hat{e}(z) = \sum_{j=1}^{w} Y_j z^{i_j}$;

calculate code polynomial $\hat{b}(z) = r(z) - \hat{e}(z)$.

Figure 2.63: Algebraic decoding algorithm for erasure correction for narrow-sense BCH codes

algebraic decoding algorithms became applicable for practical applications (Berlekamp, 1984; Massey, 1969). Further details about important applications of these classical codes, as well as codes discussed in the following chapters, can be found elsewhere (Costello et al., 1998).

3

Convolutional Codes

In this chapter we discuss binary convolutional codes. Convolutional codes belong, like the most practical block codes, to the class of linear codes. Similarly to linear block codes, a convolutional code is defined by a linear mapping of k information symbols to n code symbols. However, in contrast to block codes, convolutional codes have a memory, i.e. the current encoder output of the convolutional encoder depends on the current k input symbols and on (possibly all) previous inputs.

Today, convolutional codes are used for example in Universal Mobile Telecommunications System (UMTS) and Global System for Mobile communications (GSM) digital cellular systems, dial-up modems, satellite communications, 802.11 wireless Local Area Networks (LANs) and many other applications. The major reason for this popularity is the existence of efficient decoding algorithms. Furthermore, those decoding methods can utilise soft-input values from the demodulator which leads to significant performance gains compared with hard-input decoding.

The widespread Viterbi algorithm is a maximum likelihood decoding procedure that can be implemented with reasonable complexity in hardware as well as in software. The Bahl, Cocke, Jelinek, Raviv (BCJR) algorithm is a method to calculate reliability information for the decoder output. This so-called soft output is essential for the decoding of concatenated convolutional codes (turbo codes), which we will discuss in Chapter 4. Both the BCJR algorithm and the Viterbi algorithm are based on the trellis representation of the code. The highly repetitive structure of the code trellis makes trellis-based decoding very suitable for pipelining hardware implementations.

We will concentrate on rate $1/n$ binary linear time-invariant convolutional codes. These are the easiest to understand and also the most useful in practical applications. In Section 3.1, we discuss the encoding of convolutional codes and some of their basic properties.

Section 3.2 is dedicated to the famous Viterbi algorithm and the trellis representation of convolutional codes. Then, we discuss distance properties and error correction capabilities. The concept of soft-output decoding and the BCJR algorithm follow in Section 3.5.

Finally, we consider an application of convolutional codes for mobile communication channels. Section 3.6.2 deals with the hybrid Automatic Repeat Request (ARQ) protocols as defined in the GSM standard for enhanced packet data services. Hybrid ARQ protocols are a combination of forward error correction and the principle of automatic repeat request, i.e. to detect and repeat corrupted data. The basic idea of hybrid ARQ is that the error correction capability increases with every retransmission. This is an excellent example of an adaptive coding system for strongly time-variant channels.

3.1 Encoding of Convolutional Codes

In this section we discuss how convolutional codes are encoded. Moreover, we establish the notation for the algebraic description and consider some structural properties of binary linear convolutional codes.

3.1.1 Convolutional Encoder

The easiest way to introduce convolutional codes is by means of a specific example. Consider the convolutional encoder depicted in Figure 3.1. Information bits are shifted into a register of length $m = 2$, i.e. a register with two binary memory elements. The output sequence results from multiplexing the two sequences denoted by $\mathbf{b}^{(1)}$ and $\mathbf{b}^{(2)}$. Each output bit is generated by modulo 2 addition of the current input bit and some symbols of the register contents. For instance, the information sequence $\mathbf{u} = (1, 1, 0, 1, 0, 0, \ldots)$ will be encoded to $\mathbf{b}^{(1)} = (1, 0, 0, 0, 1, 1, \ldots)$ and $\mathbf{b}^{(2)} = (1, 1, 1, 0, 0, 1, \ldots)$. The generated code sequence after multiplexing of $\mathbf{b}^{(1)}$ and $\mathbf{b}^{(2)}$ is $\mathbf{b} = (1, 1, 0, 1, 0, 1, 0, 0, 1, 0, 1, 1, \ldots)$.

A convolutional encoder is a linear sequential circuit and therefore a Linear Time-Invariant (LTI) system. It is well known that an LTI system is completely characterized

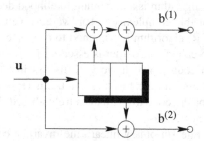

A rate $R = 1/2$ convolutional encoder with memory $m = 2$

- Each output block is calculated by modulo 2 addition of the current input block and some symbols of the register contents:

$$b_i^{(1)} = u_i + u_{i-1} + u_{i-2},$$
$$b_i^{(2)} = u_i + u_{i-2}.$$

Figure 3.1: Example of a convolutional encoder

CONVOLUTIONAL CODES

by its impulse response. Let us therefore investigate the two impulse responses of this particular encoder. The information sequence $\mathbf{u} = 1, 0, 0, 0, \ldots$ results in the output $\mathbf{b}^{(1)} = (1, 1, 1, 0, \ldots)$ and $\mathbf{b}^{(2)} = (1, 0, 1, 0, \ldots)$, i.e. we obtain the generator impulse responses $\mathbf{g}^{(1)} = (1, 1, 1, 0, \ldots)$ and $\mathbf{g}^{(2)} = (1, 0, 1, 0, \ldots)$ respectively. These generator impulse responses are helpful for calculating the output sequences for an arbitrary input sequence

$$b_i^{(1)} = \sum_{l=0}^{m} u_{i-l} g_l^{(1)} \leftrightarrow \mathbf{b}^{(1)} = \mathbf{u} * \mathbf{g}^{(1)},$$
$$b_i^{(2)} = \sum_{l=0}^{m} u_{i-l} g_l^{(2)} \leftrightarrow \mathbf{b}^{(2)} = \mathbf{u} * \mathbf{g}^{(2)}.$$

The generating equations for $\mathbf{b}^{(1)}$ and $\mathbf{b}^{(2)}$ can be regarded as convolutions of the input sequence with the generator impulse responses $\mathbf{g}^{(1)}$ and $\mathbf{g}^{(2)}$. The code $\mathbb{B}$ generated by this encoder is the set of all output sequences $\mathbf{b}$ that can be produced by convolution of arbitrary input sequence $\mathbf{u}$ with the generator impulse responses. This explains the name *convolutional codes*.

The general encoder of a rate $R = k/n$ convolutional code is depicted in Figure 3.2. Each input corresponds to a shift register, i.e. each information sequence is shifted into its own register. In contrast to block codes, the ith code block $\mathbf{b}_i = b_i^1, b_i^2, \ldots, b_i^n$ of a convolutional code sequence $\mathbf{b} = \mathbf{b}_0, \mathbf{b}_1, \ldots$ is a linear function of several information blocks $\mathbf{u}_j = u_j^1, u_j^2, \ldots, u_j^k$ with $j \in \{i - m, \ldots, i\}$ and not only of $\mathbf{b}_i$. The integer m

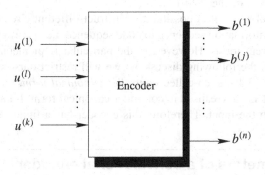

Convolutional encoder with k inputs and n outputs

- k and n denote the number of encoder inputs and outputs respectively. Thus, the code rate is $R = \frac{k}{n}$.

- The current n outputs are linear combinations of the present k input bits and the previous $k \times m$ input bits, where m is called the memory of the convolutional code.

- A binary convolutional code is often denoted by a three-tuple (n, k, m).

Figure 3.2: Convolutional encoder with k inputs and n outputs

denotes the *memory* of the encoder. The k encoder registers do not necessarily have the same length. We denote the number of memory elements of the lth register by v_l. The n output sequences may depend on any of the k registers. Thus, we require $k \times n$ impulse responses to characterise the encoding. If the shift registers are feedforward registers, i.e. they have no feedback, than the corresponding generator impulse responses $\mathbf{g}_l^{(j)}$ are limited to length $v_l + 1$. For this reason, v_l is often called the *constraint length* of the lth input sequence. The memory m of the encoder is

$$m = \max_l v_l.$$

The memory parameters of a convolutional encoder are summarised in Figure 3.3. A binary convolutional code is often denoted by a three-tuple $\mathbb{B}(n, k, m)$. For instance, $\mathbb{B}(2, 1, 2)$ represents the code corresponding to the encoder in Figure 3.1.

The reader is probably familiar with the theory of discrete-time LTI systems over the real or the complex field, which are sometimes called discrete-time real or complex filters. Similarly, convolutional encoders can be regarded as filters over finite fields – in our case the field $\mathbb{F}_2$. The theory of discrete-time LTI systems over the field $\mathbb{F}_2$ is similar to the theory for real or complex fields, except that over a finite field there is no notion of convergence of an infinite sum. We will observe that a little linear system theory is helpful in the context of convolutional codes. For example, we know different methods to construct filters for a given transfer function from system theory. The structure depicted in Figure 3.4 is a canonical form of a transversal filter, where usually the forward coefficients p_l and the backward coefficients q_l are real numbers. In case of a convolutional encoder, these coefficients are elements of the field $\mathbb{F}_2$.

With convolutional codes it is possible to construct different circuits that result in the same mapping from information sequence to code sequence, i.e. a particular encoder can be implemented in different forms. However, as the particular implementation of an encoder plays a certain role in the following discussion, we will restrict ourselves to the canonical construction in Figure 3.4, the so-called *controller canonical form*. Our example encoder depicted in Figure 3.1 is also realised in controller canonical form. It has no feedback from the register outputs to the input. Therefore, this encoder has a finite impulse response. In

---

Memory parameters of a convolutional encoder

As the numbers of memory elements of the k encoder registers may differ, we define

- the *memory* $m = \max_l v_l$ of the encoder
- the *minimum constraint length* $v_{\min} = \min_l v_l$,
- and the *overall constraint length* $v = \sum_{l=1}^{k} v_l$.

Figure 3.3: Memory parameters of a convolutional encoder

CONVOLUTIONAL CODES

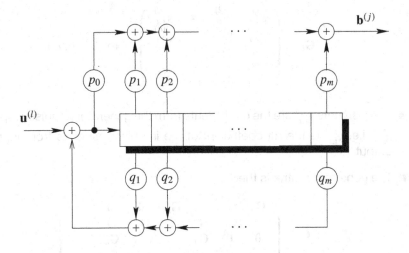

Figure 3.4: Rational transfer functions

general, an encoder may have feedback according to Figure 3.4 and therefore an infinite impulse response.

3.1.2 Generator Matrix in the Time Domain

Similarly to linear block codes, the encoding procedure can be described as a multiplication of the information sequence with a generator matrix $\mathbf{b} = \mathbf{uG}$. However, the information sequence $\mathbf{u}$ and the code sequence $\mathbf{b}$ are semi-infinite. Therefore, the generator matrix of a convolutional code also has a semi-infinite structure. It is constructed from $k \times n$ submatrices $\mathbf{G}_i$ according to Figure 3.5, where the elements of the submatrices are the coefficients from the generator impulse responses.

For instance, for the encoder in Figure 3.1 we obtain

$$\mathbf{G}_0 = (g_0^{(1)}, g_0^{(2)}) = (1\,1),$$
$$\mathbf{G}_1 = (g_1^{(1)}, g_1^{(2)}) = (1\,0),$$
$$\mathbf{G}_2 = (g_2^{(1)}, g_2^{(2)}) = (1\,1).$$

Generator matrix in the time domain

- The generator matrix is constructed from $k \times n$ submatrices

$$G_i = \begin{pmatrix} g_{1,i}^{(1)} & g_{1,i}^{(2)} & \cdots & g_{1,i}^{(n)} \\ g_{2,i}^{(1)} & g_{2,i}^{(2)} & \cdots & g_{2,i}^{(n)} \\ \vdots & \vdots & & \vdots \\ g_{k,i}^{(1)} & g_{k,i}^{(2)} & \cdots & g_{k,i}^{(n)} \end{pmatrix} \quad \text{for } i \in [0, m].$$

- The elements $g_{l,i}^{(j)}$ are the coefficients from the generator impulse responses $\mathbf{g}_l^{(j)}$, i.e. $g_{l,i}^{(j)}$ is the ith coefficient of the impulse response from input l to output j.

- The generator matrix is then

$$\mathbf{G} = \begin{pmatrix} \mathbf{G}_0 & \mathbf{G}_1 & \cdots & \mathbf{G}_m & \mathbf{0} & \mathbf{0} & \cdots \\ \mathbf{0} & \mathbf{G}_0 & \mathbf{G}_1 & \cdots & \mathbf{G}_m & \mathbf{0} & \cdots \\ \mathbf{0} & \mathbf{0} & \mathbf{G}_0 & \mathbf{G}_1 & \cdots & \mathbf{G}_m & \cdots \\ \mathbf{0} & \mathbf{0} & \mathbf{0} & \ddots & \ddots & & \ddots \end{pmatrix},$$

where a bold zero indicates an all-zero matrix.

Figure 3.5: Generator matrix in the time domain

Finally, the generator matrix is

$$\mathbf{G} = \begin{pmatrix} \mathbf{G}_0 & \mathbf{G}_1 & \cdots & \mathbf{G}_m & \mathbf{0} & \mathbf{0} & \cdots \\ \mathbf{0} & \mathbf{G}_0 & \mathbf{G}_1 & \cdots & \mathbf{G}_m & \mathbf{0} & \cdots \\ \mathbf{0} & \mathbf{0} & \mathbf{G}_0 & \mathbf{G}_1 & \cdots & \mathbf{G}_m & \cdots \\ \mathbf{0} & \mathbf{0} & \mathbf{0} & \ddots & \ddots & & \ddots \end{pmatrix} = \begin{pmatrix} 11 & 10 & 11 & 00 & 00 & \cdots \\ 00 & 11 & 10 & 11 & 00 & \cdots \\ 00 & 00 & 11 & 10 & 11 & \cdots \\ 00 & 00 & 00 & \ddots & & \ddots \end{pmatrix}.$$

With this generator matrix we can express the encoding of an information sequence, for instance $\mathbf{u} = (1, 1, 0, 1, 0, 0, \ldots)$, by a matrix multiplication

$$\mathbf{b} = \mathbf{u}\mathbf{G} = (1, 1, 0, 1, 0, 0, \ldots) \begin{pmatrix} 11 & 10 & 11 & 00 & 00 & \cdots \\ 00 & 11 & 10 & 11 & 00 & \cdots \\ 00 & 00 & 11 & 10 & 11 & \cdots \\ 00 & 00 & 00 & \ddots & & \ddots \end{pmatrix}$$

$$= (1\,1, 0\,1, 0\,1, 0\,0, 1\,0, 1\,1, 0\,0, \ldots).$$

CONVOLUTIONAL CODES

3.1.3 State Diagram of a Convolutional Encoder

Up to now we have considered two methods to describe the encoding of a convolutional code, i.e. encoding with a linear sequential circuit, a method that is probably most suitable for hardware implementations, and a formal description with the generator matrix. Now we consider a graphical representation, the so-called *state diagram*. The state diagram will be helpful later on when we consider distance properties of convolutional codes and decoding algorithms.

The state diagram of a convolutional encoder describes the operation of the encoder. From this graphical representation we observe that the encoder is a finite-state machine. For the construction of the state diagram we consider the contents of the encoder registers as *encoder state* σ. The set $\$$ of encoder states is called the *encoder state space*.

Each memory element contains only 1 bit of information. Therefore, the number of encoder states is 2^ν. We will use the symbol σ_i to denote the encoder state at time i. The state diagram is a graph that has 2^ν nodes representing the encoder states. An example of the state diagram is given in Figure 3.6. The branches in the state diagram represent possible state transitions, e.g. if the encoder in Figure 3.1 has the register contents

State diagram of a convolutional encoder

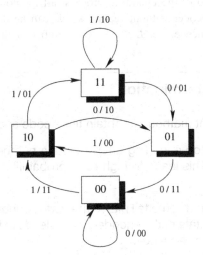

- Each node represents an encoder state, i.e. the binary contents of the memory elements.
- Each branch represents a state transition and is labelled by the corresponding k input and n output bits.

Figure 3.6: State diagram of the encoder in Figure 3.1

$\sigma_i = (00)$ and the input bit u_i at time i is a 1, then the state of the encoder changes from $\sigma_i = (00)$ to $\sigma_{i+1} = (10)$. Along with this transition, the two output bits $\mathbf{b}_i = (11)$ are generated. Similarly, the information sequence $\mathbf{u} = (1, 1, 0, 1, 0, 0, \ldots)$ corresponds to the state sequence $\sigma = (00, 10, 11, 01, 10, 01, 00, \ldots)$, subject to the encoder starting in the all-zero state. The code sequence is again $\mathbf{b} = (1\,1, 0\,1, 0\,1, 0\,0, 1\,0, 1\,1, 0\,0, \ldots)$. In general, the output bits only depend on the current input and the encoder state. Therefore, we label each transition with the k input bits and the n output bits (input/output).

3.1.4 Code Termination

In theory, the code sequences of convolutional codes are of infinite length, but for practical applications we usually employ finite sequences. Figure 3.7 gives an overview of the three different methods to obtain finite code sequences that will be discussed below.

Assume we would like to encode exactly L information blocks. The easiest procedure to obtain a finite code sequence is code truncation. With this method we stop the encoder after L information blocks and also truncate the code sequence after L code blocks. However, this straightforward approach leads to a substantial degradation of the error protection for the last encoded information bits, because the last encoded information bits influence only a small number of code bits. For instance, the last k information bits determine the last n code bits. Therefore, termination or tail-biting is usually preferred over truncation.

In order to obtain a terminated code sequence, we start encoding in the all-zero encoder state and we ensure that, after the encoding process, all memory elements contain zeros again. In the case of an encoder without feedback this can be done by adding $k \cdot m$ zero bits to the information sequence. Let L denote the number of information blocks, i.e. we

---

Methods for code termination

There are three different methods to obtain finite code sequences:

- **Truncation:** We stop encoding after a certain number of bits without any additional effort. This leads to high error probabilities for the last bits in a sequence.

- **Termination:** We add some tail bits to the code sequence in order to ensure a predefined end state of the encoder, which leads to low error probabilities for the last bits in a sequence.

- **Tail-biting:** We choose a starting state that ensures that the starting and end states are the same. This leads to equal error protection.

- Note, tail-biting increases the decoding complexity, and for termination additional redundancy is required.

Figure 3.7: Methods for code termination

CONVOLUTIONAL CODES

encode kL information bits. Adding $k \cdot m$ tail bits decreases the code rate to

$$R_{\text{terminated}} = \frac{kL}{n(L+m)} = R\frac{L}{L+m},$$

where $L/(L+m)$ is the so-called *fractional rate loss*. But now the last k information bits have code constraints over $n(m+1)$ code bits. The $kL \times n(L+m)$ generator matrix now has a finite structure

$$\mathbf{G} = \begin{pmatrix} \mathbf{G}_0 & \mathbf{G}_1 & \cdots & \mathbf{G}_m & 0 & \cdots & 0 \\ 0 & \mathbf{G}_0 & \mathbf{G}_1 & \cdots & \mathbf{G}_m & \cdots & 0 \\ 0 & 0 & \ddots & \ddots & & \ddots & \\ 0 & 0 & \cdots & \mathbf{G}_0 & \mathbf{G}_1 & \cdots & \mathbf{G}_m \end{pmatrix}.$$

The basic idea of tail-biting is that we start the encoder in the same state in which it will stop after the input of L information blocks. For an encoder without feedback, this means that we first encode the last m information blocks in order to determine the starting state of the encoder. Keeping this encoder state, we restart the encoding at the beginning of the information sequence. With this method the last m information blocks influence the first code symbols, which leads to an equal protection of all information bits. The influence of the last information bits on the first code bits is illustrated by the generator matrix of the tail-biting code. On account of the tail-biting, the generator matrix now has a finite structure. It is a $kL \times nL$ matrix

$$\mathbf{G} = \begin{pmatrix} \mathbf{G}_0 & \mathbf{G}_1 & \cdots & \mathbf{G}_m & 0 & \cdots & 0 \\ 0 & \mathbf{G}_0 & \mathbf{G}_1 & \cdots & \mathbf{G}_m & \cdots & 0 \\ 0 & 0 & \ddots & \ddots & & \ddots & \\ 0 & 0 & \cdots & \mathbf{G}_0 & \mathbf{G}_1 & \cdots & \mathbf{G}_m \\ \mathbf{G}_m & 0 & 0 & 0 & \ddots & & \vdots \\ \vdots & \ddots & 0 & \cdots & 0 & \mathbf{G}_0 & \mathbf{G}_1 \\ \mathbf{G}_1 & \cdots & \mathbf{G}_m & 0 & \cdots & 0 & \mathbf{G}_0 \end{pmatrix}.$$

For instance, the tail-biting generator matrix for the specific code $\mathbb{B}(2, 1, 2)$ is

$$\mathbf{G} = \begin{pmatrix} 11 & 10 & 11 & 00 & \cdots & 00 \\ 00 & 11 & 10 & 11 & \cdots & 00 \\ 00 & 00 & \ddots & \ddots & & \vdots \\ 00 & 00 & \cdots & 11 & 10 & 11 \\ 11 & 00 & 00 & 00 & \ddots & \vdots \\ 10 & 11 & 00 & \cdots & 00 & 11 \end{pmatrix}.$$

From this matrix, the influence of the last two information bits on the first four code bits becomes obvious.

Code termination and tail-biting, as discussed above, can only be applied to convolutional encoders without feedback. Such encoders are commonly used when the forward error correction is solely based on the convolutional code. In concatenated systems, as discussed

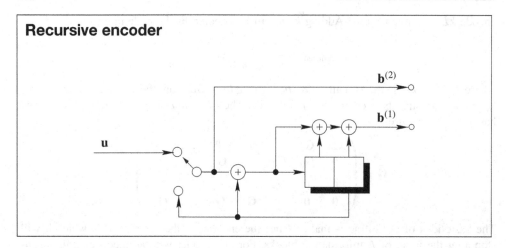

Figure 3.8: Recursive encoder for the code $\mathbb{B}(2,1,2)$

in Chapter 4, usually recursive encoders, i.e. encoders with feedback, are employed. In the case of recursive encoders, we cannot simply add m zeros to the information sequence. The termination sequence for a recursive encoder depends on the encoder state. However, the termination sequence can be easily determined. Consider, for instance, the encoder in Figure 3.8. To encode the information bits, the switch is in the upper position. The tail bits are then generated by shifting the switch into the down position. This forces the encoder state back to the all-zero state in m transitions at most.

The method of tail-biting for recursive encoders is more complicated, because in this case the start state depends on the complete information sequence. An algorithm to calculate the encoder start state for tail-biting with recursive encoders is presented elsewhere (Weiss et al., 1998).

3.1.5 Puncturing

In Figure 3.2 we have introduced the general convolutional encoder with k inputs and n outputs. In practical applications we will usually only find encoders for rate $1/n$ convolutional codes, i.e. with one input. One reason for this is that there exists a simple method to construct high-rate codes from rate $R = 1/n$ codes. This method is called puncturing, because a punctured code sequence is obtained by periodically deleting a part of the code bits of rate $R = 1/n$ code (cf. Section 2.2.7). However, the main reason is that the decoding of high-rate convolutional codes can be significantly simplified by using punctured codes. This will be discussed in more detail in Section 3.4.2. Tables of punctured convolutional codes with a large free distance can be found elsewhere (Lee, 1997).

Consider again the convolutional encoder shown in Figure 3.1. For each encoding step the encoder produces two code bits $b_i^{(1)}$ and $b_i^{(2)}$. Now let us delete all bits of the sequence $b^{(2)}$ for odd times i, i.e. we delete $b_{2i+1}^{(2)}$ from each 4-bit output block. We obtain the encoder output

$$\mathbf{b} = (b_0^{(1)}, b_0^{(2)}, b_1^{(1)}, b_1^{(2)}, b_2^{(1)}, b_2^{(2)}, \ldots) \Rightarrow \mathbf{b}_{\text{punctured}} = (b_0^{(1)}, b_0^{(2)}, b_1^{(1)}, b_2^{(1)}, b_2^{(2)}, \ldots).$$

CONVOLUTIONAL CODES

Code puncturing

- A punctured code sequence is obtained by periodically deleting a part of the code bits of rate $R = 1/n$ code.

- The puncturing pattern is determined by an $n \times T$ puncturing matrix **P**, where the jth row corresponds to the jth encoder output and T is the period of the puncturing pattern. A zero element means that the corresponding code bit will be deleted, a code bit corresponding to a 1 will be submitted.

- Example: Puncturing of the code $\mathbb{B}(2, 1, 2)$ according to $\mathbf{P} = \begin{pmatrix} 1 & 1 \\ 1 & 0 \end{pmatrix}$ results in a code $\mathbb{B}_{\text{punctured}}(3, 2, 1)$.

Figure 3.9: Code puncturing

We can express the periodic deletion pattern by an $n \times T$ puncturing matrix, where the jth row corresponds to the jth encoder output and T is the period of the puncturing pattern. For our example we have

$$\mathbf{P} = \begin{pmatrix} 1 & 1 \\ 1 & 0 \end{pmatrix},$$

where '0' means that the corresponding code bit will be deleted; a code bit corresponding to a '1' will be submitted. The puncturing procedure is summarised in Figure 3.9. Obviously, the original rate $R = 1/2$ mother code has become a rate $R = 2/3$ code after puncturing according to **P**. But what is the generator matrix of this code and what does the corresponding encoder look like?

To answer these questions, we consider the semi-infinite generator matrix of the mother code

$$\mathbf{G} = \begin{pmatrix} 11 & 10 & 11 & 00 & 00 & \ldots \\ 00 & 11 & 10 & 11 & 00 & \ldots \\ 00 & 00 & 11 & 10 & 11 & \ldots \\ 00 & 00 & 00 & \ddots & & \ddots \end{pmatrix}.$$

This matrix was constructed from the submatrices

$$\mathbf{G}_0 = (11), \mathbf{G}_1 = (10) \text{ and } \mathbf{G}_2 = (11).$$

However, we can also find another interpretation of the generator matrix **G**. Considering the submatrices

$$\mathbf{G}'_0 = \begin{pmatrix} 1 & 1 & 1 & 0 \\ 0 & 0 & 1 & 1 \end{pmatrix} \text{ and } \mathbf{G}'_1 = \begin{pmatrix} 1 & 1 & 0 & 0 \\ 1 & 0 & 1 & 1 \end{pmatrix}$$

we can interpret the matrix **G** as the generator matrix of a code with the parameters $k = 2$, $n = 4$ and $m = 1$. Note that other interpretations with $k = 3, 4, \ldots$ would also be possible. The generator matrix of the punctured code is obtained by deleting every column that corresponds to a deleted code bit. In our example, this is every fourth column, resulting in

$$\mathbf{G}_{\text{punctured}} = \begin{pmatrix} 11 & 1 & 11 & 0 & 00 & \cdots \\ 00 & 1 & 10 & 1 & 00 & \cdots \\ 00 & 0 & 11 & 1 & 11 & \cdots \\ 00 & 0 & 00 & & \ddots & \ddots \end{pmatrix}$$

and the submatrices

$$\mathbf{G}'_{0,\text{punctured}} = \begin{pmatrix} 1 & 1 & 1 \\ 0 & 0 & 1 \end{pmatrix} \text{ and } \mathbf{G}'_{1,\text{punctured}} = \begin{pmatrix} 1 & 1 & 0 \\ 1 & 0 & 1 \end{pmatrix}.$$

From these two matrices we could deduce the six generator impulse responses of the corresponding encoder. However, the construction of the encoder is simplified when we consider the generator matrix in the D-domain, which we will discuss in the next section.

3.1.6 Generator Matrix in the D-Domain

As mentioned above, an LTI system is completely characterised by its impulse response. However, it is sometimes more convenient to specify an LTI system by its transfer function, in particular if the impulse response is infinite. Moreover, the fact that the input/output relation of an LTI system may be written as a convolution in the time domain or as a multiplication in a transformed domain suggests the use of a transformation in the context of convolutional codes. We use the D-transform

$$\mathbf{x} = x_i, x_{i+1}, x_{i+2}, \ldots \circ\!\!-\!\!\bullet X(D) = \sum_{i=j}^{+\infty} x_i D^i, \; j \in \mathbb{Z}.$$

Using the D-transform, the sequences of information and code blocks can be expressed in terms of the delay operator D as follows

$$\mathbf{u}(D) = \mathbf{u}_0 + \mathbf{u}_1 D + \mathbf{u}_2 D^2 + \cdots,$$
$$\mathbf{b}(D) = \mathbf{b}_0 + \mathbf{b}_1 D + \mathbf{b}_2 D^2 + \cdots.$$

Moreover, infinite impulse responses can be represented by rational transfer functions

$$G_{l,j}(D) = \frac{P_{l,j}(D)}{Q_{l,j}(D)} = \frac{p_0 + p_1 D + \cdots + p_m D^m}{1 + q_1 D + \cdots + q_m D^m}.$$

The encoding process that is described as a convolution in the time domain can be expressed by a simple multiplication in the D-domain

$$\mathbf{b}(D) = \mathbf{U}(D)\mathbf{G}(D),$$

where $\mathbf{G}(D)$ is the encoding (or generator) matrix of the convolutional code. In general, a generator matrix $\mathbf{G}(D)$ is a $k \times n$ matrix. The elements $G_{l,j}(D)$ of this matrix are realisable rational functions. The term *realisable* reflects the fact that a linear sequential circuit always

CONVOLUTIONAL CODES

has a causal impulse response. Hence, we can only realise transfer functions that result in a causal impulse response. For the encoder in Figure 3.1 we obtain

$$G(D) = (1 + D + D^2, 1 + D^2).$$

The generator matrix of the punctured convolutional code from Section 3.1.5 is

$$\begin{aligned}
\mathbf{G}_{\text{punctured}}(D) &= \mathbf{G}'_{0,\text{punctured}} + D\mathbf{G}'_{1,\text{punctured}} \\
&= \begin{pmatrix} 1 & 1 & 1 \\ 0 & 0 & 1 \end{pmatrix} + D \begin{pmatrix} 1 & 1 & 0 \\ 1 & 0 & 1 \end{pmatrix} \\
&= \begin{pmatrix} 1+D & 1+D & 1 \\ D & 0 & 1+D \end{pmatrix}.
\end{aligned}$$

Considering the general controller canonical form of a shift register in Figure 3.4, it is now easy to construct the encoder of the punctured code as given in Figure 3.10.

Usually an *octal notation* is used to specify the generator polynomials of a generator matrix. For example, the polynomial $g(D) = 1 + D + D^2 + D^4$ can be represented by a binary vector $\mathbf{g} = (11101)$, where the elements are the binary coefficients of the polynomial

---

Encoder of a punctured code

- Puncturing of the code $\mathbb{B}(2, 1, 2)$ according to $\mathbf{P} = \begin{pmatrix} 1 & 1 \\ 1 & 0 \end{pmatrix}$ results in a code $\mathbb{B}_{\text{punctured}}(3, 2, 1)$ with the following encoder

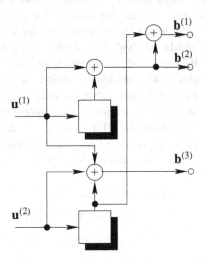

Figure 3.10: Encoder of the punctured code $\mathbb{B}_{\text{punctured}}(3, 2, 1)$

$g(D)$ with the coefficient of the lowest order in the leftmost position. To convert this vector to octal numbers, we use right justification, i.e. $\mathbf{g} = (11101)$ becomes $\mathbf{g} = (011101)$. Now, the binary vector $\mathbf{g} = (011101)$ is equivalent to $(35)_8$ in octal notation. For example, the generator matrix of our canonical example code $\mathbb{B}(2, 1, 2)$ can be stated as $(7\,5)_8$, whereas the generator matrix $(35\,23)_8$ defines the code $\mathbb{B}(2, 1, 4)$.

3.1.7 Encoder Properties

We have already mentioned that a particular encoder is just one possible realisation of a generator matrix. Moreover, there exist a number of different generator matrices that produce the same set of output sequences. Therefore, it is important to distinguish between properties of encoders, generator matrices and codes.

A convolutional code is the set of all possible output sequences of a convolutional encoder. Two encoders are called *equivalent encoders* if they encode the same code. Two encoding matrices $\mathbf{G}(D)$ and $\mathbf{G}'(D)$ are called *equivalent generator matrices* if they encode the same code. For equivalent generator matrices $\mathbf{G}(D)$ and $\mathbf{G}'(D)$ we have

$$\mathbf{G}'(D) = \mathbf{T}(D)\mathbf{G}(D),$$

with $\mathbf{T}(D)$ a non-singular $k \times k$ matrix.

We call $\mathbf{G}(D)$ a *polynomial generator matrix* if $Q_{l,j}(D) = 1$ for all submatrices $G_{l,j}(D)$. Some polynomial generator matrices lead to an undesired mapping from information sequence to code sequence. That is, an information sequence containing many (possibly infinite) 1s is mapped to a code sequence with only a few (finite) number of 1s. As a consequence, a small number of transmission errors can lead to a large (possibly infinite) number of errors in the estimated information sequence. Such a generator matrix is called a *catastrophic generator matrix*. Note that the catastrophic behaviour is not a property of the code but results from the mapping of information sequence to code sequence. Hence, it is a property of the generator matrix. Methods to test whether a generator matrix is catastrophic can be found elsewhere (Bossert, 1999; Johannesson and Zigangirov, 1999). The state diagram of the encoder provides a rather obvious condition for a catastrophic mapping. If there exists a loop (a sequence of state transitions) in the state diagram that produces zero output for a non-zero input, this is a clear indication of catastrophic mapping.

For example, the generator matrix $\mathbf{G}(D) = (1 + D, 1 + D^2)$ is catastrophic. The corresponding encoder and the state diagram are depicted in Figure 3.11, where the critical loop in the state diagram is indicated by a dashed line. The generator matrix results from

$$\begin{aligned}\mathbf{G}(D) &= T(D)\mathbf{G}'(D) \\ &= (1 + D)(1, 1 + D) \\ &= (1 + D, 1 + D^2).\end{aligned}$$

Note that in our example all elements of $\mathbf{G}(D)$ have a common factor $(1 + D)$. In general, a generator matrix $\mathbf{G}(D)$ of a rate $1/n$ code is catastrophic if and only if all elements of $\mathbf{G}(D)$ have a common factor other than D.

CONVOLUTIONAL CODES

Catastrophic encoder with state diagram

- Catastrophic encoder:

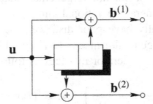

- State diagram with zero loop for non-zero input:

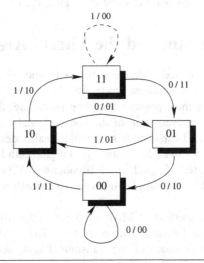

Figure 3.11: Encoder and state diagram corresponding to the catastrophic generator matrix $\mathbf{G}(D) = (1 + D, 1 + D^2)$

A *systematic encoding* results in a mapping where the code sequence contains the unchanged information sequence. We call $\mathbf{G}(D)$ a *systematic generator matrix* if it satisfies

$$\mathbf{G}(D) = \begin{pmatrix} 1 & & & G_{1,k+1}(D) & \cdots & G_{1,n}(D) \\ & 1 & & \vdots & & \vdots \\ & & \ddots & & & \\ & & & 1 & G_{k,k+1}(D) & \cdots & G_{k,n}(D) \end{pmatrix}$$

$$= (\mathbf{I}_k \mathbf{A}(D)).$$

That is, the systematic generator matrix contains an $k \times k$ identity matrix $\mathbf{I}_k$. In general, the k unit vectors may be arbitrarily distributed over the n columns of the generator matrix. Systematic generator matrices will be of special interest in Chapter 4, because they are used to construct powerful concatenated codes. A systematic generator matrix is never catastrophic, because a non-zero input leads automatically to a non-zero encoder output. Every convolutional code has a systematic generator matrix. The elements of a systematic generator matrix will in general be rational functions. Codes with polynomial systematic generator matrices usually have poor distance- and error-correcting capabilities. For instance, for the polynomial generator matrix $\mathbf{G}(D) = (1 + D + D^2, 1 + D^2)$ we can write the two equivalent systematic generator matrices as follows

$$\mathbf{G}'(D) = \left(1, \frac{1+D^2}{1+D+D^2}\right) \text{ and } \mathbf{G}''(D) = \left(\frac{1+D+D^2}{1+D^2}, 1\right).$$

The encoder for the generator matrix $\mathbf{G}''(D)$ is depicted in Figure 3.8, i.e. the encoders in Figure 3.1 and Figure 3.8 encode the same code $\mathbb{B}(2, 1, 2)$.

3.2 Trellis Diagram and the Viterbi Algorithm

Up to now we have considered different methods to encode convolutional codes. In this section we discuss how convolutional codes can be used to correct transmission errors. We consider possibly the most popular decoding procedure, the *Viterbi algorithm*, which is based on a graphical representation of the convolutional code, the *trellis diagram*. The Viterbi algorithm is applied for decoding convolutional codes in Code Division Multiple Access (CDMA) (e.g. IS-95 and UMTS) and GSM digital cellular systems, dial-up modems, satellite communications (e.g. Digital Video Broadcast (DVB) and 802.11 wireless LANs. It is also commonly used in other applications such as speech recognition, keyword spotting and bioinformatics.

In coding theory, the method of MAP decoding is usually considered to be the optimal decoding strategy for forward error correction. The MAP decision rule can also be used either to obtain an estimate of the transmitted code sequence on the whole or to perform bitwise decisions for the corresponding information bits. The decoding is based on the received sequence and an a-priori distribution over the information bits. Figure 3.12 provides an overview over different decoding strategies for convolutional codes. All four formulae define decision rules. The first rule considers MAP sequence estimation, i.e. we are looking for the code sequence $\hat{\mathbf{b}}$ that maximizes the a-posteriori probability $\Pr\{\mathbf{b}|\mathbf{r}\}$. MAP decoding is closely related to the method of ML decoding. The difference is that MAP decoding exploits an a-priori distribution over the information bits, and with ML decoding we assume equally likely information symbols.

Considering Bayes law $\Pr\{\mathbf{r}\}\Pr\{\mathbf{b}|\mathbf{r}\} = \Pr\{\mathbf{r}|\mathbf{b}\}\Pr\{\mathbf{b}\}$, we can formulate the MAP rule as

$$\hat{\mathbf{b}} = \operatorname*{argmax}_{\mathbf{b}} \{\Pr\{\mathbf{b}|\mathbf{r}\}\} = \operatorname*{argmax}_{\mathbf{b}} \left\{\frac{\Pr\{\mathbf{r}|\mathbf{b}\}\Pr\{\mathbf{b}\}}{\Pr\{\mathbf{r}\}}\right\}.$$

The constant factor $\Pr\{\mathbf{r}\}$ is the same for all code sequences. Hence, it can be neglected and we obtain

$$\hat{\mathbf{b}} = \operatorname*{argmax}_{\mathbf{b}} \{\Pr\{\mathbf{r}|\mathbf{b}\}\Pr\{\mathbf{b}\}\}.$$

CONVOLUTIONAL CODES

Decoding strategies

- Maximum A-Posteriori (MAP) sequence estimation:

$$\hat{\mathbf{b}} = \underset{\mathbf{b}}{\operatorname{argmax}}\, \{\Pr\{\mathbf{b}|\mathbf{r}\}\} \qquad (3.1)$$

- Maximum Likelihood (ML) sequence estimation:

$$\hat{\mathbf{b}} = \underset{\mathbf{b}}{\operatorname{argmax}}\, \{\Pr\{\mathbf{r}|\mathbf{b}\}\} \qquad (3.2)$$

- Symbol-by-symbol MAP decoding:

$$\hat{u}_t = \underset{u_t}{\operatorname{argmax}}\, \{\Pr\{u_t|\mathbf{r}\}\} \quad \forall\, t \qquad (3.3)$$

- Symbol-by-symbol ML decoding:

$$\hat{u}_t = \underset{u_t}{\operatorname{argmax}}\, \{\Pr\{\mathbf{r}|u_t\}\} \quad \forall\, t \qquad (3.4)$$

Figure 3.12: Decoding rules for convolutional codes

The term $\Pr\{\mathbf{b}\}$ is the a-priori probability of the code sequence $\mathbf{b}$. If all information sequences and therefore all code sequences are equally likely, then $\Pr\{\mathbf{b}\}$ is also a constant. In this case we can neglect $\Pr\{\mathbf{b}\}$ in the maximisation and have

$$\hat{\mathbf{b}} = \underset{\mathbf{b}}{\operatorname{argmax}}\, \{\Pr\{\mathbf{r}|\mathbf{b}\}\},$$

the so-called ML sequence estimation rule. Hence, ML decoding is equivalent to MAP decoding if no a-priori information is available.

The Viterbi algorithm finds the best sequence $\hat{\mathbf{b}}$ according to the ML sequence estimation rule, i.e. it selects the code sequence $\hat{\mathbf{b}}$ that maximizes the conditional probability $\Pr\{\mathbf{r}|\mathbf{b}\}$

$$\hat{\mathbf{b}} = \underset{\mathbf{b}}{\operatorname{argmax}}\, \{\Pr\{\mathbf{r}|\mathbf{b}\}\}.$$

Therefore, implementations of the Viterbi algorithm are also called maximum likelihood decoders. In Section 3.5 we consider the BCJR algorithm that performs symbol-by-symbol decoding.

3.2.1 Minimum Distance Decoding

In Section 2.1.2 we have already considered the concept of maximum likelihood decoding and observed that for the Binary Symmetric Channel (BSC) we can actually apply minimum distance decoding in order to minimise the word error probability. To simplify the following discussion, we also consider transmission over the BSC. In later sections we will extend these considerations to channels with continuous output alphabets.

It is convenient to revise some results concerning minimum distance decoding. Remember that the binary symmetric channel has the input and output alphabet $\mathbb{F}_2 = \{0, 1\}$. The BSC complements the input symbol with probability $\varepsilon < \frac{1}{2}$. When an error occurs, a 0 is received as a 1, and vice versa. The received symbols do not reveal where the errors have occurred.

Let us first consider a binary block code $\mathbb{B}$ of length n and rate $R = \frac{k}{n}$. Let $\mathbf{r}$ denote the received sequence. In general, if $\mathbb{B}$ is used for transmission over a noisy channel and all code words $\mathbf{b} \in \mathbb{B}$ are a priori equally likely, then the optimum decoder must compute the 2^k probabilities $\Pr\{\mathbf{r}|\mathbf{b}\}$. In order to minimise the probability of a decoding error, the decoder selects the code word corresponding to the greatest value of $\Pr\{\mathbf{r}|\mathbf{b}\}$.

For the BSC, the error process is independent of the channel input, thus we have $\Pr\{\mathbf{r}|\mathbf{b}\} = \Pr\{\mathbf{r} - \mathbf{b}\}$. Furthermore, the error process is memoryless which yields

$$\Pr\{\mathbf{r}|\mathbf{b}\} = \Pr\{\mathbf{r} - \mathbf{b}\} = \prod_{i=1}^{n} \Pr\{r_i|b_i\}.$$

Note that we have $\Pr\{r_i|b_i\} = \varepsilon$ if r_i and b_i differ. Similarly, we have $\Pr\{r_i|b_i\} = 1 - \varepsilon$ for $r_i = b_i$. Now, consider the Hamming distance $\text{dist}(\mathbf{r}, \mathbf{b})$ between the received sequence $\mathbf{r}$ and a code word $\mathbf{b} \in \mathbb{B}$, i.e. the number of symbols where the two sequences differ. Using the Hamming distance, we can express the likelihood $\Pr\{\mathbf{r}|\mathbf{b}\}$ as

$$\Pr\{\mathbf{r}|\mathbf{b}\} = \prod_{i=1}^{n} \Pr\{r_i|b_i\} = \varepsilon^{\text{dist}(\mathbf{r},\mathbf{b})}(1 - \varepsilon)^{n - \text{dist}(\mathbf{r},\mathbf{b})},$$

because we have $\text{dist}(\mathbf{r}, \mathbf{b})$ bit positions where $\mathbf{r}$ and $\mathbf{b}$ differ and $n - \text{dist}(\mathbf{r}, \mathbf{b})$ positions where the two sequences match. Hence, for the likelihood function we obtain

$$\begin{aligned} \Pr\{\mathbf{r}|\mathbf{b}\} &= \varepsilon^{\text{dist}(\mathbf{r},\mathbf{b})}(1 - \varepsilon)^{n - \text{dist}(\mathbf{r},\mathbf{b})} \\ &= (1 - \varepsilon)^n \left(\frac{\varepsilon}{1-\varepsilon}\right)^{\text{dist}(\mathbf{r},\mathbf{b})}. \end{aligned}$$

The term $(1 - \varepsilon)^n$ is a constant and is therefore independent of $\mathbf{r}$ and $\mathbf{b}$. For $\varepsilon < 0.5$, the term $\left(\frac{\varepsilon}{1-\varepsilon}\right)$ satisfies $\left(\frac{\varepsilon}{1-\varepsilon}\right) < 1$. Thus, the maximum of $\Pr\{\mathbf{r}|\mathbf{b}\}$ is attained for the code word $\mathbf{b}$ that minimizes the distance $\text{dist}(\mathbf{r}, \mathbf{b})$. Consequently, the problem of finding the most likely code word can be solved for the BSC by minimum distance decoding. A minimum distance decoder selects the code word $\hat{\mathbf{b}}$ that minimizes the Hamming distance to the received sequence

$$\hat{\mathbf{b}} = \underset{\mathbf{b} \in \mathbb{B}}{\arg\min}\, \text{dist}(\mathbf{r}, \mathbf{b}).$$

The derivation of the minimum distance decoding rule is summarized in Figure 3.13. Viterbi's celebrated algorithm is an efficient implementation of minimum distance decoding. This algorithm is based on the trellis representation of the code, which we will discuss next.

CONVOLUTIONAL CODES

> **Minimum distance decoding**
>
> - In order to minimise the probability of a decoding error, the maximum likelihood decoder selects the code word corresponding to the greatest value of $\Pr\{\mathbf{r}|\mathbf{b}\}$.
>
> - The error process of the BSC is memoryless and independent of the channel input
>
> $$\Pr\{\mathbf{r}|\mathbf{b}\} = \Pr\{\mathbf{r} - \mathbf{b}\} = \prod_{i=1}^{n} \Pr\{r_i|b_i\} = \varepsilon^{\mathrm{dist}(\mathbf{r},\mathbf{b})}(1-\varepsilon)^{n-\mathrm{dist}(\mathbf{r},\mathbf{b})}.$$
>
> - We obtain the likelihood function
>
> $$\Pr\{\mathbf{r}|\mathbf{b}\} = (1-\varepsilon)^n \left(\frac{\varepsilon}{1-\varepsilon}\right)^{\mathrm{dist}(\mathbf{r},\mathbf{b})}.$$
>
> - For $\varepsilon < 0.5$, the maximum of $\Pr\{\mathbf{r}|\mathbf{b}\}$ is attained for the code word $\mathbf{b}$ that minimizes the distance $\mathrm{dist}(\mathbf{r},\mathbf{b})$.
>
> - A minimum distance decoder selects the code word that minimizes the Hamming distance to the received sequence
>
> $$\hat{\mathbf{b}} = \underset{\mathbf{b} \in \mathbb{B}}{\mathrm{argmin}}\, \mathrm{dist}(\mathbf{r},\mathbf{b}) = \underset{\mathbf{b}}{\mathrm{argmax}}\, \Pr\{\mathbf{r}|\mathbf{b}\}$$

Figure 3.13: Minimum distance decoding

It is particularly efficient for convolutional codes, because the trellises of these codes have a very regular structure.

3.2.2 Trellises

In Section 3.1.3 we have seen that the encoder of a convolutional code can be considered as a finite-state machine that is characterized by a state diagram. Alternatively, a convolutional encoder may be represented by a trellis diagram. The concept of the trellis diagram was invented by Forney (Forney, Jr, 1973b, 1974). The trellis for the encoder in Figure 3.1 is depicted in Figure 3.14. A trellis is a directed graph, where the nodes represent encoder states. In contrast to the state diagram, the trellis has a time axis, i.e. the ith level of the trellis corresponds to all possible encoder states at time i. The trellis of a convolutional code has a very regular structure. It can be constructed from the trellis module, which is simply a state diagram, where the states σ_i at each time i and σ_{i+1} at time $i+1$ are

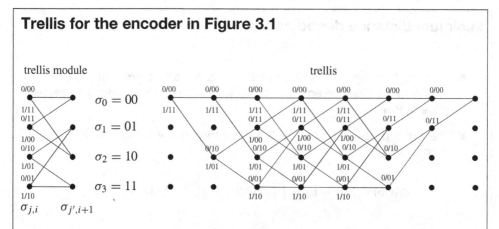

Figure 3.14: Trellis for the encoder in Figure 3.1

depicted separately. The transitions leaving a state are labelled with the corresponding input and output bits (input/output).

As we usually assume that the encoder starts in the all-zero state, the trellis also starts in the zero state. Hence, the first m levels differ from the trellis module. There are only initial state transitions that depart from the all-zero state. In Figure 3.14 we consider a terminated convolutional code with five information bits and two bits for termination. Therefore, the final encoder state is always the all-zero state. All code words begin and end with the encoder state $\sigma = 00$. Moreover, each non-zero code word corresponds to a sequence of state transitions that depart from the all-zero state some number of times and return to the all-zero state. The trellis is a representation of the terminated convolutional codes, i.e. there is a one-to-one correspondence between the set of all paths through the trellis and the set of code sequences, i.e. the code $\mathbb{B}$. Each path from the initial all-zero state to the final all-zero state corresponds to a code word, and vice versa. Moreover, there is a one-to-one correspondence between the set of all paths through the trellis and the set of information sequences. Note that this one-to-one correspondence exists only if the encoder is non-catastrophic.

3.2.3 Viterbi Algorithm

We have seen in Section 3.2.1 that the problem of maximum likelihood decoding for the BSC may be reduced to minimum distance decoding. In 1967, Viterbi introduced an

CONVOLUTIONAL CODES

efficient algorithm to find the code word with minimum distance to the received sequence in a trellis (Viterbi, 1967). This algorithm is currently the most popular decoding procedure for convolutional codes. The Viterbi algorithm is also applied to numerous other applications such as speech recognition and equalization for transmission over channels with memory. We will, however, concentrate on Viterbi decoding.

A brute-force approach to minimum distance decoding would be to calculate all the distances dist($\mathbf{r}$, $\mathbf{b}$) for all possible code words. The efficiency of Viterbi's procedure comes from the fact that we can eliminate many code words from the list of possible candidates without calculating all the distances. Therefore, we will not immediately consider complete code sequences. Instead, we pass the trellis from the initial node to the terminating node and thereby calculate distances for partial code sequences.

The Viterbi algorithm is a *dynamic programming* algorithm. In computer science, dynamic programming is a method for reducing the runtime of algorithms exhibiting the properties of optimal substructures, which means that optimal solutions of subproblems can be used to find the optimal solution of the overall problem. The subproblems of the Viterbi algorithm are finding the most likely path from the starting node to each node in the trellis. The optimal solution for the nodes of level $i + 1$ is, however, simple if we know the solutions for all nodes of level i. Therefore, the Viterbi algorithm traverses the trellis from left to right, finding the overall solution, the maximum likelihood estimate, when the terminating node is reached.

In this section we only consider terminated convolutional codes with L information blocks, i.e. kL information bits, and m blocks for termination. Let us first consider an example of Viterbi decoding. Assume that we use the terminated convolutional code with the trellis depicted in Figure 3.14 for transmission over the BSC. We have sent the code word $\mathbf{b} = (11\,01\,01\,00\,01\,01\,11)$ and received the sequence $\mathbf{r} = (10\,01\,01\,10\,01\,01\,01)$. How should we decode this sequence in order to correct possible transmission errors and determine the transmitted code sequence? The Viterbi algorithm traverses the trellis from left to right and searches for the optimum path that is closest to the received sequence $\mathbf{r}$, with the term *closest* referring to the smallest Hamming distance.

The decoding process for the first three trellis levels is given in Figure 3.15. At the first stage we only have two code segments 00 and 11; both segments differ from the received block 10 in one position. In the process of the Viterbi algorithm we store the corresponding distance values as the so-called *node metric* $\Lambda(\sigma_{j,i})$, where $\sigma_{j,i}$ denotes the jth state at time i. We initialise the first node $\sigma_{0,0}$ with the metric $\Lambda(\sigma_{0,0}) = 0$. For all other nodes the node metric is calculated from the node metric of the predecessor node and the distance of the currently processed code block. Hence, for the first transition we have $\Lambda(\sigma_{0,1}) = \Lambda(\sigma_{0,0}) + 1 = 1$ and $\Lambda(\sigma_{2,1}) = \Lambda(\sigma_{0,0}) + 1 = 1$. At the second stage we consider the received block 01 and obtain the node metrics

$$\Lambda(\sigma_{0,2}) = \Lambda(\sigma_{0,1}) + 1 = 2,$$
$$\Lambda(\sigma_{1,2}) = \Lambda(\sigma_{2,1}) + 2 = 3,$$
$$\Lambda(\sigma_{2,2}) = \Lambda(\sigma_{0,1}) + 1 = 2,$$
$$\Lambda(\sigma_{3,2}) = \Lambda(\sigma_{2,1}) + 0 = 1.$$

At the third stage the situation changes. Now, two transitions arrive at each state. As we are interested in the path closest to the received sequence, we take the minimum of the

CONVOLUTIONAL CODES

Example of the Viterbi algorithm – forward pass (1)

- Received sequence $\mathbf{r} = (10\,01\,01\,10\,01\,01\,01)$.
- For the first $m = 2$ transitions, the state metrics are the accumulated branch metrics.

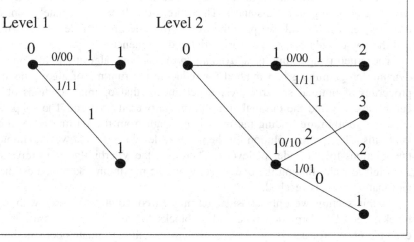

Figure 3.15: Example of the Viterbi algorithm – forward pass (1)

metric values (distances)

$$\Lambda(\sigma_{0,3}) = \min\left\{\Lambda(\sigma_{0,2}) + 1, \Lambda(\sigma_{1,2}) + 1\right\} = \min\{3, 4\} = 3,$$

$$\Lambda(\sigma_{1,3}) = \min\left\{\Lambda(\sigma_{2,2}) + 2, \Lambda(\sigma_{3,2}) + 0\right\} = \min\{4, 1\} = 1,$$

$$\Lambda(\sigma_{2,3}) = \min\left\{\Lambda(\sigma_{0,2}) + 1, \Lambda(\sigma_{1,2}) + 1\right\} = \min\{3, 4\} = 3,$$

$$\Lambda(\sigma_{3,3}) = \min\left\{\Lambda(\sigma_{2,2}) + 0, \Lambda(\sigma_{3,2}) + 2\right\} = \min\{2, 3\} = 2.$$

Obviously, the path with the larger metric value has a larger distance to the received sequence. This path can now be eliminated, because it cannot be part of the optimum code sequence. The decoder discards all transitions that do not correspond to the minimum metric value. The one remaining path for each state is called the *survivor*. It has to be stored. In Figure 3.16 the survivors are labelled by arrows. Note that the node metric is the smallest distance of any partial code sequence that starts at the initial node and ends at this particular node.

The above procedure is now continued until the terminating node of the trellis is reached. The resulting trellis is depicted in Figure 3.17. At the terminating node we will also have a single survivor. This final survivor determines the best path through the trellis. The corresponding code word can be found by passing backwards through the trellis. Now, we traverse the trellis from right to left, following the stored survivors. In our example we

CONVOLUTIONAL CODES

Example of the Viterbi algorithm – forward pass (2)

- For each node, calculate the metric of the paths that enter the node

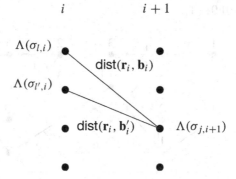

$$\Lambda(\sigma_{j,i+1}) = \min\left\{\Lambda(\sigma_{l,i}) + \text{dist}(\mathbf{r}_i, \mathbf{b}_i), \Lambda(\sigma_{l',i}) + \text{dist}(\mathbf{r}_i, \mathbf{b}'_i)\right\}.$$

- Select the path with the smallest metric (survivor).
- The third transition (survivors are labelled by arrows):

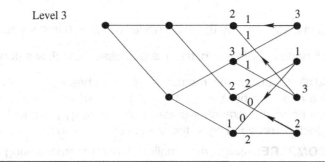

Figure 3.16: Example of the Viterbi algorithm – forward pass (2)

obtain the estimated code word $\hat{\mathbf{b}} = (11\,01\,01\,00\,01\,01\,11)$. From the node metric of the final node $\Lambda(\sigma_{0,7}) = 3$ we conclude that the distance to the received sequence is 3. Hence, we have corrected three errors.

A general description of the Viterbi algorithm is summarised in Figure 3.18. We will consider a version for channels with continuous alphabets and some implementation issues later in Section 3.4. Step 2 defines the so-called add–compare–select recursion of the Viterbi algorithm. Owing to the highly repetitive structure of the trellis, this recursion can be efficiently implemented in hardware. Further implementation issues will be discussed in Section 3.4, too.

Example of the Viterbi algorithm – backward pass

- Starting from the terminating node and going back to the initial node using the labelled survivors, we obtain the estimated code word $\hat{\mathbf{b}} = (11\,01\,01\,00\,01\,01\,11)$.

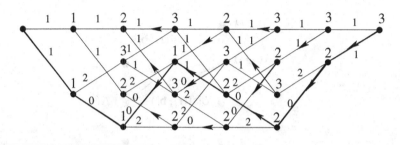

Figure 3.17: Example of the Viterbi algorithm – backward pass

Viterbi algorithm

Step 1: Assign metric zero to the initial node $\Lambda(\sigma_{0,0}) = 0$ and set time $i = 1$.

Step 2: Each node $\sigma_{j,i}$ of time instant i is processed as follows (forward pass):

  a. **ADD**: Calculate the metrics of all branches $\sigma_{j',i-1} \to \sigma_{j,i}$ that enter the node $\sigma_{j,i}$ by $\Lambda(\sigma_{j',i-1}) + \mathrm{dist}(\mathbf{r}_i, \mathbf{b}'_i)$, where $\Lambda(\sigma_{j',i-1})$ is the node metric of this branch's predecessor node $\sigma_{j',i-1}$ and $\mathbf{b}'_i$ is the code block corresponding to the branch $\sigma_{j',i-1} \to \sigma_{j,i}$.

  b. **COMPARE**: Assign the smallest branch metric among all branches merging at node $\sigma_{j,i}$ as the node metric $\Lambda(\sigma_{j,i})$.

  c. **SELECT**: Store the branch corresponding to the smallest branch metric as the survivor.

Step 3: If $i \leq L + m$, then continue with the next level, i.e. increment i by 1 and go to step 2. Otherwise go to step 4.

Step 4: To find the best path, we have to start from the terminating node $\sigma_{0,L+m}$ and go to the initial node $\sigma_{0,0}$, following the survivors (backward pass).

Figure 3.18: Viterbi algorithm

CONVOLUTIONAL CODES

3.3 Distance Properties and Error Bounds

In this section we consider some performance measures for convolutional codes. The analysis of the decoding performance of convolutional codes is based on the notion of an *error event*. Consider, for example, a decoding error with the Viterbi algorithm, i.e. the transmitted code sequence is $\mathbf{b}$ and the estimated code sequence is $\mathbf{b}' \neq \mathbf{b}$. Typically, these code sequences will match for long periods of time but will differ for some code sequence segments. An error event is a code sequence segment where the transmitted and the estimated code sequences differ. It is convenient to define an error event as a path through the trellis. Both sequences $\mathbf{b}$ and $\mathbf{b}'$ are represented by unique paths through the trellis. An error event is a code segment (path) that begins when the path for $\mathbf{b}'$ diverges from the path $\mathbf{b}$, and ends when these two paths merge again.

Now, remember that a convolutional code is a linear code. Hence, the error event $\mathbf{b} - \mathbf{b}'$ is a code sequence. It is zero for all times where $\mathbf{b}$ coincides with $\mathbf{b}'$. Therefore, we may define an error event as a code sequence segment that diverges from the all-zero code sequence and merges with the all-zero sequence at some later time. Furthermore, an error event can only start at times when $\mathbf{b}$ coincides with $\mathbf{b}'$. That is, at times when the decoder is in the correct state. For the BSC, transmission errors occur statistically independently. Consequently, different error events are also statistically independent.

3.3.1 Free Distance

We would like to determine the minimum number of channel errors that could lead to a decoding error. Generally, we would have to consider all possible pairs of code sequences. However, with the notion of an error event, we can restrict ourselves to possible error events. Let the received sequence be $\mathbf{r} = \mathbf{b} + \mathbf{e}$, i.e. $\mathbf{e}$ is the error sequence. With minimum distance decoding an error only occurs if

$$\text{dist}(\mathbf{r}, \mathbf{b}) = \text{wt}(\mathbf{r} - \mathbf{b}) \geq \text{dist}(\mathbf{r}, \mathbf{b}') = \text{wt}(\mathbf{r} - \mathbf{b}')$$

or

$$\text{wt}(\mathbf{e}) \geq \text{wt}(\mathbf{b} - \mathbf{b}' + \mathbf{e}) \geq \text{wt}(\mathbf{b} - \mathbf{b}') - \text{wt}(\mathbf{e}).$$

Therefore, the error event $\mathbf{b} - \mathbf{b}'$ can only occur if $\text{wt}(\mathbf{e}) \geq \text{wt}(\mathbf{b} - \mathbf{b}')/2$.

Remember that the error event $\mathbf{b} - \mathbf{b}'$ is a code sequence. By analogy with the minimum Hamming distance of linear binary block codes, we define the free distance of a linear binary convolutional code

$$d_{\text{free}} = \min_{\mathbf{b},\mathbf{b}' \in \mathbb{B}, \mathbf{b} \neq \mathbf{b}'} \text{dist}(\mathbf{b}, \mathbf{b}')$$

as the minimum Hamming distance between two code sequences. Owing to the linearity, we have $\text{dist}(\mathbf{b}, \mathbf{b}') = \text{wt}(\mathbf{b} - \mathbf{b}')$, where $\mathbf{b} - \mathbf{b}'$ is a code sequence. We assume without loss of generality that an error event is a path that diverges from and remerges with the all-zero sequence. Therefore, we can determine the free distance by considering the Hamming weights of all non-zero code sequences

$$d_{\text{free}} = \min_{\mathbf{b} \in \mathbb{B}, \mathbf{b} \neq \mathbf{0}} \text{wt}(\mathbf{b}).$$

> **Free distance**
>
> - By analogy with the minimum Hamming distance of linear binary block codes, we define the *free distance* of a linear binary convolutional code
>
> $$d_{\text{free}} = \min_{\mathbf{b},\mathbf{b}'\in\mathbb{B}, \mathbf{b}\neq\mathbf{b}'} \text{dist}(\mathbf{b},\mathbf{b}') = \min_{\mathbf{b}\in\mathbb{B}, \mathbf{b}\neq 0} \text{wt}(\mathbf{b}) \qquad (3.5)$$
>
> as the minimum Hamming distance between two code sequences.
>
> - A convolutional code can correct all error patterns with weight up to
>
> $$e = \lfloor \frac{d_{\text{free}}-1}{2} \rfloor.$$
>
> - The free distance is a code property. Convolutional codes with the best known free distance for given rate and memory are called *optimum free distance codes* (OFDs).

Figure 3.19: Free distance

Consequently, the minimum number of transmission errors that could lead to a decoding error is $e+1$ with

$$e = \lfloor \frac{d_{\text{free}}-1}{2} \rfloor.$$

The definition of the free distance and our first result on the error-correcting capabilities of convolutional codes are summarized in Figure 3.19. It is easy to verify that the code $\mathbb{B}(2,1,2)$ represented by the trellis in Figure 3.14 has free distance 5 and can correct all error patterns with weight up to 2.

The free distance of a convolutional code is an important performance measure. Convolutional codes with the best known free distance for given rate and memory are called *optimum free distance codes* (OFDs). Tables of OFD codes for different rates can be found elsewhere (Lee, 1997).

3.3.2 Active Distances

A convolutional code with Viterbi decoding will in general correct much more than $e = \lfloor \frac{d_{\text{free}}-1}{2} \rfloor$ errors. In Section 3.2.3 we have seen that the code $\mathbb{B}(2,1,2)$ with free distance $d_{\text{free}} = 5$ corrected three errors. Actually, this code could correct hundreds of errors if larger code sequences were considered. With block codes, we observed that $e+1$ channel errors with $e = \lfloor \frac{d-1}{2} \rfloor$ could lead to a decoding error. In principle, these $e+1$ errors could be arbitrarily distributed over the n code symbols. With convolutional codes, $e+1$ channel errors could also lead to an error event. Yet these $e+1$ errors have to occur in close proximity, or, in other words, an error event caused by $e+1$ channel errors will be rather limited in length. In this section we consider a distance measure that takes the length of error events into account. We will see that the error correction capability of a convolutional

CONVOLUTIONAL CODES

code increases with growing error event length. The rate of growth, the so-called slope α, will be an important measure when we consider the error correction capability of a concatenated convolutional code in Chapter 4.

We define the correct path trough a trellis to be the path determined by the encoded information sequence. This information sequence also determines the code sequence as well as a sequence of encoder states. The active distance measures are defined as the minimal weight of a set of code sequence segments $\mathbf{b}_{[i_1,i_2]} = \mathbf{b}_{i_1}\mathbf{b}_{i_1+1}\cdots\mathbf{b}_{i_2}$ which is given by a set of encoder state sequences according to Figure 3.20 (Höst et al., 1999).[1] The set $S_{[i_1,i_2]}^{0,0}$ formally defines the set of state sequences that correspond to error events as discussed at the beginning of this section. That is, we define an error event as a code segment starting in a correct state σ_{i_1} and terminating in a correct state σ_{i_2+1}. As the code is linear and an error event is a code sequence segment, we can assume without loss of generality that the correct state is the all-zero state. The error event differs at some, but not necessarily

---

Active burst distance

- Let $S_{[i_1,i_2]}^{\sigma_s,\sigma_e}$ denote the set of encoder state sequences $\sigma_{[i_1,i_2]} = \sigma_{i_1}\sigma_{i_1+1}\cdots\sigma_{i_2}$ that start at depth i_1 in some state $\sigma_{i_1} \in \sigma_s$ and terminate at depth i_2 in some state $\sigma_{i_2} \in \sigma_e$ and do not have all-zero state transitions along with all-zero information block weight in between:

$$S_{[i_1,i_2]}^{\sigma_s,\sigma_e} = \{\sigma_{[i_1,i_2]} : \sigma_{i_1} \in \sigma_s, \sigma_{i_2} \in \sigma_e \text{ and not}$$
$$\sigma_i = 0, \sigma_{i+1} = 0 \text{ with } \mathbf{u}_i = 0, i_1 \leq i \leq i_2 - 1\} \quad (3.6)$$

where σ_s and σ_e denote the sets of possible starting and ending states.

- The jth-order *active burst distance* is

$$a_j^b \stackrel{\text{def}}{=} \min_{S_{[0,j+1]}^{0,0}} \{\text{wt}(\mathbf{b}_{[0,j]})\} \quad (3.7)$$

where $j \geq \nu_{\min}$.

- An error event $\mathbf{b}_{[i_1,i_2]}$ can only occur with minimum distance decoding if the channel error pattern $\mathbf{e}_{[i_1,i_2]}$ has weight

$$e \geq \frac{a_{i_2-i_1}^b}{2}.$$

Figure 3.20: Active burst distance

---

[1]This definition of state sequences was presented elsewhere (Jordan et al., 1999). It differs slightly from the original definition (Höst et al., 1999). Here, all-zero to all-zero state transitions that are not generated by all-zero information blocks are included in order to consider partial (unit) memory codes.

at all, states within the interval $[i_1, i_2]$ from the correct path. The active burst distance a_j^b as defined in Figure 3.20 is the minimum weight of an error event of $j+1$ code blocks. Therefore, we have

$$d_{\text{free}} = \min_j \{a^b(j)\}.$$

Let $\mathbf{e}_{[i_1,i_2]}$ denote an error pattern and e its weight, i.e. $\mathbf{e}_{[i_1,i_2]}$ is a binary sequence with e 1s distributed over the interval $[i_1, i_2]$. According to the discussion in Section 3.3.1, a decoding error with minimum distance decoding could only occur if $e = \text{wt}(\mathbf{e}_{[i_1,i_2]}) \geq \text{wt}(\mathbf{b})/2$ for some code sequence $\mathbf{b}$. Consequently, a minimum distance decoder of a convolutional code $\mathbb{B}$ can only output an error event $\mathbf{b}_{[i_1,i_2]}$ if the corresponding error pattern $\mathbf{e}_{[i_1,i_2]}$ has weight

$$e \geq \frac{a_{i_2-i_1}^b}{2}.$$

There exist some more active distance measures that are defined in Figure 3.21. Those distances will be required in Chapter 4. They differ from the burst distance in the definition of the sets of starting and ending states. Generally, all active distances can be lower bounded

---

Other active distances

- The jth-order *active column distance* is

$$a_j^c \stackrel{\text{def}}{=} \min_{S_{[0,j+1]}^{0,\sigma}} \{\text{wt}(\mathbf{b}_{[0,j]})\} \qquad (3.8)$$

where σ denotes any encoder state.

- The jth-order *active reverse column distance* is

$$a_j^{rc} \stackrel{\text{def}}{=} \min_{S_{[0,j+1]}^{\sigma,0}} \{\text{wt}(\mathbf{b}_{[0,j]})\} \qquad (3.9)$$

where σ denotes any encoder state.

- The jth-order *active segment distance* is

$$a_j^s \stackrel{\text{def}}{=} \min_{S_{[m,m+j+1]}^{\sigma_1,\sigma_2}} \{\text{wt}(\mathbf{b}_{[0,j]})\} \qquad (3.10)$$

where σ_1 and σ_2 denote any encoder state.

Figure 3.21: Other active distances

CONVOLUTIONAL CODES

by linear functions with the same slope α. Therefore, we can write

$$a_j^b \geq \alpha \cdot j + \beta^b,$$
$$a_j^c \geq \alpha \cdot j + \beta^c,$$
$$a_j^{rc} \geq \alpha \cdot j + \beta^{rc},$$
$$a_j^s \geq \alpha \cdot j + \beta^s,$$

where $\beta^b, \beta^c, \beta^{rc}$ and β^s are rational constants. For example, the code $\mathbb{B}(2,1,2)$ has free distance $d_{\text{free}} = 5$. With the encoder $(7\,5)_8$, we obtain the parameters $\alpha = 1/2$, $\beta^b = 9/2$, $\beta^c = 3/2$, $\beta^{rc} = 3/2$ and $\beta^s = -1/2$. With this code, three channel errors could lead to an error event. However, this error event would be limited to length $j = 3$, i.e. $n(j+1) = 8$ code bits. An error event of length $j = 11$ with 24 code bits would require at least

$$e \geq \frac{a_j^b}{2} \geq \frac{\alpha j + \beta^b}{2} = \frac{11 \cdot \frac{1}{2} + \frac{9}{2}}{2} = 5$$

channel errors.

From these definitions it follows that the active distances are encoder properties, not code properties.[2] As a catastrophic encoder leads to a zero-loop in the state diagram, it is easy to see that a catastrophic encoding would lead to the slope $\alpha = 0$.

Both parameters, α and d_{free}, characterise the distance properties of convolutional codes. The free distance determines the code performance for low channel error rates, whereas codes with good active distances, and high slope α, achieve better results for very high channel error rates. The latter case is important for code concatenation which we will consider in Chapter 4. However, convolutional codes with large slopes also yield better tail-biting codes (Bocharova et al., 2002). A table with the free distances and slopes of the active distances for rate $R = 1/2$ OFD convolutional codes is given in Figure 3.22. We

---

Table of code parameters

■ The following table presents the free distances and slopes of the active distances for rate $R = 1/2$ OFD convolutional codes with different memories.

Memory	Polynomial	Free distance d_{free}	Slope α
2	(7 5)	5	0.5
3	(17 15)	6	0.5
4	(35 23)	7	0.36
5	(75 53)	8	0.33
6	(171 133)	10	0.31

Figure 3.22: Table of code parameters

---

[2] In fact the active distances are invariant over the set of minimal encoding matrices. For the definition of a minimal encoding matrix, see elsewhere (Bossert, 1999; Johannesson and Zigangirov, 1999).

notice that with increasing memory the free distance improves while the slope α decreases. Tables of codes with good slope can be found elsewhere (Jordan et al., 2004b, 2000).

3.3.3 Weight Enumerators for Terminated Codes

Up to now we have only considered code sequences or code sequence segments of minimum weight. However, the decoding performance with maximum likelihood decoding is also influenced by code sequences with higher weight and other code parameters such as the number of code paths with minimum weight. The complete error correction performance of a block code is determined by its *weight distribution*. Consider, for example, the weight distribution of the Hamming code $\mathbb{B}(7, 4)$. This code has a total of $2^k = 16$ code words: the all-zero code word, seven code words of weight $d = 3$, seven code words of weight 4 and one code word of weight 7. In order to write this weight distribution in a compact way, we specify it as a polynomial of the dummy variable W, i.e. we define the Weight Enumerating Function (WEF) of a block code $\mathbb{B}(n, k)$ as

$$A_{\text{WEF}}(W) = \sum_{w=0}^{n} a_w W^w,$$

where a_w is the number of code words of weight w in $\mathbb{B}$. The numbers $a_0, \ldots, a_n$ are the weight distribution of $\mathbb{B}$. For the Hamming code $\mathbb{B}(7, 4)$ we have

$$A_{\text{WEF}}(W) = 1 + 7W^3 + 7W^4 + W^7.$$

With convolutional codes we have basically code sequences of infinite length. Therefore, we cannot directly state numbers of code words of a particular weight. To solve this problem, we will introduce the concept of path enumerators in the next section. Now, we discuss a method to evaluate weight distributions of terminated convolutional codes which are in fact block codes. Such weight enumerators are, for instance, required to calculate the expected weight distributions of code concatenations with convolutional component codes, as will be discussed in Chapter 4.

Consider the trellis of the convolutional code $\mathbb{B}(2, 1, 2)$ as given in Figure 3.14, where we labelled the branches with input and output bits corresponding to a particular state transition. Similarly, we can label the trellis with weight enumerators as in Figure 3.23, i.e. every branch is labelled by a variable W^w with w being the number of non-zero code bits corresponding to this state transition. Using such a labelled trellis, we can now employ the forward pass of the Viterbi algorithm to compute the WEF of a terminated convolutional code. Instead of a branch metric, we compute a weight enumerator $A_i^{(j)}(W)$ for each node in the trellis, where $A_i^{(j)}(W)$ denotes the enumerator for state σ_j at level i. Initialising the enumerator of the first node with $A_0^{(0)}(W) = 1$, we could now traverse the trellis from left to right, iteratively computing the WEF for all other nodes. In each step of the forward pass, we compute the enumerators $A_{i+1}^{(j)}(W)$ at level $i + 1$ on the basis of the enumerators of level i and the labels of the corresponding transitions as indicated in Figure 3.23. That is, we multiply the enumerators of level i with the corresponding transition label and sum over all products corresponding to paths entering the same node. The enumerator $A_{L+m}^{(0)}(W)$ of the final node is equal to the desired overall WEF.

CONVOLUTIONAL CODES

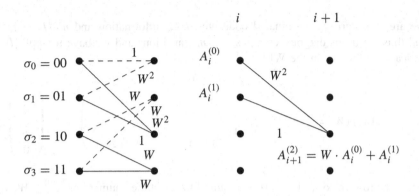

- A trellis module of the $(7\ 5)_8$ convolutional code. Each branch is labelled with a branch enumerator.

- We define a $2^\nu \times 2^\nu$ transition matrix $\mathbf{T}$. The coefficient $\tau_{l,j}$ of $\mathbf{T}$ is the weight enumerator of the transition from state σ_l to state σ_j, where we set $\tau_{l,j} = 0$ for impossible transitions.

- The weight enumerator function is evaluated iteratively

$$\mathbf{A}_0(W) = (1\ 0\ \cdots\ 0)^\mathrm{T},$$

$$\mathbf{A}_{i+1}(W) = \mathbf{T} \cdot \mathbf{A}_i(W),$$

$$A_{\mathrm{WEF}}(W) = (1\ 0\ \cdots\ 0) \cdot \mathbf{A}_{L+m}(W).$$

Figure 3.23: Calculating the weight enumerator function

Using a computer algebra system, it is usually more convenient to represent the trellis module by a transition matrix $\mathbf{T}$ and to calculate the WEF with matrix operations. The trellis of a convolutional encoder with overall constraint length ν has the physical state space $\mathbb{S} = \{\sigma_0, \ldots, \sigma_{2^\nu - 1}\}$. Therefore, we define a $2^\nu \times 2^\nu$ transition matrix $\mathbf{T}$ so that the coefficients $\tau_{l,j}$ of $\mathbf{T}$ are the weight enumerators of the transition from state σ_l to state σ_j, where we set $\tau_{l,j} = 0$ for impossible transitions. The weight enumerators of level i can now be represented by a vector

$$\mathbf{A}_i(W) = (A_i^{(0)}\ A_i^{(1)}\ \cdots\ A_i^{(2^\nu - 1)})^\mathrm{T}$$

with the special case

$$\mathbf{A}_0(W) = (1\ 0\ \cdots\ 0)^\mathrm{T}$$

for the starting node. Every step of the Viterbi algorithm can now be expressed by a multiplication

$$\mathbf{A}_{i+1}(W) = \mathbf{T} \cdot \mathbf{A}_i(W).$$

We are considering a terminated code with $k \cdot L$ information and $n \cdot (L+m)$ code bits, and thus the trellis diagram contains $L + m$ transitions and we have to apply $L + m$ multiplications. We obtain the WEF

$$A_{\text{WEF}}(W) = (1\ 0\ \cdots\ 0) \cdot \begin{pmatrix} \tau_{0,0} & \cdots & \tau_{1,1} \\ \vdots & & \vdots \\ \tau_{0,2^\nu-1} & \cdots & \tau_{2^\nu-1,2^\nu-1} \end{pmatrix}^{L+m} \cdot \begin{pmatrix} 1 \\ 0 \\ \vdots \\ 0 \end{pmatrix},$$

where the row vector $(1\ 0\ \cdots\ 0)$ is required to select the enumerator $A_{L+m}^{(0)}(W)$ of the final node. The WEF may also be evaluated iteratively as indicated in Figure 3.23. An example of iterative calculation is given in Figure 3.24.

The concept of the weight enumerator function can be generalized to other enumerator functions, e.g. the Input–Output Weight Enumerating Function (IOWEF)

$$A_{\text{IOWEF}}(I, W) = \sum_{i=0}^{k} \sum_{w=0}^{n} a_{i,w} I^i W^w,$$

where $a_{i,w}$ represents the number of code words with weight w generated by information words of weight i. The input–output weight enumerating function considers not only the weight of the code words but also the mapping from information word to code words. Therefore, it also depends on the particular generator matrix. For instance, the Hamming code $\mathbb{B}(7, 4)$ with systematic encoding has the IOWEF

$$A_{\text{IOWEF}}(I, W) = 1 + I(3W^3 + W^4) + I^2(3W^3 + 3W^4) + I^3(W^3 + 3W^4) + I^4 W^7.$$

Note that by substituting $I = 1$ we obtain the WEF from $A_{\text{IOWEF}}(I, W)$

$$A_{\text{WEF}}(W) = A_{\text{IOWEF}}(I, W)|_{I=1}.$$

To evaluate the input–output weight enumerating function for a convolutional encoder, two transition matrices $\mathbf{T}$ and $\mathbf{T}'$ are required. The coefficients $\tau_{i,j}$ of $\mathbf{T}$ are input–output enumerators, for example $\tau_{0,2} = IW^2$ for the transition from state σ_0 to state σ_2 with the $(7\,5)_8$ encoder. The matrix $\mathbf{T}'$ regards the tailbits necessary for termination and therefore only contains enumerators for code bits, e.g. $\tau'_{0,2} = W^2$. We obtain

$$A_{\text{IOWEF}}(I, W) = (1\ 0\ \cdots\ 0) \cdot T^L T'^m \cdot (1\ 0\ \cdots\ 0)^{\text{T}}.$$

CONVOLUTIONAL CODES

Calculating the weight enumerator function – example

- For the $(7\ 5)_8$ convolutional code we obtain

$$\mathbf{A}_{i+1}(W) = \begin{pmatrix} A_{i+1}^{(0)}(W) \\ A_{i+1}^{(1)}(W) \\ A_{i+1}^{(2)}(W) \\ A_{i+1}^{(3)}(W) \end{pmatrix} = \begin{pmatrix} 1 & W^2 & 0 & 0 \\ 0 & 0 & W & W \\ W^2 & 1 & 0 & 0 \\ 0 & 0 & W & W \end{pmatrix} \cdot \begin{pmatrix} A_i^{(0)}(W) \\ A_i^{(1)}(W) \\ A_i^{(2)}(W) \\ A_i^{(3)}(W) \end{pmatrix}.$$

- Consider the procedure for $L = 2$:

$$\begin{pmatrix} 1 \\ 0 \\ 0 \\ 0 \end{pmatrix} \xrightarrow{i=1} \begin{pmatrix} 1 \\ 0 \\ W^2 \\ 0 \end{pmatrix} \xrightarrow{i=2} \begin{pmatrix} 1 \\ W^3 \\ W^2 \\ W^3 \end{pmatrix} \xrightarrow{i=3} \begin{pmatrix} 1+W^5 \\ W^3+W^4 \\ W^2+W^3 \\ W^3+W^4 \end{pmatrix}$$

$$\xrightarrow{i=4} \begin{pmatrix} 1+2W^5+W^6 \\ W^3+2W^4 \\ W^2+W^3+W^4+W^7 \\ W^3+2W^4+W^5 \end{pmatrix}.$$

- This results in the weight enumerating function

$$A_{\text{WEF}}(W) = A_{L+m}^0(W) = A_4^0(W) = 1 + 2W^5 + W^6.$$

Figure 3.24: Calculating the weight enumerator function – example

3.3.4 Path Enumerators

At the beginning of this section we have introduced the notion of an error event, i.e. if an error occurs, the correct sequence and the estimated sequence will typically match for long periods of time but will differ for some code sequence segments. An error event is a code sequence segment where the transmitted and the estimated code sequence differ. Without loss of generality we can assume that the all-zero sequence was transmitted. Therefore, we can restrict the analysis of error events to code sequence segments that diverge from the all-zero sequence and remerge at some later time. Such a code sequence segment corresponds to a path in the trellis that leaves the all-zero state and remerges with the all-zero state. A weight distribution for such code sequence segments is typically called *the path enumerator*.

In this section we will discuss a method for calculating such path enumerators. Later on, we will see how the path enumerator can be used to estimate the decoding performance for maximum likelihood decoding.

CONVOLUTIONAL CODES

We will assume that the error event starts at time zero. Hence, all code sequence segments under consideration correspond to state sequences $(0, \sigma_1, \sigma_2, \ldots, \sigma_j, 0, \ldots)$. Note that j is greater than or equal to m, because a path that diverges from the all-zero state requires at least $m+1$ transitions to reach the zero state again. Once more, we will discuss the procedure to derive the path enumerator for a particular example using the $(7\,5)_8$ convolutional code. Consider, therefore, the signal flow chart in Figure 3.25. This is basically the state diagram from Figure 3.6. The state transitions are now labelled with enumerators

Path enumerator function

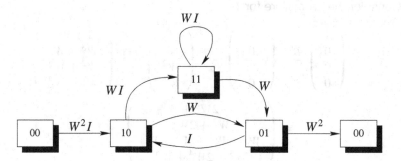

- Path enumerator function can be derived from a signal flow chart similar to the state diagram. State transitions are labelled with weight enumerators, and the loop from the all-zero state to the all-zero state is removed.

- We introduce a dummy weight enumerator function for each non-zero state and obtain a set of linear equations

$$A^{(1)}(I, W) = WA^{(2)}(I, W) + WA^{(3)}(I, W),$$
$$A^{(2)}(I, W) = IA^{(1)}(I, W) + IW^2,$$
$$A^{(3)}(I, W) = WIA^{(2)}(I, W) + WIA^{(3)}(I, W),$$
$$A_{\text{IOPEF}}(I, W) = A^{(1)}(I, W)W^2.$$

- Solving this set of equations yields the path enumerator functions

$$A_{\text{IOPEF}}(I, W) = \frac{W^5 I}{(1 - 2WI)}$$

and

$$A_{\text{PEF}}(I, W) = A_{\text{IOPEF}}(I, W)|_{I=1} = \frac{W^5}{(1 - 2W)}.$$

Figure 3.25: Path enumerator function

CONVOLUTIONAL CODES

for the number of code bits (exponent of W) and the number of information bits (exponent of I) that correspond to the particular transition. Furthermore, the loop from the all-zero state to the all-zero state is removed, because we only consider the state sequence $(0, \sigma_1, \sigma_2, \ldots, \sigma_j, 0, \ldots)$ where only the first transition starts in the all-zero state the last transition terminates in the all-zero state and there are no other transitions to the all-zero state in between. In order to calculate the Input–Output Path Enumerator Function (IOPEF), we introduce an enumerator for each non-zero state, comprising polynomials of the dummy variables W and I. For instance, $A^{(2)}(I, W)$ denotes the enumerator for state $\sigma_2 = (10)$. From the signal flow chart we derive the relation $A^{(2)}(I, W) = IA^{(1)}(I, W) + IW^2$. Here, $A^{(2)}(I, W)$ is the label of the initial transition IW^2 plus the enumerator of state σ_1 multiplied by I, because the transition from state σ_1 to state σ_2 has one non-zero information bit and only zero code bits. Similarly, we can derive the four linear equations in Figure 3.25, which can be solved for $A_{\text{IOPEF}}(I, W)$, resulting in

$$A_{\text{IOPEF}}(I, W) = \frac{W^5 I}{(1 - 2WI)}.$$

As with the IOWEF, we can derive the Path Enumerator Function (PEF) from the input–output path enumerator by substituting $I = 1$ and obtain

$$A_{\text{PEF}}(W) = A_{\text{IOPEF}}(I, W)|_{I=1} = \frac{W^5}{(1 - 2W)}.$$

3.3.5 Pairwise Error Probability

In Section 3.3.1 and Section 3.3.2 we have considered error patterns that could lead to a decoding error. In the following, we will investigate the probability of a decoding error with minimum distance decoding. In this section we derive a bound on the so-called *pairwise error probability*, i.e. we consider a block code $\mathbb{B} = \{\mathbf{b}, \mathbf{b}'\}$ that has only two code words and estimate the probability that the minimum distance decoder decides on the code word $\mathbf{b}'$ when actually the code word $\mathbf{b}$ was transmitted. The result concerning the pairwise error probability will be helpful when we consider codes with more code words or possible error events of convolutional codes.

Again, we consider transmission over the BSC. For the code $\mathbb{B} = \{\mathbf{b}, \mathbf{b}'\}$, we can characterise the behaviour of the minimum distance decoder with two decision regions (cf. Section 2.1.2). We define the set $\mathbb{D}$ as the decision region of the code word $\mathbf{b}$, i.e. $\mathbb{D}$ is the set of all received words $\mathbf{r}$ so that the minimum distance decoder decides on $\mathbf{b}$. Similarly, we define the decision region $\mathbb{D}'$ for the code word $\mathbf{b}'$. Note that for the code with two code words we have $\mathbb{D}' = \mathbb{F}^n \setminus \mathbb{D}$.

Assume that the code word $\mathbf{b}$ was transmitted over the BSC. In this case, we can define the conditional pairwise error probability as

$$P_{e|\mathbf{b}} = \sum_{\mathbf{r} \notin \mathbb{D}} \Pr\{\mathbf{r}|\mathbf{b}\}.$$

Similarly, we define

$$P_{e|\mathbf{b}'} = \sum_{\mathbf{r} \notin \mathbb{D}'} \Pr\{\mathbf{r}|\mathbf{b}'\}$$

and obtain the average pairwise error probability

$$P_e = P_{e|b}\Pr\{\mathbf{b}\} + P_{e|b'}\Pr\{\mathbf{b'}\}.$$

We proceed with the estimation of $P_{e|b}$. Summing over all received vectors $\mathbf{r}$ with $\mathbf{r} \notin \mathbb{D}$ is usually not feasible. Therefore, it is desirable to sum over all possible received vectors $\mathbf{r} \in \mathbb{F}^n$. In order to obtain a reasonable estimate of $P_{e|b}$ we multiply the term $\Pr\{\mathbf{r}|\mathbf{b}\}$ with the factor

$$\sqrt{\frac{\Pr\{\mathbf{r}|\mathbf{b'}\}}{\Pr\{\mathbf{r}|\mathbf{b}\}}} \begin{cases} \geq 1 \text{ for } \mathbf{r} \notin \mathbb{D} \\ \leq 1 \text{ for } \mathbf{r} \in \mathbb{D} \end{cases}$$

This factor is greater than or equal to 1 for all received vectors $\mathbf{r}$ that lead to a decoding error, and less than or equal to 1 for all others. We have

$$P_{e|b} \leq \sum_{\mathbf{r} \notin \mathbb{D}} \Pr\{\mathbf{r}|\mathbf{b}\} \sqrt{\frac{\Pr\{\mathbf{r}|\mathbf{b'}\}}{\Pr\{\mathbf{r}|\mathbf{b}\}}} = \sum_{\mathbf{r} \notin \mathbb{D}} \sqrt{\Pr\{\mathbf{r}|\mathbf{b}\}\Pr\{\mathbf{r}|\mathbf{b'}\}}.$$

Now, we can sum over all possible received vectors $\mathbf{r} \in \mathbb{F}^n$

$$P_{e|b} \leq \sum_{\mathbf{r}} \sqrt{\Pr\{\mathbf{r}|\mathbf{b}\}\Pr\{\mathbf{r}|\mathbf{b'}\}}.$$

The BSC is memoryless, and we can therefore write this estimate as

$$P_{e|b} \leq \sum_{r_1} \cdots \sum_{r_n} \sqrt{\prod_{i=1}^{n} \Pr\{r_i|b_i\}\Pr\{r_i|b_i'\}} = \prod_{i=1}^{n} \sum_{r} \sqrt{\Pr\{r|b_i\}\Pr\{r|b_i'\}}.$$

For the BSC, the term $\sum_r \sqrt{\Pr\{r|b_i\}\Pr\{r|b_i'\}}$ is simply

$$\sqrt{\Pr\{0|b_i\}\Pr\{0|b_i'\}} + \sqrt{\Pr\{1|b_i\}\Pr\{1|b_i'\}} = \begin{cases} \sqrt{\varepsilon^2} + \sqrt{(1-\varepsilon)^2} = 1 \text{ for } b_i = b_i' \\ 2\sqrt{\varepsilon(1-\varepsilon)} \text{ for } b_i \neq b_i' \end{cases}.$$

Hence, we have

$$P_{e|b} \leq \prod_{i=1}^{n} \sum_{r} \sqrt{\Pr\{r|b_i\}\Pr\{r|b_i'\}} = \left(2\sqrt{\varepsilon(1-\varepsilon)}\right)^{\text{dist}(\mathbf{b},\mathbf{b'})}.$$

Similarly, we obtain $P_{e|b'} \leq \left(2\sqrt{\varepsilon(1-\varepsilon)}\right)^{\text{dist}(\mathbf{b},\mathbf{b'})}$. Thus, we can conclude that the pairwise error probability is bounded by

$$P_e \leq \left(2\sqrt{\varepsilon(1-\varepsilon)}\right)^{\text{dist}(\mathbf{b},\mathbf{b'})}.$$

This bound is usually called the *Bhattacharyya bound*, and the term $2\sqrt{\varepsilon(1-\varepsilon)}$ the *Bhattacharyya parameter*. Note, that the Bhattacharyya bound is independent of the total number of code bits. It only depends on the Hamming distance between two code words. Therefore,

CONVOLUTIONAL CODES

Bhattacharyya bound

- The pairwise error probability for the two code sequences **b** and **b'** is defined as
$$P_e = P_{e|b}\Pr\{\mathbf{b}\} + P_{e|b'}\Pr\{\mathbf{b'}\}$$
with conditional pairwise error probabilities
$$P_{e|b} = \sum_{\mathbf{r}\notin \mathbb{D}}\Pr\{\mathbf{r}|\mathbf{b}\} \text{ and } P_{e|b'} = \sum_{\mathbf{r}\notin \mathbb{D}'}\Pr\{\mathbf{r}|\mathbf{b'}\}$$
and the decision regions $\mathbb{D}$ and $\mathbb{D}'$.

- To estimate $P_{e|b}$, we multiply the term $\Pr\{\mathbf{r}|\mathbf{b}\}$ with the factor
$$\sqrt{\frac{\Pr\{\mathbf{r}|\mathbf{b'}\}}{\Pr\{\mathbf{r}|\mathbf{b}\}}} \begin{cases} \geq 1 \text{ for } \mathbf{r}\notin \mathbb{D} \\ \leq 1 \text{ for } \mathbf{r}\in \mathbb{D} \end{cases}$$
and sum over all possible received vectors $\mathbf{r}\in \mathbb{F}^n$
$$P_{e|b} \leq \sum_{\mathbf{r}} \sqrt{\Pr\{\mathbf{r}|\mathbf{b}\}\Pr\{\mathbf{r}|\mathbf{b'}\}}.$$

- For the BSC this leads to the Bhattacharyya bound
$$P_e \leq \left(2\sqrt{\varepsilon(1-\varepsilon)}\right)^{\text{dist}(\mathbf{b},\mathbf{b'})}.$$

Figure 3.26: Bhattacharyya bound

it can also be used to bound the pairwise error probability for two convolutional code sequences. The derivation of the Bhattacharyya bound is summarised in Figure 3.26.

For instance, consider the convolutional code $\mathbb{B}(2,1,2)$ and assume that the all-zero code word is transmitted over the BSC with crossover probability $\varepsilon = 0.01$. What is the probability that the Viterbi algorithm will result in the estimated code sequence $\hat{\mathbf{b}} = (11\,10\,11\,00\,00\,\ldots)$? We can estimate this pairwise error probability with the Bhattacharyya bound. We obtain the Bhattacharyya parameter $2\sqrt{\varepsilon(1-\varepsilon)} \approx 0.2$ and the bound $P_e \leq 3.2\cdot 10^{-4}$. In this particular case we can also calculate the pairwise error probability. The Viterbi decoder will select the sequence $(11\,10\,11\,00\,00\,\ldots)$ if at least three channel errors occur in any of the five non-zero positions. This event has the probability
$$\sum_{e=3}^{5}\binom{5}{e}\varepsilon^e(1-\varepsilon)^{5-e} \approx 1\cdot 10^{-5}.$$

3.3.6 Viterbi Bound

We will now use the pairwise error probability for the BSC to derive a performance bound for convolutional codes with Viterbi decoding. A good measure for the performance of a convolutional code is the *bit error probability* P_b, i.e. the probability that an encoded information bit will be estimated erroneously in the decoder. However, it is easier to derive bounds on the *burst error probability* P_B, which is the probability that an error event will occur at a given node. Therefore, we start our discussion with the burst error probability.[3]

We have already mentioned that different error events are statistically independent for the BSC. However, the burst error probability is not the same for all nodes along the correct path. An error event can only start at times when the estimated code sequence coincides with the correct one. Therefore, the burst error probability for the initial node (time $i = 0$) is greater than for times $i > 0$. We will derive a bound on P_B assuming that the error event starts at time zero. This yields a bound that holds for all nodes along the correct path.

Remember that an error event is a path through the trellis, that is to say, a code sequence segment. Let $\mathbf{b}$ denote the correct code sequence and $\mathcal{E}(\mathbf{b}')$ denote the event that the code sequence segment $\mathbf{b}'$ causes a decoding error starting at time zero. A necessary condition for an error event starting at time zero is that the corresponding code sequence segment has a distance to the received sequence that is less than the distance between the correct path and $\mathbf{r}$. This condition is not sufficient, because there might exist another path with an even smaller distance. Therefore, we have

$$P_B \leq \Pr\{\cup_{\mathbf{b}'} \mathcal{E}(\mathbf{b}')\},$$

where the union is over all possible error events diverging from the initial trellis node. We can now use the union bound to estimate the union of events $\Pr\{\cup_{\mathbf{b}'} \mathcal{E}(\mathbf{b}')\} \leq \sum_{\mathbf{b}'} \Pr\{\mathcal{E}(\mathbf{b}')\}$ and obtain

$$P_B \leq \sum_{\mathbf{b}'} \Pr\{\mathcal{E}(\mathbf{b}')\}.$$

Assume that $\text{dist}(\mathbf{b}, \mathbf{b}') = w$. We can use the Bhattacharyya bound to estimate $\Pr\{\mathcal{E}(\mathbf{b}')\}$

$$\Pr\{\mathcal{E}(\mathbf{b}')\} \leq \left(2\sqrt{\varepsilon(1-\varepsilon)}\right)^w.$$

Let a_w be the number of possible error events of weight w starting at the initial node. We have

$$P_B \leq \sum_{w=d_{\text{free}}}^{\infty} a_w \left(2\sqrt{\varepsilon(1-\varepsilon)}\right)^w.$$

Note, that the set of possible error events is the set of all paths that diverge from the all-zero path and remerge with the all-zero path. Hence, a_w is the weight distribution of the convolutional code, and we can express our bound in terms of the path enumerator function $A(W)$

$$P_B \leq \sum_{w=d_{\text{free}}}^{\infty} a_w \left(2\sqrt{\varepsilon(1-\varepsilon)}\right)^w = A_{\text{PEF}}(W)|_{W=2\sqrt{\varepsilon(1-\varepsilon)}}.$$

This bound is called the *Viterbi bound* (Viterbi, 1971).

---
[3] The burst error probability is sometimes also called the *first event error probability*.

CONVOLUTIONAL CODES

> **Viterbi bound**
>
> ■ The *burst error probability* P_B is the probability that an error event will occur at a given node.
>
> ■ For transmission over the BSC with maximum likelihood decoding the burst error probability is bounded by
>
> $$P_B \leq \sum_{w=d_{\text{free}}}^{\infty} a_w \left(2\sqrt{\varepsilon(1-\varepsilon)}\right)^w = A_{\text{PEF}}(W)|_{W=2\sqrt{\varepsilon(1-\varepsilon)}}.$$
>
> ■ The *bit error probability* P_b is the probability that an encoded information bit will be estimated erroneously in the decoder.
>
> ■ The bit error probability for transmission over the BSC with maximum likelihood decoding is bounded by
>
> $$P_b \leq \frac{1}{k} \frac{\partial A_{\text{IOPEF}}(I,W)}{\partial I}\bigg|_{I=1;\, W=2\sqrt{\varepsilon(1-\varepsilon)}}.$$

Figure 3.27: Viterbi bound

Consider, for example, the code $\mathbb{B}(2,1,2)$ with path enumerator $A_{\text{PEF}}(W) = \frac{W^5}{1-2W}$. For the BSC with crossover probability $\varepsilon = 0.01$, the Viterbi bound results in

$$P_B \leq A_{\text{PEF}}(W)|_{W\approx 0.2} \approx 5 \cdot 10^{-4}.$$

In Section 3.3.5 we calculated the Bhattacharyya bound $P_e \leq 3.2 \cdot 10^{-4}$ on the pairwise error probability for the error event $(11\,10\,11\,00\,00\,\ldots)$. We observe that for $\varepsilon = 0.01$ this path, which is the path with the lowest weight $w = d_{\text{free}} = 5$, determines the overall decoding error performance with Viterbi decoding.

Based on the concept of error events, it is also possible to derive an upper bound on the bit error probability P_b (Viterbi, 1971). We will only state the result without proof and discuss the basic idea. A proof can be found elsewhere (Johannesson and Zigangirov, 1999). The bit error probability for transmission over the BSC with maximum likelihood decoding is bounded by

$$P_b \leq \frac{1}{k} \frac{\partial A_{\text{IOPEF}}(I,W)}{\partial I}\bigg|_{I=1;\, W=2\sqrt{\varepsilon(1-\varepsilon)}}.$$

The bound on the burst error probability and the bound on the bit error probability are summarized in Figure 3.27. In addition to the bound for the burst error probability, we require an estimate of the expected number of information bits that occur in the event of a decoding error. Here, the input–output path enumerator is required. Remember that, by substituting $W = 2\sqrt{\varepsilon(1-\varepsilon)}$ in the path enumerator, we evaluate the pairwise error probability for all possible Hamming weights w, multiply each pairwise error probability by the number of paths of the particular weight and finally sum over all those terms. In order to calculate the bit error probability, we have to take the expected number of information bits into account. Therefore, we derive the input–output path enumerator with respect to the variable I. Consequently, the pairwise error probabilities are weighted with the number of information bits corresponding to the particular error event.

Again, consider the code $\mathbb{B}(2, 1, 2)$ with the generator matrix $(7\,5)_8$ in octal notation. With this generator matrix we have the input–output path enumerator

$$A_{\text{IOPEF}}(I, W) = \frac{W^5 I}{(1 - 2WI)}$$

and the derivative

$$\frac{\partial A_{\text{IOPEF}}(I, W)}{\partial I} = \frac{W^5}{(1 - 2WI)^2}.$$

For the BSC with crossover probability $\varepsilon = 0.01$, the bound results in

$$P_{\text{b}} \leq \left. \frac{W^5}{(1 - 2WI)^2} \right|_{W \approx 0.2} \approx 9 \cdot 10^{-4}.$$

3.4 Soft-input Decoding

Up to now, we have only considered so called hard-input decoding, i.e. in Section 3.2 we have assumed that the channel is a BSC which has only two output values. In this case we can employ minimum distance decoding with the Hamming metric as the distance measure. In general, the transmission channel may have a continuous output alphabet like, for example, the Additive White Gaussian Noise (AWGN) channel or fading channels in mobile communications. In this section we will generalise the concept of minimum distance decoding to channels with a continuous output alphabet. We will observe that we can still use the Viterbi algorithm, but with a different distance measure.

Later on in this section we consider some implementation issues. In particular, we discuss the basic architecture of a hardware implementation of the Viterbi algorithm.

3.4.1 Euclidean Metric

As an example of a channel with a continuous output alphabet, we consider the AWGN channel, where we assume that Binary Phase Shift Keying (BPSK) is used for modulation. Hence, the code bits $b_i \in \mathbb{F}_2$ are mapped to the transmission symbols $x_i \in \{-1, +1\}$ according to

$$x_i = 2b_i - 1.$$

If the receiver performs coherent demodulation, then the received symbols at the decoder input are

$$r_i = x_i + n_i,$$

CONVOLUTIONAL CODES

where we have assumed that the energy of the transmitted signal is normalised to 1. The Gaussian random variable n_i represents the additive noise. Furthermore, we assume that the channel is memoryless, i.e. the noise samples n_i are statistically independent. In this case the channel can be characterised by the following probability density function

$$p(r_i|b_i) = \frac{1}{\sqrt{2\pi\sigma^2}} \exp\left(-\frac{(r_i - x_i)^2}{2\sigma^2}\right),$$

where σ^2 is the variance of the additive Gaussian noise.

Again, we would like to perform ML sequence estimation according to Figure 3.12. Note that, using Bayes' rule, the MAP criterion can be expressed as

$$\hat{\mathbf{b}} = \underset{\mathbf{b}}{\operatorname{argmax}} \left\{ \Pr\{\mathbf{b}|\mathbf{r}\} \right\} = \underset{\mathbf{b}}{\operatorname{argmax}} \left\{ \frac{p(\mathbf{r}|\mathbf{b})\Pr\{\mathbf{b}\}}{p(\mathbf{r})} \right\},$$

where $p(\mathbf{r})$ is the probability density function of the received sequence and $p(\mathbf{r}|\mathbf{b})$ is the conditional probability density function given the code sequence $\mathbf{b}$. Assuming, again, that the information bits are statistically independent and equally likely, we obtain the ML rule

$$\hat{\mathbf{b}} = \underset{\mathbf{b}}{\operatorname{argmax}} \left\{ \Pr\{\mathbf{r}|\mathbf{b}\} \right\} = \underset{\mathbf{b}}{\operatorname{argmax}} \left\{ \prod_i p(r_i|b_i) \right\}.$$

For decoding we can neglect constant factors. Thus, we have

$$\hat{\mathbf{b}} = \underset{\mathbf{b}}{\operatorname{argmax}} \left\{ \prod_i \exp\left(-\frac{(r_i - x_i)^2}{2\sigma^2}\right) \right\}.$$

Taking the logarithm and again neglecting constant factors, we obtain

$$\hat{\mathbf{b}} = \underset{\mathbf{b}}{\operatorname{argmax}} \left\{ \sum_i -(r_i - x_i)^2 \right\} = \underset{\mathbf{b}}{\operatorname{argmin}} \left\{ \sum_i (r_i - x_i)^2 \right\}.$$

Note that the term $\sum_i (r_i - x_i)^2$ is the square of the *Euclidean metric*

$$\operatorname{dist}_E(\mathbf{r}, \mathbf{x}) = \sqrt{\sum_i (r_i - x_i)^2}, \quad r_i, x_i \in \mathbb{R}.$$

Therefore, it is called the *squared Euclidean distance*. Consequently, we can express the ML decision criterion in terms of a minimum distance decoding rule, but now with the squared Euclidean distance as the distance measure

$$\hat{\mathbf{b}} = \underset{\mathbf{b}}{\operatorname{argmin}} \left\{ \operatorname{dist}_E^2(\mathbf{r}, \mathbf{x}) \right\}.$$

3.4.2 Support of Punctured Codes

Punctured convolutional codes as discussed in Section 3.1.5 are derived from a rate $R = 1/n$ mother code by periodically deleting a part of the code bits. Utilising an appropriate metric, like the squared Euclidean distance, we can decode the punctured code using the trellis of the mother code.

The trick is that we compute the branch metric so that for each deleted bit the contribution to all branch metric values is constant. For example, with the squared Euclidean distance we can use a simple depuncturing unit before the Viterbi decoder that inserts the value zero for each punctured bit into the sequence of received values. Consequently, the missing information does not influence the decisions of the Viterbi algorithm, because the zero values do not alter the metric calculation. Yet, we can use the trellis of the original mother code for decoding.

The application of puncturing is motivated by the fact that Viterbi decoding of high-rate convolutional codes can be significantly simplified by using punctured codes. The trellis of a convolutional code has 2^ν nodes in each trellis section, where ν is the overall constraint length of the code. Furthermore, each node (accept for the m terminating trellis sections) has 2^k outgoing branches. Hence, there are $2^{\nu+k}$ branches per trellis section and roughly $(L+m)2^{\nu+k}$ branches in the terminated trellis. The number of branches determines the number of operations of the Viterbi algorithm and therefore the computational requirements. The complexity increases linearly with the length L of the trellis, but exponentially with the constraint length ν and with the dimension k.

With punctured codes we always use the trellis of the rate $R = 1/n$ mother code for decoding. Hence, each trellis section has only $2^{\nu+1}$ branches. Of course, the addition of the zero values causes some computational overhead. However, the total number of additions is much smaller if we use puncturing and depuncturing. For instance, with a rate $R = 2/3$ code of overall constraint length $\nu = 6$, we have to compute $3 \cdot 2^{\nu+2} = 768$ additions for each trellis section or 384 additions per information bit. Using the trellis of the mother code, we have to compute $2 \cdot 2^{\nu+1} = 256$ additions per information bit.

3.4.3 Implementation Issues

For applications with low data rates, the Viterbi algorithm can be implemented in software on a Digital Signal Processor (DSP). However, for high data rates the high computational requirements of maximum likelihood decoding demand hardware implementations in Very Large-Scale Integration (VLSI) technology or hybrid DSP architectures, where the Viterbi algorithm runs in a dedicated processor part or coprocessor.

Figure 3.28 provides the block diagram of a receiver structure with Viterbi decoding. The first block is the so-called Automatic Gain Control (AGC). The AGC is an adaptive device that adjusts the gain of the received signal to an appropriate level for the analogue-to-digital (A/D) conversion. For instance, in Section 3.4.1 we have assumed that the energy of the signal is normalised to 1. In this case the AGC should measure the average signal level and implement a scaling by $\frac{1}{\sqrt{E_s}}$.

The A/D conversion provides a quantised channel output. It is important that the A/D conversion reduces the word length to the required minimum, because the complexity of the Viterbi decoder depends strongly on the word length of the branch metrics. Massey presented a procedure to calculate the quantisation thresholds for the AWGN channel (Massey, 1974). Using this procedure, it was shown in (Johannesson and Zigangirov, 1999) that two quantisation levels would lead to roughly 2 dB loss in signal-to-noise ratio compared with the unquantised AWGN channel. This is essentially the difference between hard- and

CONVOLUTIONAL CODES

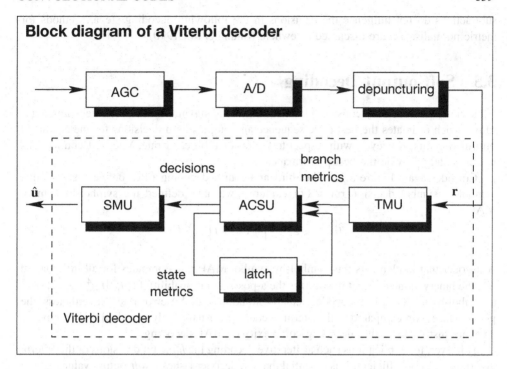

Figure 3.28: Block diagram of a receiver structure with Viterbi decoding

soft-input decoding for the AWGN channel. In Section 1.2.4 we have already seen that, with respect to the channel capacity, the difference is actually up to 3 dB (cf. Figure 1.6 on page 8). A 2-bit quantisation, i.e. four levels, would lead to approximately 0.57 dB loss, whereas with eight levels the loss is only about 0.16 dB. Hence, under ideal conditions a 3-bit quantisation should already provide sufficient accuracy. However, the computed quantisation thresholds depend on the actual signal-to-noise ratio and require an ideal gain control. Therefore, in practice a larger word length for the channel values might be required.

The block depuncturing in Figure 3.28 represents the depuncturing unit discussed in the previous section and is only required if punctured codes are used.

The Viterbi decoder in Figure 3.28 depicts the structure of a hardware implementation. The Transition Metric Unit (TMU) calculates all possible branch metrics in the Viterbi algorithms. For a rate $R = k/n$ code, 2^n different branch metric values have to be computed for each trellis section. The Add Compare Select Unit (ACSU) performs the forward recursion for all trellis nodes, i.e. the ACSU updates the state metric values. The ACSU is recursive which is indicated by the latch. Finally, the decisions for each trellis node, i.e. the local survivors, are stored in the Survivor Memory Unit (SMU) which also performs the backward recursion.

The ACSU performs the actual add–compare–select operation and is therefore the most complex part of the Viterbi decoder. In order to use the same unit for all trellis sections, a metric normalisation is required that ensures a fixed word length. Note that we can subtract a common value from all state metric values in any level of the trellis, because these subtractions do not influence the decisions in the following decoding steps. Methods for metric normalisation are discussed elsewhere (Shung et al., 1990).

3.5 Soft-output Decoding

Up to now, we have only considered Viterbi's famous maximum likelihood decoding algorithm which estimates the best code sequence and outputs hard decisions for the estimated information bits. However, with respect to the residual bit error rate, Viterbi decoding does not ultimately provide the best performance.

Consider again Figure 3.12. The third and fourth decoding rules define the so-called symbol-by-symbol decision rules. For example, we have defined the symbol-by-symbol MAP rule

$$\hat{u}_t^{(l)} = \underset{u_t^{(l)}}{\operatorname{argmax}} \left\{ \Pr\{u_t^{(l)}|\mathbf{r}\} \right\} \quad \forall \, t.$$

A decoder that implements this symbol-by-symbol MAP rule computes for all information bits the binary value $\hat{u}_t^{(l)}$ that maximizes the a-posteriori probability $\Pr\{u_t^{(l)}|\mathbf{r}\}$.[4] Hence, such a symbol-by-symbol MAP decoder actually minimizes the bit error rate. Nevertheless, the performance gain compared with Viterbi decoding is usually only marginal and commonly does not justify the higher decoding complexity of MAP decoding.

However, for applications such as iterative decoding the *hard* bit decisions of the Viterbi algorithm are not sufficient. It is essential that the decoder issues a *soft* output value and not merely provides the binary value of the estimated information or code bits (*hard output*). The term *soft-output decoding* refers to decoding procedures that additionally calculate a reliability information for each estimated bit, the so-called *soft output*. Consequently, the Viterbi algorithm is a hard-output decoder.

There are basically two types of Soft-Input Soft-Output (SISO) decoding methods. The first type are soft-output extensions of the Viterbi algorithm. The most popular representative of this type is the Soft-Output Viterbi Algorithm (SOVA) proposed elsewhere (Hagenauer and Hoeher, 1989). The SOVA selects the survivor path as the Viterbi algorithm. To calculate reliabilities, it utilises the fact that the difference in the branch metrics between the survivor and the discarded branches indicates the reliability of each decision in the Viterbi algorithm. A similar reliability output algorithm was recently presented (Freudenberger and Stender, 2004).

The second class of soft-output algorithms are based on the so-called *BCJR algorithm* which is named after its inventors Bahl, Cocke, Jelinek and Raviv (Bahl et al., 1974). The BCJR algorithm is a symbol-by-symbol MAP decoding algorithm. A similar symbol-by-symbol MAP algorithm was also independently developed by McAdam, Welch and Weber (McAdam et al., 1972). The soft outputs of the BCJR algorithm are the symbol-by-symbol a-posteriori probabilities $\Pr\{u_t^{(l)}|\mathbf{r}\}$. Therefore, this kind of soft-output decoding is also called A-Posteriori Probability (APP) decoding.

APP decoding provides the best performance in iterative decoding procedures. However, the SOVA has a significantly lower complexity, with only slight degradation in the

---

[4]Note that if all information bits are a priori equally likely, the symbol-by-symbol MAP rule results in the ML symbol-by-symbol rule.

CONVOLUTIONAL CODES

decoding performance. Therefore, the SOVA is also popular for practical decoder implementations.

In this section we discuss APP decoding based on the BCJR algorithm for convolutional codes. We first derive the original algorithm. Then, we consider a second version that is more suitable for implementations as it solves some numerical issues of the former version.

3.5.1 Derivation of APP Decoding

The BCJR algorithm is a symbol-by-symbol APP decoder, i.e. it calculates the a-posteriori probability for each information or code symbol. In the literature there exist many modifications of the original BCJR algorithm (see, for example, Hagenauer et al., 1996; Robertson et al., 1997; Schnug, 2002). We will discuss a realisation with soft output as log-likelihood values (cf. Section 4.1.3). For the sake of simplicity, we will assume a terminated convolutional code.

The log-likelihood ratio (L-value) of the binary random variable x is defined as

$$L(x) = \ln \frac{\Pr\{x=0\}}{\Pr\{x=1\}}.$$

From this L-value we can calculate the probabilities

$$\Pr\{x=0\} = \frac{1}{1+e^{-L(x)}} \text{ and } \Pr\{x=1\} = \frac{1}{1+e^{L(x)}}.$$

Hence, we can formulate the symbol-by-symbol MAP decoding rule in terms of the log-likelihood ratio

$$L(\hat{u}_t^{(l)}) = \ln \frac{\Pr\{u_t^{(l)}=0|\mathbf{r}\}}{\Pr\{u_t^{(l)}=1|\mathbf{r}\}}$$

and obtain

$$\hat{u}_t^{(l)} = \begin{cases} 0 & \text{if } L(\hat{u}_t^{(l)}) \geq 0 \\ 1 & \text{if } L(\hat{u}_t^{(l)}) < 0 \end{cases}.$$

We observe that the hard MAP decision of the symbol can be based on the sign of this L-value, and the quantity $|L(\hat{u}_t^{(l)})|$ is the reliability of this decision. The problem of APP decoding is therefore equivalent to calculating the log-likelihood ratios $L(\hat{u}_t^{(l)})$ for all t.

In the following we discuss a method to compute the L-values $L(\hat{u}_t^{(l)})$ on the basis of the trellis representation of a convolutional code. First of all, note that the probability $\Pr\{u_t^{(l)}=0|\mathbf{r}\}$ can be calculated from the a-posteriori probabilities $\Pr\{\mathbf{b}|\mathbf{r}\}$ of all code sequences $\mathbf{b}$ which correspond to an information bit $u_t^{(l)}=0$ at position t. We have

$$\Pr\{u_t^{(l)}=0|\mathbf{r}\} = \sum_{\mathbf{b}\in\mathbb{B}, u_t^{(l)}=0} \Pr\{\mathbf{b}|\mathbf{r}\}.$$

Similarly, $\Pr\{u_t^{(l)}=1|\mathbf{r}\}$ can be calculated from the APP of all code sequences $\mathbf{b}$

$$\Pr\{u_t^{(l)}=1|\mathbf{r}\} = \sum_{\mathbf{b}\in\mathbb{B}, u_t^{(l)}=1} \Pr\{\mathbf{b}|\mathbf{r}\}.$$

This yields

$$L(\hat{u}_t^{(l)}) = \ln \frac{\Pr\{u_t^{(l)} = 0 | \mathbf{r}\}}{\Pr\{u_t^{(l)} = 1 | \mathbf{r}\}} = \ln \frac{\sum_{\mathbf{b} \in B, u_t^{(l)}=0} \Pr\{\mathbf{b}|\mathbf{r}\}}{\sum_{\mathbf{b} \in B, u_t^{(l)}=1} \Pr\{\mathbf{b}|\mathbf{r}\}}.$$

Now, assume that the channel is memoryless and the information bits are statistically independent. Then we have

$$\Pr\{\mathbf{b}|\mathbf{r}\} = \frac{\Pr\{\mathbf{r}|\mathbf{b}\}\Pr\{\mathbf{b}\}}{\Pr\{\mathbf{r}\}} = \prod_i \frac{\Pr\{\mathbf{r}_i|\mathbf{b}_i\}\Pr\{\mathbf{b}_i\}}{\Pr\{\mathbf{r}_i\}}.$$

Cancelling the terms $\Pr\{\mathbf{r}_i\}$, we obtain

$$L(\hat{u}_t^{(l)}) = \ln \frac{\sum_{\mathbf{b} \in B, u_t^{(l)}=0} \prod_i \Pr\{\mathbf{r}_i|\mathbf{b}_i\}\Pr\{\mathbf{b}_i\}}{\sum_{\mathbf{b} \in B, u_t^{(l)}=1} \prod_i \Pr\{\mathbf{r}_i|\mathbf{b}_i\}\Pr\{\mathbf{b}_i\}}.$$

In particular, for depth t we obtain

$$\prod_i \Pr\{\mathbf{r}_i|\mathbf{b}_i\} = \prod_{i<t} \Pr\{\mathbf{r}_i|\mathbf{b}_i\} \cdot \Pr\{\mathbf{r}_t|\mathbf{b}_t\}\Pr\{\mathbf{b}_t\} \cdot \prod_{i>t} \Pr\{\mathbf{r}_i|\mathbf{b}_i\}.$$

Consider now Figure 3.29 which illustrates a segment of a code trellis. Each state transition $\sigma_t' \to \sigma_{t+1}$ from a state σ_t' at depth t to a state σ_{t+1} at depth $t+1$ corresponds to k

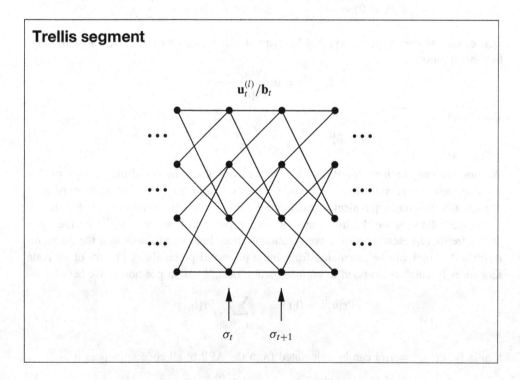

Figure 3.29: Trellis segment of a convolutional code

CONVOLUTIONAL CODES

APP decoding

- Initialisation: $\alpha(\sigma_0) = \beta(\sigma_{L+m}) = 1$.
- Transition probabilities:
$$\gamma(\sigma_t, \sigma'_{t+1}) = \Pr\{\mathbf{r}_t | \mathbf{b}_t\} \Pr\{\mathbf{b}_t\} \qquad (3.11)$$

- Forward recursion: starting from σ_0, calculate
$$\alpha(\sigma'_t) = \sum_{\sigma_{t-1}} \gamma(\sigma_{t-1}, \sigma'_t) \alpha(\sigma_{t-1}) \qquad (3.12)$$

- Backward recursion: starting from σ_{L+m}, calculate
$$\beta(\sigma'_t) = \sum_{\sigma_{t+1}} \gamma(\sigma'_t, \sigma_{t+1}) \beta(\sigma_{t+1}) \qquad (3.13)$$

- Output:
$$L(\hat{u}_t) = \ln \frac{\sum_{\sigma'_t \to \sigma_{t+1}, u_t^{(l)} = 0} \alpha(\sigma'_t) \cdot \gamma(\sigma'_t, \sigma_{t+1}) \cdot \beta(\sigma_{t+1})}{\sum_{\sigma'_t \to \sigma_{t+1}, u_t^{(l)} = 1} \alpha(\sigma'_t) \cdot \gamma(\sigma'_t, \sigma_{t+1}) \cdot \beta(\sigma_{t+1})} \qquad (3.14)$$

Figure 3.30: APP decoding with probabilities

information bits $\mathbf{u}_t$ and n code bits $\mathbf{b}_t$. Using the trellis to represent the convolutional code, we can write the L-value $L(\hat{u}_t^{(l)})$ as

$$L(\hat{u}_t^{(l)}) = \ln \frac{\sum_{\sigma'_t \to \sigma_{t+1}, u_t^{(l)}=0} \Pr\{\sigma'_t \to \sigma_{t+1} | \mathbf{r}\}}{\sum_{\sigma'_t \to \sigma_{t+1}, u_t^{(l)}=1} \Pr\{\sigma'_t \to \sigma_{t+1} | \mathbf{r}\}} = \ln \frac{\sum_{\sigma'_t \to \sigma_{t+1}, u_t^{(l)}=0} \Pr\{\sigma'_t \to \sigma_{t+1}, \mathbf{r}\}}{\sum_{\sigma'_t \to \sigma_{t+1}, u_t^{(l)}=1} \Pr\{\sigma'_t \to \sigma_{t+1}, \mathbf{r}\}}.$$

Again, proceeding from the assumption of a memoryless channel and statistically independent information symbols, we can rewrite the transition probability $\Pr\{\sigma'_t \to \sigma_{t+1}, \mathbf{r}\}$. In particular, for depth t we obtain

$$\Pr\{\sigma'_t \to \sigma_{t+1}, \mathbf{r}\} = \underbrace{\Pr\{\sigma'_t, \mathbf{r}_{[0,t-1]}\}}_{\alpha(\sigma'_t)} \cdot \underbrace{\Pr\{\sigma'_t \to \sigma_{t+1}, \mathbf{r}_t\}}_{\gamma(\sigma'_t, \sigma_{t+1})} \cdot \underbrace{\Pr\{\sigma_{t+1}, \mathbf{r}_{[t+1, L+m-1]}\}}_{\beta(\sigma_{t+1})}$$

and

$$L(\hat{u}_t^{(l)}) = \ln \frac{\sum_{\sigma'_t \to \sigma_{t+1}, u_t^{(l)}=0} \alpha(\sigma'_t) \cdot \gamma(\sigma'_t, \sigma_{t+1}) \cdot \beta(\sigma_{t+1})}{\sum_{\sigma'_t \to \sigma_{t+1}, u_t^{(l)}=1} \alpha(\sigma'_t) \cdot \gamma(\sigma'_t, \sigma_{t+1}) \cdot \beta(\sigma_{t+1})}.$$

$\Pr\{\sigma'_t, \mathbf{r}_{[0,t-1]}\}$ is the joint probability of the event that the received sequence is $\mathbf{r}_{[0,t-1]}$ up to time $t - 1$ and the transmitted sequence passes through state σ'_t. We use the abbreviated notation $\alpha(\sigma'_t)$ to denote this value. Similarly, we use $\beta(\sigma_{t+1})$ and $\gamma(\sigma'_t, \sigma_{t+1}, \mathbf{r}_t)$ to denote the probabilities $\Pr\{\sigma_{t+1}, \mathbf{r}_{[t+1,L+m-1]}\}$ and $\Pr\{\sigma'_t \to \sigma_{t+1}, \mathbf{r}_t\}$ respectively. Hence, $\beta(\sigma_{t+1})$ is the joint probability of the event that the transmitted sequence passes through state σ_{t+1} and we receive $\mathbf{r}_{[t+1,L+m-1]}$ for the tail of the code sequence. Finally, $\gamma(\sigma'_t, \sigma_{t+1})$ is the joint probability of the state transition $\sigma'_t \to \sigma_{t+1}$ and the n-tuple $\mathbf{r}_t$.

The BCJR algorithm is an efficient method for calculating the joint probabilities $\alpha(\sigma'_t)$ and $\beta(\sigma_t)$ recursively. During the forward recursion, starting with the initial node σ_0, we compute

$$\alpha(\sigma'_t) = \sum_{\sigma_{t-1}} \gamma(\sigma_{t-1}, \sigma'_t) \alpha(\sigma_{t-1})$$

with the initial condition $\alpha(\sigma_0) = 1$ and with the transition probabilities

$$\gamma(\sigma_t, \sigma'_{t+1}) = \Pr\{\mathbf{r}_t | \mathbf{b}_t\} \Pr\{\mathbf{b}_t\} = \Pr\{\mathbf{r}_t, \mathbf{b}_t\}.$$

Note that this forward recursion is similar to the forward recursion of the Viterbi algorithm, i.e. for each state we calculate a joint probability based on the previous states and state transitions.

Then, we start the backward recursion at the terminating node σ_{L+m} with the condition $\beta(\sigma_{L+m}) = 1$. We compute

$$\beta(\sigma'_t) = \sum_{\sigma_{t+1}} \gamma(\sigma'_t, \sigma_{t+1}) \beta(\sigma_{t+1}).$$

This is essentially the same procedure as the forward recursion starting from the terminated end of the trellis. During the backward recursion we also calculate the soft-output L-values according to Equation (3.14). All steps of the BCJR algorithm are summarised in Figure 3.30. Consider now the calculation of the values $\gamma(\sigma_t, \sigma'_{t+1}) = \Pr\{\mathbf{r}_t | \mathbf{b}_t\} \Pr\{\mathbf{b}_t\} = \Pr\{\mathbf{r}_t | \mathbf{b}_t\} \Pr\{\mathbf{u}_t\}$. As the channel is memoryless and the information bits are statistically independent, we have

$$\gamma(\sigma_t, \sigma'_{t+1}) = \Pr\{\mathbf{r}_t | \mathbf{b}_t\} \Pr\{\mathbf{u}_t\} = \prod_{j=1}^{n} \Pr\{r_t^{(j)} | b_t^{(j)}\} \prod_{l=1}^{k} \Pr\{u_t^{(l)}\}.$$

In order to calculate $\gamma(\sigma_t, \sigma'_{t+1})$, we can use the a-priori log-likelihood ratios for information bits and the L-values for the received symbols. Note that for an a-priori probability we have

$$\Pr\{u_t^{(l)} = 0\} = \frac{1}{1 + e^{-L(u_t^{(l)})}} \quad \text{and} \quad \Pr\{u_t^{(l)} = 1\} = \frac{1}{1 + e^{L(u_t^{(l)})}} = \frac{e^{-L(u_t^{(l)})}}{1 + e^{-L(u_t^{(l)})}}.$$

To calculate $\gamma(\sigma_t, \sigma'_{t+1})$ it is sufficient to use the term

$$e^{-L(u_t^{(l)}) u_t^{(l)}} = \begin{cases} 1 & \text{, if } u_t^{(l)} = 0 \\ e^{-L(u_t^{(l)})} & \text{, if } u_t^{(l)} = 1 \end{cases}$$

CONVOLUTIONAL CODES

instead of the probabilities $\Pr\{u_t^{(l)} = 0\}$ or $\Pr\{u_t^{(l)} = 1\}$, because the numerator terms $1 + e^{-L(u_t^{(l)})}$ will be cancelled in the calculation of the final a-posteriori L-values. Similarly, we can use $e^{-L(r_t^{(j)}|b_t^{(j)})b_t^{(j)}}$ instead of the probabilities $\Pr\{r_t^{(j)}|b_t^{(j)}\}$. Certainly, the $\gamma(\sigma_t, \sigma'_{t+1})$ values no longer represent probabilities. Nevertheless, we can use

$$\gamma(\sigma_t, \sigma'_{t+1}) = \prod_{j=1}^{n} e^{-L(r_t^{(j)}|b_t^{(j)})b_t^{(j)}} \prod_{l=1}^{k} e^{-L(u_t^{(l)})u_t^{(l)}}$$

in the BCJR algorithm and obtain the correct soft-output values.

3.5.2 APP Decoding in the Log Domain

The implementation of APP decoding as discussed in the previous section leads to some numerical issues, in particular for good channel conditions where the input L-values are large. In this section we consider another version of APP decoding, where we calculate logarithms of probabilities instead of the probabilities. Therefore, we call this procedure APP decoding in the log domain.

Basically, we will use the notation as introduced in the previous section. In particular, we use

$$\overline{\alpha}(\sigma_t) = \ln(\alpha(\sigma_t)),$$
$$\overline{\beta}(\sigma_t) = \ln(\beta(\sigma_t)),$$
$$\overline{\gamma}(\sigma'_t, \sigma_{t+1}) = \ln(\gamma(\sigma'_t, \sigma_{t+1})).$$

On account of the logarithm, the multiplications of probabilities in the update equations in Figure 3.30 correspond to sums in the log domain. For instance, the term $\gamma(\sigma_{t-1}, \sigma'_t) \alpha(\sigma_{t-1})$ corresponds to $\overline{\gamma}(\sigma_{t-1}, \sigma'_t) + \overline{\alpha}(\sigma_{t-1})$ in the log domain. To calculate $\overline{\gamma}(\sigma_{t-1}, \sigma'_t)$, we can now use

$$\overline{\gamma}(\sigma_{t-1}, \sigma'_t) = -\sum_{j=1}^{n} L(r_t^{(j)}|b_t^{(j)})b_t^{(j)} - \sum_{l=1}^{k} L(u_t^{(l)})u_t^{(l)}$$

which follows from

$$\overline{\gamma}(\sigma_t, \sigma'_{t+1}) = \ln\left(\prod_{j=1}^{n} e^{-L(r_t^{(j)}|b_t^{(j)})b_t^{(j)}} \prod_{l=1}^{k} e^{-L(u_t^{(l)})u_t^{(l)}}\right).$$

However, instead of the sum $x_1 + x_2$ of two probabilities x_1 and x_2, we have to calculate $\ln(e^{\bar{x}_1} + e^{\bar{x}_2})$ in the log domain. Consequently, the update equation for the forward recursion is now

$$\overline{\alpha}(\sigma'_t) = \ln\left(\sum_{\sigma_{t-1}} e^{(\overline{\gamma}(\sigma_{t-1}, \sigma'_t) + \overline{\alpha}(\sigma_{t-1}))}\right).$$

Similarly, we obtain the update equation

$$\overline{\beta}(\sigma'_t) = \ln\left(\sum_{\sigma_{t+1}} e^{(\overline{\gamma}(\sigma'_t, \sigma_{t+1}) + \overline{\beta}(\sigma_{t+1}))}\right)$$

for the backward recursion and equation

$$L(\hat{u}_t^{(l)}) = \ln \frac{\sum_{\sigma_t' \to \sigma_{t+1}, u_t^{(l)}=0} e^{\overline{\alpha}(\sigma_t') + \overline{\gamma}(\sigma_t', \sigma_{t+1}) + \overline{\beta}(\sigma_{t+1})}}{\sum_{\sigma_t' \to \sigma_{t+1}, u_t^{(l)}=1} e^{\overline{\alpha}(\sigma_t') + \overline{\gamma}(\sigma_t', \sigma_{t+1}) + \overline{\beta}(\sigma_{t+1})}}$$

for the soft output. On account of $\ln(1) = 0$ the initialisation is now $\overline{\alpha}(\sigma_0) = \overline{\beta}(\sigma_{L+m}) = 0$.

The most complex operation in the log domain version of the APP decoding algorithm corresponds to the sum of two probabilities x_1 and x_2. In this case, we have to calculate $\ln\left(e^{\overline{x}_1} + e^{\overline{x}_2}\right)$. For this calculation we can also use the *Jacobian logarithm* which yields

$$\ln\left(e^{\overline{x}_1} + e^{\overline{x}_2}\right) = \max\{\overline{x}_1, \overline{x}_2\} + \ln\left(1 + e^{-|\overline{x}_1 - \overline{x}_2|}\right).$$

APP decoding in the log domain

- Initialisation: $\overline{\alpha}(\sigma_0) = \overline{\beta}(\sigma_{L+m}) = 0$.
- State transitions:

$$\overline{\gamma}(\sigma_t, \sigma_{t+1}') = \ln(\Pr\{\mathbf{r}_t|\mathbf{b}_t\}) + \ln(\Pr\{\mathbf{b}_t\}) \qquad (3.15)$$

or

$$\overline{\gamma}(\sigma_{t-1}, \sigma_t') = -\sum_{j=1}^{n} L(r_t^{(j)}|b_t^{(j)})b_t^{(j)} - \sum_{l=1}^{k} L(u_t^{(l)})u_t^{(l)} \qquad (3.16)$$

- Forward recursion: starting from σ_0, calculate

$$\overline{\alpha}(\sigma_t') = \max_{\sigma_{t-1}} \left(\overline{\gamma}(\sigma_{t-1}, \sigma_t') + \overline{\alpha}(\sigma_{t-1})\right) + f_c(\cdot) \qquad (3.17)$$

- Backward recursion: starting from σ_{L+m}, calculate

$$\overline{\beta}(\sigma_t') = \max_{\sigma_{t+1}} \left(\overline{\gamma}(\sigma_t', \sigma_{t+1}) + \overline{\beta}(\sigma_{t+1})\right) + f_c(\cdot) \qquad (3.18)$$

- Output:

$$L(\hat{u}_t^{(l)}) = \max_{\sigma_t' \to \sigma_{t+1}, u_t^{(l)}=0} \left(\overline{\alpha}(\sigma_t') + \overline{\gamma}(\sigma_t', \sigma_{t+1}) + \overline{\beta}(\sigma_{t+1})\right) + f_c(\cdot)$$
$$- \max_{\sigma_t' \to \sigma_{t+1}, u_t^{(l)}=1} \left(\overline{\alpha}(\sigma_t') + \overline{\gamma}(\sigma_t', \sigma_{t+1}) + \overline{\beta}(\sigma_{t+1})\right) - f_c(\cdot) \qquad (3.19)$$

Figure 3.31: APP decoding in the log domain

To reduce the decoding complexity, this expression is usually implemented as

$$\ln\left(e^{\bar{x}_1} + e^{\bar{x}_2}\right) \approx \max\{\bar{x}_1, \bar{x}_2\} + f_c(\bar{x}_1, \bar{x}_2),$$

where $f_c(\bar{x}_1, \bar{x}_2)$ is a correction function that approximates the term $\ln\left(1 + e^{-|\bar{x}_1 - \bar{x}_2|}\right)$. This function depends only on the absolute value of the difference $\bar{x}_1 - \bar{x}_2$ and can be implemented with a look-up table, but the approximation with $\ln\left(e^{\bar{x}_1} + e^{\bar{x}_2}\right) \approx \max\{\bar{x}_1, \bar{x}_2\}$ can also be employed and usually leads only to a minor performance degradation. The latter approach is called the *max-log approximation*. Besides the significantly lower complexity, the max-log approximation does not require exact input L-values. A constant factor common to all input values does not influence the maximum operation. Hence, the max-log approximation has no need for a signal-to-noise ratio (SNR) estimation.

Note that the max-log approximation provides the same hard-decision output as the Viterbi algorithm (Robertson *et al.*, 1997; Schnug, 2002). Hence, it no longer minimises the bit error rate.

The complete algorithm is summarised in Figure 3.31. The term $f_c(\cdot)$ denotes the correction function for all values included in the corresponding maximisation. If there are more than two terms, the algorithm can be implemented recursively where we always consider value pairs. For instance, in order to evaluate $\ln(e^{x_1} + e^{x_2} + e^{x_3} + e^{x_4})$, we first calculate $x' = \ln(e^{x_1} + e^{x_2})$ and $x'' = \ln(e^{x_3} + e^{x_4})$. Then we have

$$\ln(e^{x'} + e^{x''}) = \ln(e^{x_1} + e^{x_2} + e^{x_3} + e^{x_4}).$$

3.6 Convolutional Coding in Mobile Communications

In this section we will consider some examples for convolutional coding in mobile communications. With mobile communication channels it is necessary to use channel coding for transmission in order to avoid losses due to transmission errors. Mobile communication standards like GSM and UMTS distinguish between speech and data communication. The transmission of coded speech is a real-time service that allows only a small latency, but is rather impervious to residual errors in the decoded data stream, i.e. residual errors can usually be efficiently masked by the speech decoder. Data services, on the other hand, usually require very low residual bit error rates, but allow for some additional delay caused by retransmissions due to transmission errors. We first consider the transmission of speech signals according to the GSM and the UMTS standards. In the subsequent sections we discuss how convolutional codes are employed in GSM to achieve very efficient retransmission schemes that yield low residual error rates.

3.6.1 Coding of Speech Data

Usually, speech-coded data require a forward error correction scheme that allows for unequal error protection, i.e. there are some very important bits in the speech data stream that should be protected by robust channel coding. Then, there is a certain portion of data bits that is impervious to transmission errors. Consequently, speech coding and forward error correction should be designed together in order to get a good overall coding result.

Both the GSM and UMTS standards define different speech coding types. We will, however, only consider the so-called full-rate and enhanced-full-rate codecs which illustrate how the speech coding and channel coding cooperate. The Full-Rate (FR) codec was the first digital speech coding standard used in the GSM digital mobile phone system developed in the late 1980s. It is still used in GSM networks, but will gradually be replaced by Enhanced-Full-Rate (EFR) and Adaptive MultiRate (AMR) standards, which provide much higher speech quality.

With the GSM full-rate codec, the analogue speech signal is usually sampled with a sampling rate of 8000 samples per second and an 8-bit resolution of the analogue-to-digital conversion. This results in a data rate of 64 kbps. The FR speech coder reduces this rate to 13 kbps. To become robust against transmission errors, these speech-coded data are encoded with a convolutional code, yielding a transmission rate of 22.8 kbps. We will now focus on this convolutional encoding.

The speech coder delivers a sequence of blocks of data to the channel encoder. One such block of data is called a speech frame. With the FR coder, 13 000 bits per second from the encoder are transmitted in frames of 260 bits, i.e. a 260-bit frame every 20 ms. These 260 bits per frame are differentiated into classes according to their importance to the speech quality. Each frame contains 182 bits of class 1 which will be protected by channel coding, and 78 bits of class 2 which will be transmitted without protection. The class 1 bits are further divided into class 1a and class 1b. Class 1a bits are protected by a cyclic code and the convolutional code, whereas class 1b bits are protected by the convolutional code only. The 50 bits of class 1a are the most important bits in each frame, and the 132 bits of class 1b are less important than the class 1a bits, but more important than the 78 bits of class 2. The FR uses a Cyclic Redundancy Check (CRC) code (cf. Section 2.3.2) with the generator polynomial $g(x) = x^3 + x + 1$ for error detection. This code is only applied to the 50 bits in class 1a. As the generator polynomial indicates, the encoding of the CRC code results in three parity bits for error detection.

For the EFR coder, each block from the speech encoder contains only 244 information bits. The block of 244 information bits passes through a preliminary stage, applied only to EFR. This precoding produces 260 bits corresponding to the 244 input bits and 16 redundancy bits. For channel coding, those 260 bits are interpreted as FR data, i.e. as 182 bits of class 1 and 78 bits of class 2. Hence, the channel coding for EFR is essentially the same as for FR.

For channel coding, the memory $m = 4$ convolutional encoder in Figure 3.32 is used. To terminate the code, four zero bits are added to the 185 class 1 bits (including the three CRC bits). Those 189 bits are then encoded with the rate $R = 1/2$ convolutional code, resulting in 378 bits. These code bits are transmitted together with the uncoded 78 less important bits from class 2, so that every speech frame corresponds to a total of 456 transmitted bits. The 50 speech frames per second yield a transmitted rate of 22.8 kbps.

With mobile communication channels, transmission errors usually occur in bursts of errors, i.e. the channel errors are correlated. This correlation occurs both in time and in frequency direction. We have already seen that convolutional codes can correct much more errors than indicated by the free distance of the code. However, this is only possible if the errors are spread over the code sequence. On the other hand, if a burst error occurs, and more than half the free distance of bits is altered, this may lead to a decoding error. The GSM standard takes different measures to cope with these correlated transmission errors.

CONVOLUTIONAL CODES

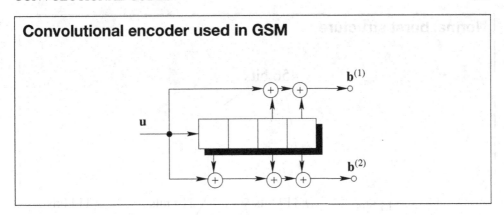

Figure 3.32: Rate $R = 1/2$ convolutional encoder memory $m = 4$

The first method is called *bit interleaving*. Bit interleaving uses the fact that a burst of errors possibly contains a large number of errors within close proximity, but the occurrence of a burst is a relatively unlikely event. Therefore, the code bits are interleaved (permuted) and then transmitted. That way, the errors of a single burst are distributed over the complete block. It is then very likely that the distributed errors will only affect a correctable number of bits. To spread the transmitted bit over a longer period of time, GSM uses a Time Division Multiple Access (TDMA) structure with eight users per frequency band, where a single time slot is smaller than a speech frame. In the GSM standard such a slot is called a *normal burst*. It has a duration of 0.577 ms corresponding to 156.25-bit periods. However, as a normal burst is used to carry both signalling and user information over the GSM air interface, only 114 bits can by used for the coded speech data. Consequently, a speech frame is distributed over four normal bursts. This is illustrated in Figure 3.33.

If the mobile subscriber is moving, e.g. in a car, the errors in different normal bursts are usually uncorrelated. Therefore, the interleaving in GSM spreads adjacent code bits over different normal bursts. In some GSM cells, *frequency hopping* is used to provide uncorrelated normal bursts for slow-moving subscribers and as a means against frequency-selective fading and interference. With frequency hopping, the carrier of the transmitting radio signals is rapidly switched to another carrier frequency, i.e. with every normal burst.

Figure 3.34 provides some simulation results for the transmission with the GSM FR codec. Obviously, the unprotected class 2 bits have a much higher bit error rate than the coded bits from class 1. The bits of class 1a have a slightly better error protection than class 1b. This results from the code termination and the positions in the interleaving. Owing to code termination, the encoder states at the start and at the end of the terminated code word are known. This leads to a slightly better protection for the information bits close to the code word ends. Moreover, the code bits corresponding to class 1a information bits are located close to the pilot sequence within every normal burst. The pilot symbols are bits known to the receiver that are used for channel estimation. This channel estimation is more reliable for nearby code bits.

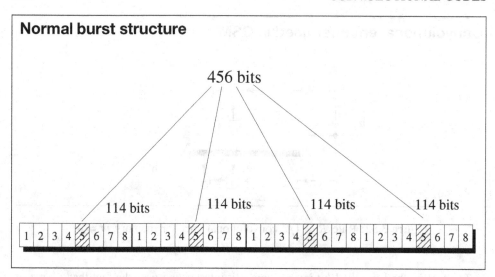

Figure 3.33: Normal burst structure of the GSM air interface

The UMTS standard provides several speech and channel coding modes. The AMR speech codec defines coding methods for data rates from 4.75 to 12.2 kbps, where the 12.2 kbps mode is equivalent to the GSM EFR codec. Moreover, the basic procedure for the convolutional coding is also similar. However, UMTS employs a powerful convolutional code with memory $m = 8$ and code rate $R = 1/3$.

The 12.2 kbps speech coding mode uses 20 ms speech frames with 244 data bits per frame. In the UMTS system, those 244 bits are also partitioned into three classes according to their relevance to the speech decoder. Class A contains the 81 most important bits, while class B with 103 bits and class C with 60 bits are less vital. Only the bits of class A are protected with a CRC code. However, this code provides more reliability as eight redundancy bits are used for error detection.

All three classes are encoded with the same rate $R = 1/3$ convolutional code, where each class is encoded separately. Hence, eight tail bits are added for each class. This results in 828 code bits. Then, puncturing is used to reduce the number of code bits to 688 for each speech frame.

3.6.2 Hybrid ARQ

In this section we discuss hybrid ARQ protocols as an application of convolutional codes for mobile communication channels. The principle of automatic repeat-request protocols is to detect and repeat corrupted data. This means that the receiver has to check whether the arriving information is correct or erroneous. Therefore, all ARQ schemes require an error detection mechanism. Usually, a small amount of redundancy is added to the information packet, e.g. with a CRC code.

If the received data packet is correct, the receiver sends an *acknowledgement* (ACK) to the transmitter, otherwise additional information for the erroneous packet is requested through a *not acknowledgement* (NACK) command. In the simplest case this retransmission

CONVOLUTIONAL CODES

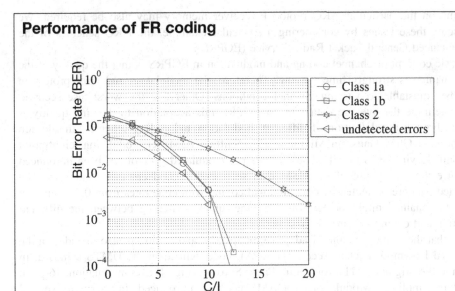

Performance of FR coding

- Channel model: Typical Urban (TU) channel with 50 km/h (no frequency hopping).
- Class 2 bits are transmitted without channel coding.
- The curve for the undetected errors considers all errors of class 1a that are not detected by the CRC code.

Figure 3.34: Simulation results for the full-rate coding scheme

request is answered by sending an exact copy of the packet. However, there exist more advanced retransmission schemes which will be discussed in this section.

Pure ARQ systems, where only error detection is used, are not suitable for time-variant mobile channels as successful transmissions become very unlikely for poor channel conditions. In order to improve the average throughput for such channels, ARQ systems are combined with additional forward error correction. This combination of ARQ and forward error correction is called *hybrid ARQ*.

There exist different types of hybrid ARQ. With *type-I hybrid ARQ* schemes the same encoded packet is sent for transmission and retransmission. Thus, the ARQ protocol does not influence the forward error correction. With *type-II hybrid ARQ* schemes, transmission and retransmission differ.

Packet data systems with ARQ schemes require some additional components compared with circuit-switched transmission systems. We have already mentioned the error detection mechanism. The transmission of the acknowledgement (ACK)/not acknowledgement (NACK) commands needs a reliable return channel. The transmitter has to store sent packets until an acknowledgement is received, thus transmitter memory is mandatory. Moreover, depending on the particular ARQ protocol, receiver memory may also be required. We will discuss these issues by considering a particular example, the ARQ schemes of the GSM/Enhanced General Packet Radio Service (EGPRS).

A basic concept of channel coding and modulation in EGPRS is that the quality of the mobile channel is strongly time varying. Therefore, the modulation and error protection should be adjustable to varying channel conditions. Otherwise the worst-case scenario would determine the average data rate. In EGPRS this adaptation to the link quality is achieved through the definition of different modulation and coding schemes. In addition to the original GSM Gaussian Minimum Key Shifting (GMSK) modulation, eight-point Phase Shift Keying (8PSK) with 3 bits of data per modulated symbol has been introduced to enhance the throughput of the GSM data services for good channel conditions. The transmitted data are protected by convolutional coding with code rates from 0.37 to nearly 1.0. As the channel might be fast time varying, quick switching between the different modulation and coding schemes is possible.

Note that the same principle of adaptive modulation and coding is also included in the High-Speed Downlink Packet Access (HSDPA) mode defined in the UMTS standard. In addition to the original UMTS four-point Phase Shift Keying (4PSK) modulation, 16-point Quadrature Amplitude Modulation (16QAM) has been introduced. Furthermore, several coding schemes are defined that make it possible to adapt the required redundancy to the channel condition.

3.6.3 EGPRS Modulation and Coding

The well-established GSM standard originally only provided circuit-switched data services with low transmission rates of up to 9.6 kbps. High-Speed Circuit-Switched Data (HSCSD) allows rates of up to 57.6 kbps (14.4 kbps/time slot), and, with the General Packet Radio Service (GPRS), packet data services with gross data rates of 182 kbps (22.8 kbps/time slot) become possible. The new Enhanced Data rates for GSM Evolution (EDGE) standard provides even higher data rates (up to 384 kbps). EDGE covers both ECSD for enhanced circuit-switched connections and EGPRS for enhanced packet data services. This section only deals with the latter standard. EGPRS is the GPRS evolutional upgrade. However, EGPRS preserves the most important GSM air interface features, such as the 200 kHz channelling and the TDMA scheme, i.e. every band of 200 kHz is subdivided into eight time slots. GSM originally used GMSK for modulation, which allowed for 1 bit of data per modulated symbol. In order to enhance the throughput of the GSM data services, a second modulation scheme, 8PSK (with 3 bits of data per modulated symbol) in addition to GMSK, has been introduced.

In order to ensure reliable packet data services with EGPRS, hybrid ARQ protocols will be employed. A *link adaptation* (type-I hybrid ARQ) algorithm adapts the modulation and coding scheme to the current channel condition. This should provide a mechanism to have a smooth degradation of the data rate for the outer cell areas. The more sophisticated *incremental redundancy* (type-II hybrid ARQ) schemes automatically adjust themselves to the channel condition by sending additional redundancy for not acknowledged data packets. Applied to time-variant channels, IR schemes allow higher throughputs compared with standard link adaptation schemes. However, for poor channel conditions the average delay may increase dramatically. Therefore, in EGPRS a combination of link adaptation

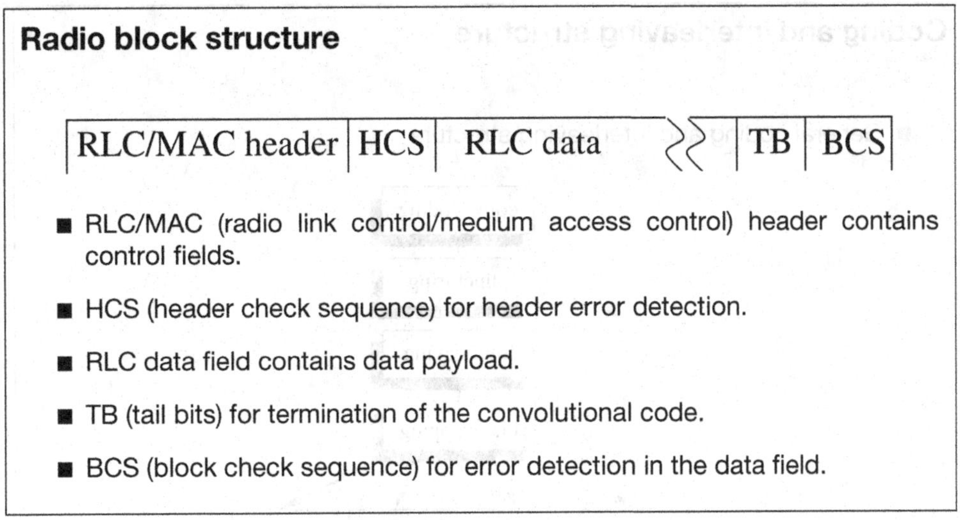

CONVOLUTIONAL CODES

> **Radio block structure**
>
> | RLC/MAC header | HCS | RLC data ⟩⟩ | TB | BCS |
>
> - RLC/MAC (radio link control/medium access control) header contains control fields.
> - HCS (header check sequence) for header error detection.
> - RLC data field contains data payload.
> - TB (tail bits) for termination of the convolutional code.
> - BCS (block check sequence) for error detection in the data field.

Figure 3.35: Radio block structure

and incremental redundancy is used. While Link Adaptation (LA) is mandatory in EGPRS, Incremental Redundancy (IR) is optional for the networks.

In Figure 3.35 the EGPRS radio block structure is presented. One radio block consists of one Radio Link Control (RLC)/Medium Access Control (MAC) header and one or two RLC data blocks. The RLC/MAC header contains control fields like a packet sequence number. The Header Check Sequence (HCS) field is used for header error detection. The payload data are contained in the RLC data field. Attached to each RLC data block there is a Tail Bits (TB) and a Block Check Sequence (BCS) field.

Figure 3.36 gives an overview of the EGPRS channel coding and interleaving structure. The first step independently encodes data and header. For headers, rate $R = 1/3$ tail-biting convolutional codes are used. For data encoding, rate $R = 1/3$ terminated convolutional codes are employed. In order to adjust the code rates, encoding is followed by puncturing. After puncturing, the remaining bits are interleaved and mapped on four consecutive bursts. For some coding schemes, bit swapping is applied after the interleaving. Figure 3.36 also provides a detailed view of the RLC structure for the example of the modulation and coding scheme Modulation and Coding Scheme (MCS)-9.

In contrast to GSM with pure GMSK modulation, EDGE will use both GMSK and 8PSK modulation. GMSK is used as a fall-back solution if 8PSK is not appropriate for the current channel condition. The combination of modulation type and code rate defines the MCS. A total of nine such modulation and coding schemes exists, partly GMSK and 8PSK modulated. These nine schemes are partitioned into three families (A,B,C), where the families correspond to different segmentations of the data stream from higher layers. Thus, switching without data resegmentation is only possible within one family.

The table in Figure 3.37 provides an overview. With coding schemes MCS 7, MCS 8, and MCS 9, two RLC blocks are transmitted with one radio block, while for all other schemes only one RLC data block per radio block is transmitted. The bit swapping

CONVOLUTIONAL CODES

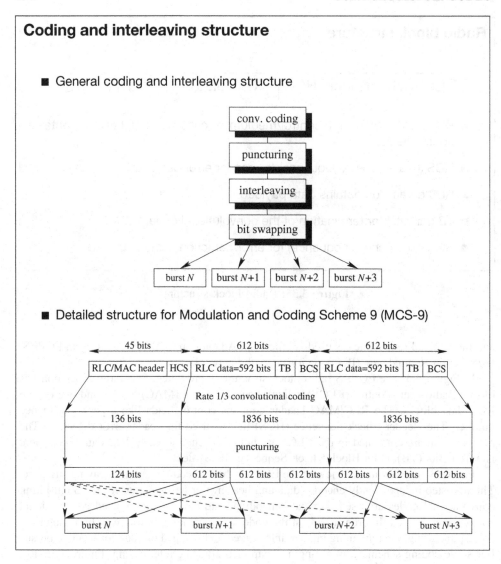

Figure 3.36: Coding and interleaving structure for EGPRS

guarantees that, with MCS 8 and MCS 9, data interleaving is done over two bursts, while with MCS 7 the two RLC data blocks are interleaved over four bursts. The RLC/MAC header is always interleaved over four normal bursts. Depending on the coding scheme, three different header types are defined: type 1 for MCS 7, 8 and 9, type 2 for MCS 5 and 6 and type 3 for MCS 1, 2, 3 and 4.

For IR the possibility of mixed retransmission schemes MCS 5–7 and MCS 6–9 exists. For instance, the first transmission is done with MCS 6 while the retransmissions are carried out with MCS 9. This is possible because MCS 6 and 9 correspond to the same mother

CONVOLUTIONAL CODES

EGPRS modulation and coding schemes

MCS	Data code rate	Header code rate	Modulation	Data within one radio block	Family	Data rate (kbps)
MCS-9	1.0	0.36	8-PSK	2 × 592	A	59.2
MCS-8	0.92	0.36		2 × 544	A	54.4
MCS-7	0.76	0.36		2 × 448	B	44.8
MCS-6	0.49	1/3		592	A	29.6
				544+48		27.2
MCS-5	0.37	1/3		448	B	22.4
MCS-4	1.0	0.53	GMSK	352	C	17.6
MCS-3	0.8	0.53		296	A	14.8
				272+24		13.6
MCS-2	0.66	0.53		224	B	11.2
MCS-1	0.53	0.53		176	C	8.8

Figure 3.37: EGPRS modulation and coding schemes

code and termination length (592 information bits). These retransmission schemes should be more efficient than pure MCS 6 schemes.

3.6.4 Retransmission Mechanism

In terms of average data rate, the selective repeat technique is the most efficient retransmission mechanism. With Selective Repeat ARQ (SR-ARQ), the transmitter continuously sends packets. If the receiver detects an erroneous packet, a NACK command requests a retransmission. Upon reception of this NACK, the transmitter stops its continuous transmission, sends the retransmission and then goes on with the next packet.

For SR-ARQ, each packet requires a unique sequence number that allows the receiver to identify arriving packets and to send requests for erroneously received packets. Moreover, the receiver needs memory in order to deliver the data in correct order to the higher layers.

In EGPRS a variation of the above selective repeat ARQ is employed. In order to use the return channel effectively, ACK/NACK messages are collected and transmitted upon request (polling). If this polling is done periodically, we obtain a block-wise ACK/NACK signalling, as indicated in Figure 3.38. Here, the polling period or Acknowledgement Period (AP) is 6, i.e. after six received packets the receiver sends a list of six ACK/NACK messages. Of course, this polling scheme introduces an additional delay, so that there is a trade-off between delay and signalling efficiency.

The number of possible packet sequence numbers has to be limited, because the transmitter and receiver memories are limited. In order to take this fact into account, the transmitter is only allowed to send packets that have sequence numbers within a certain interval (window). If an ACK is received, the window can be shifted to the next not

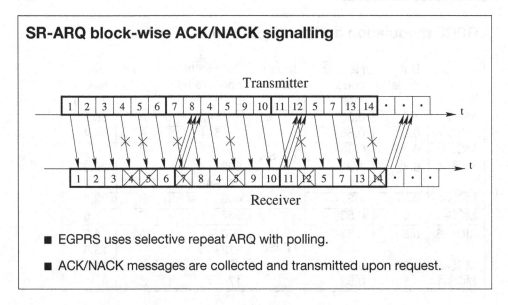

Figure 3.38: SR-ARQ block-wise ACK/NACK signalling

acknowledged sequence number. For the example in Figure 3.38, assume the *window size* (WS) is 10, i.e. for the first transmissions sequence numbers within $[1, \ldots, 10]$ are allowed. After the first ACK/NACK signalling, the window can be shifted and we have $[4, \ldots, 13]$. After the second ACK/NACK signalling we obtain the window $[5, \ldots, 14]$. A third erroneous reception of packet 5 would lead to a so-called *stalled* condition, i.e. the receiver would not be allowed to send new data packets until packet 5 is received successfully. To prevent stalled conditions, window size and acknowledgement period have to be chosen carefully. In EGPRS the maximum window size depends on the multislot class, i.e. on the number of time slots that are assigned to one user. The window size is limited to 192 if only one time slot per carrier is used.

3.6.5 Link Adaptation

The ARQ scheme should be adjustable to varying channel conditions. Otherwise the worst-case scenario would determine the average throughput. For this link adaptation (LA), different coding schemes with different code rates and a switching mechanism are required. In Figure 3.39, simulation results for BPSK modulation and a Gaussian channel are depicted. The ordinate is the average throughput normalised to the maximum throughput. Three different code rates, $R = 1$, $R = 1/2$ and $R = 1/3$, have been used. For good channel conditions the code rate limits the throughput, i.e. pure ARQ ($R = 1$) outperforms the hybrid schemes. However, for signal-to-noise ratios below 12.5 dB, successful transmissions become very unlikely with pure ARQ. At this point the hybrid scheme with code rate $R = 1/2$ still provides optimum performance. For very poor channel conditions the rate $R = 1/3$ scheme performs best. Each scheme in Figure 3.39 outperforms the others over a certain E_s/N_0 region. Outside this range, either the throughput falls rapidly or the

CONVOLUTIONAL CODES

Throughput of type-I ARQ

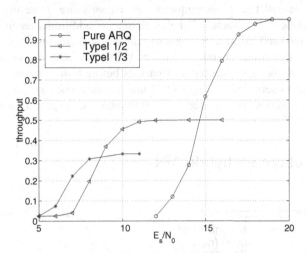

- Each coding scheme outperforms the others over a certain E_s/N_0 region.
- A link adaptation scheme should choose the coding scheme with best possible average throughput for the current channel condition.
- The maximum possible throughput with link adaptation is the envelope of all throughput curves.

Figure 3.39: Throughput of type-I ARQ for a Gaussian channel

coding overhead limits the maximum throughput. Hence, as long as the channel is static around a specific operational point, type-I schemes perform well.

In mobile channels the signal-to-noise ratio is time variant. Fast power control can provide nearly static channel conditions. However, there are limits to power control, e.g. transmit power limits or for high mobile velocity. A link adaptation scheme should choose the coding scheme with best possible average throughput for the current channel condition. Therefore, we can state the maximum possible throughput with link adaptation, i.e. the envelope of all throughput curves, without actually considering the switching algorithm.

3.6.6 Incremental Redundancy

The basic property of type-II hybrid ARQ is that the error correction capability increases with every retransmission. This can be done by storing erroneous packets (soft values) to use them as additional redundancy for later versions of the same packet. The amount of redundancy is automatically adjusted to the current channel condition. As the redundancy increases with each retransmission, this technique is called incremental redundancy (IR).

If the retransmission request is answered by sending an exact copy of the packet (as with pure ARQ), we call the IR technique *diversity combining*. Here the soft values of different transmissions are combined. For this technique, additional receiver memory for the soft values is required. Moreover, the soft combining increases receiver complexity. However, the error correction capability increases with every retransmission.

Another possibility is to encode the data packets before transmission, e.g. with a convolutional code, but to send only a segment of the complete code sequence with the first transmission. If retransmissions are necessary, additional code segments are sent.

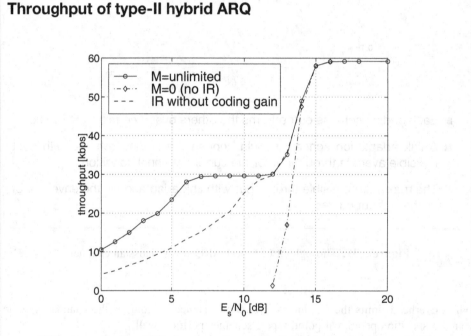

■ With diversity combining, we notice remarkable gains compared with pure ARQ.

■ Forward error correction with maximum ratio combining achieves an additional coding gain.

■ The amount of redundancy is automatically adjusted to the current channel condition.

Figure 3.40: Throughput of type-II hybrid ARQ for a Gaussian channel

CONVOLUTIONAL CODES

Consider, for example, the coding scheme MCS-9 which is presented in Figure 3.36. Two data packets share a common packet header which contains a unique identifier for each data packet. The header and the data packets are encoded with a rate 1/3 convolutional code with memory $m = 6$. For the data part, code termination is applied. The shorter header block is encoded with a tail-biting code. The code word for the header is punctured and the resulting 124 bits are equally distributed over one frame, i.e. four time slots. The code bits of the two data blocks are subdivided into three equally large code segments of 612 bits. Such a code segment is interleaved over two bursts. For every data block of 592 bits, only one code segment of 612 bits is transmitted per frame. Additional code segments are transmitted on request.

Of course, the number of different code segments is limited. If more retransmissions are necessary, copies of already sent code segments will be transmitted. If these copies are only exchanged for the already received versions and not combined, we call the receiver technique *pure code combining*. If copies are combined, we have *maximum ratio combining*.

In Figure 3.40, simulation results for a Gaussian channel with the EGPRS coding scheme MCS-9 are depicted. We compare pure ARQ (no IR) with IR. For the incremental redundancy schemes we have no memory constraints. Here, we use diversity combining and maximum ratio combining.

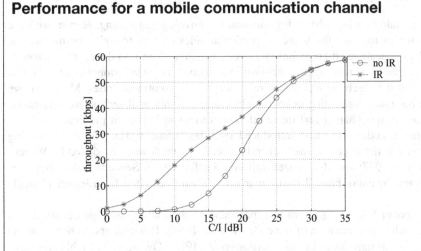

Figure 3.41: Throughput of type-II hybrid ARQ for a mobile communication channel

With diversity combining, we already notice remarkable gains compared with pure ARQ. However, maximum ratio combining utilises an additional coding gain. This coding gain is the reason for the plateau within the curve. In this region of signal-to-noise ratios it becomes very unlikely that the first transmission is successful. Yet, the first retransmission increases the redundancy, so that the need for further retransmission also becomes unlikely.

Figure 3.41 presents the performance of coding scheme MCS-9 for a mobile communication channel. For the mobile channel we used the typical urban (TU) channel model as defined in the GSM standard with a mobile velocity of 3 km/h and a single interfere. Owing to the time-varying nature of the mobile channel, the curve for the IR scheme shows no plateau as for the stationary gaussian channel. Compared with the transmission without code combining, the performance is now improved for all data rates below 50 kbps, where the gain in terms of carrier-to-interference ratio increases for low data rates up to 10 dB.

Incremental redundancy schemes with convolutional coding are usually based on *rate-compatible punctured convolutional codes*. With these punctured codes, a rate compatibility restriction on the puncturing tables ensures that all code bits of high rate codes are used by the lower-rate codes. These codes are almost as good as the best known general convolutional codes of the respective rates. Tables of good rate-compatible punctured codes are given elsewhere (Hagenauer, 1988).

3.7 Summary

This chapter should provide a basic introduction to convolutional coding. Hence, we have selected the topics that we think are of particular relevance to today's communication systems and to concatenated convolutional codes which will be introduced in Chapter 4. However, this chapter is not a comprehensive introduction to convolutional codes. We had to omit much of the algebraic and structural theory of convolutional codes. Moreover, we had to leave out many interesting decoding algorithms. In this final section we summarise the main issues of this chapter and make references to some important publications.

Convolutional codes were first introduced by Elias (Elias, 1955). The first decoding method for convolutional codes was sequential decoding which was introduced by Wozencraft (Wozencraft 1957; see also Wozencraft and Reiffen 1961). Sequential decoding was further developed by Fano (Fano, 1963), Zigangirov (Zigangirov, 1966) and Jelinek (Jelinek, 1969).

The widespread Viterbi algorithm is a maximum likelihood decoding procedure that is based on the trellis representation of the code (Viterbi, 1967). This concept, to represent the code by a graph, was introduced by Forney (Forney, Jr, 1974). Owing to the highly repetitive structure of the code trellis, trellis-based decoding is very suitable for pipelining hardware implementations. Consequently, maximum likelihood decoding has become much more popular for practical applications than sequential decoding, although the latter decoding method has a longer history.

Moreover, the Viterbi algorithm is very impervious to imperfect channel identification. On the other hand, the complexity of Viterbi decoding grows exponentially with the overall constraint length of the code. Today, Viterbi decoders with a constraint length of up to 9 are found in practical applications. Decoding of convolutional codes with a larger constraint length is the natural domain of sequential decoding, because its decoding complexity is determined by the channel condition and not by the constraint length. Sequential decoding is therefore applied when very low bit error rates are required. Recently, sequential decoding has attracted some research interest for applications in hybrid ARQ protocols (Kallel, 1992; Orten, 1999). A good introduction to sequential decoding can be found in the literature (Bossert, 1999; Johannesson and Zigangirov, 1999).

The BCJR algorithm is a method for calculating reliability information for the decoder output (Bahl *et al.*, 1974). This so-called soft output is essential for the decoding of concatenated convolutional codes (turbo codes), which we will discuss in Chapter 4. Like the Viterbi algorithm, the BCJR algorithm is also based on the code trellis which makes it suitable for hardware as well as for software implementations.

Early contributions to the algebraic and structural theory of convolutional codes were made by Massey and Sain (Massey and Sain, 1967, 1968) and by Forney (Forney, Jr, 1970, 1973a), and more recently by Forney again (Forney, Jr, 1991; Forney, Jr *et al.*, 1996). Available books (Bossert, 1999; Johannesson and Zigangirov, 1999) are good introductory texts to this subject. McEliece provides a comprehensive exposition of the algebraic theory of convolutional codes (McEliece, 1998).

As mentioned above, this chapter has focused on the application of convolutional codes in today's communication systems. In Section 3.6 we have considered some examples for convolutional coding in mobile communications. In particular, we have discussed the speech coding and the hybrid ARQ protocols as defined in the GSM standard. Similar concepts are also applied in UMTS mobile communications. Some further results about the GSM link adaptation and incremental redundancy schemes can be found in the literature (Ball *et al.*, 2004a,b). We would like to acknowledge that the simulation results in Section 3.6 are published by courtesy of Nokia Siemens Networks.[5]

---

[5]Nokia Siemens Networks, COO RA RD System Architecture, RRM, Radio and Simulations, GSM/EDGE & OFDM Mobile Radio, Sankt-Martinstrasse 76, 81617 Munich, Germany.

4

Turbo Codes

In this chapter we will discuss the construction of long powerful codes based on the concatenation of simple component codes. The first published concatenated codes were the *product codes* introduced by Elias, (Elias, 1954). The concatenation scheme according to Figure 4.1 was introduced and investigated by Forney (Forney, Jr, 1966) in his PhD thesis. With this serial concatenation scheme the data are first encoded with a so-called *outer code*, e.g. a Reed–Solomon code. The code words of this outer code are then encoded with a second, so-called *inner code*, for instance a binary convolutional code.

After transmission over the noisy channel, first the inner code is decoded, usually using soft-input decoding. The inner decoding results in smaller bit error rates at the output of the inner decoder. Therefore, we can consider the chain of inner encoder, channel and inner decoder as a superchannel with a much smaller error rate than the original channel. Then, an outer, usually algebraic, decoder is used for correcting the residual errors. This two-stage decoding procedure has a much smaller decoding complexity compared with the decoding of a single code of the same overall length. Such *classical* concatenation schemes with an outer Reed–Solomon code and an inner convolutional code (Justesen et al., 1988) are used in satellite communications as well as in digital cellular systems such as the Global System for Mobile communications (GSM).

During recent years, a great deal of research has been devoted to the concatenation of convolutional codes. This research was initiated by the invention of the so-called *turbo codes* (Berrou et al., 1993). Turbo codes are a class of error-correcting codes based on a parallel concatenated coding scheme, where at least two systematic encoders are linked by an interleaver. In the original paper, Berrou, Glavieux and Thitimasjshima showed by simulation that turbo codes employing convolutional component codes are capable of achieving bit error rates as small as 10^{-5} at a code rate of $R = 1/2$ and a signal-to-noise ratio E_b/N_0 of just 0.7 dB above the theoretical Shannon limit.

However, in order to achieve this level of performance, large block sizes of several thousand code bits are required. The name *turbo* reflects a property of the employed iterative decoding algorithm: the decoder output of one iteration is used as the decoder input of the next iteration. This cooperation of the outer and inner decoder is indicated by the feedback link in Figure 4.1.

Coding Theory – Algorithms, Architectures, and Applications André Neubauer, Jürgen Freudenberger, Volker Kühn
© 2007 John Wiley & Sons, Ltd

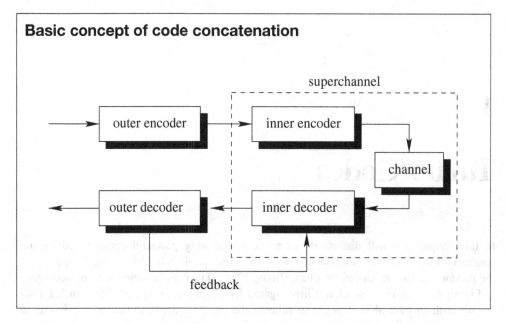

Figure 4.1: Basic concept of code concatenation

The concept of iterative decoding is, however, much older than turbo codes. It was introduced by Gallager (Gallager, 1963), who also invented the class of Low-Density Parity Check (LDPC) codes. In fact, turbo codes can be considered as a particular construction of low-density parity-check codes. Yet, LDPC codes were largely forgotten soon after their invention. There are several reasons why LDPC codes were initially neglected. First of all, the computational power to exploit iterative decoding schemes was not available until recently. In the early years of coding theory, Reed–Solomon codes with algebraic decoding were much more practical. Furthermore, the concatenated Reed–Solomon and convolutional codes were considered perfectly suitable for error control coding.

The reinvention of iterative decoding by Berrou, Glavieux and Thitimasjshima also led to the rediscovery of LDPC codes (see, for example, (MacKay, 1999). LDPC codes have a remarkably good performance; for example, LDPC codes with a code length of 1 million were constructed with a signal-to-noise ratio of only 0.3 dB above the Shannon limit (Richardson et al., 2001).

Today, there exist numerous publications on the construction of LDPC codes and turbo codes that also show the excellent performance of these code constructions. It is not possible to cover even a modest fraction of these publications in a single book chapter. Moreover, we have to leave out much of the interesting theory. Hence, this text is not a survey. Nevertheless, we hope that this chapter provides an interesting and useful introduction to the subject. Our discussion focuses on the basic properties of concatenated convolutional codes.

We start with an introduction to LDPC codes and Tanner graphs in Section 4.1. In Section 4.2 we introduce product codes as a first example of code concatenation that also illustrates the connections between concatenated codes and LDPC codes. We discuss the encoding and decoding of turbo-like codes in Section 4.3. Then, we present three different methods to analyse the code properties of concatenated convolutional codes.

In Section 4.4 we consider the analysis of the iterative decoding algorithm for concatenated convolutional codes. Therefore, we utilise the extrinsic information transfer characteristics as proposed by ten Brink (ten Brink, 2000).

One of the first published methods to analyse the performance of turbo codes and serial concatenations was proposed by Benedetto and Montorsi (Benedetto and Montorsi, 1996, 1998). With this method we calculate the average weight distribution of an ensemble of codes. Based on this weight distribution it is possible to bound the average maximum likelihood performance of the considered code ensemble. This method will be discussed in Section 4.5. Later on, in Section 4.6, we will focus our attention on the minimum Hamming distance of the concatenated codes. By extending earlier concepts (Höst et al., 1999), we are able to derive lower bounds on the minimum Hamming distance. Therefore, we discuss the class of woven convolutional codes that makes it possible to construct codes with large minimum Hamming distances.

4.1 LDPC Codes

LDPC codes were originally invented by Robert Gallager in his PhD thesis (Gallager, 1963). They were basically forgotten shortly after their invention, but today LDPC codes are among the most popular topics in coding theory.

Any linear block code can be defined by its parity-check matrix. If this matrix is sparse, i.e. it contains only a small number of 1s per row or column, then the code is called a low-density parity-check code. The sparsity of the parity-check matrix is a key property of this class of codes. If the parity-check matrix is sparse, we can apply an efficient iterative decoding algorithm.

4.1.1 Codes Based on Sparse Graphs

Today, LDPC codes are usually defined in terms of a sparse bipartite graph, the so-called *Tanner graph* (Tanner, 1981).[1] Such an undirected graph has two types of node, the *message nodes* and *check nodes*.[2]

Figure 4.2 provides the Tanner graph of the Hamming code $\mathbb{B}(7, 4, 3)$ (cf. Section 2.2.8 where an equivalent Hamming code is used). The n nodes on the left are the message nodes. These nodes are associated with the n symbols $b_1, \ldots, b_n$ of a code word. The r nodes $c_1, \ldots, c_r$ on the right are the so-called check nodes and represent parity-check equations. For instance, the check node c_1 represents the equation $b_1 \oplus b_4 \oplus b_5 \oplus b_7 = 0$. Note that in this section we use the symbol $\oplus$ to denote the addition modulo 2 in order to distinguish it from the addition of real numbers. The equation $b_1 \oplus b_4 \oplus b_5 \oplus b_7 = 0$ is determined by the edges connecting the check node c_1 with the message nodes b_1, b_4, b_5

---

[1] A graph is the basic object of study in a mathematical discipline called graph theory. Informally speaking, a graph is a set of objects called nodes connected by links called edges. A sparse graph is a graph with few edges, and a bipartite graph is a special graph where the set of all nodes can be divided into two disjoint sets $\mathbb{A}$ and $\mathbb{B}$ such that every edge has one end-node in $\mathbb{A}$ and one end-node in $\mathbb{B}$.

[2] In an undirected graph the connections between nodes have no particular direction, i.e. an edge from node i to node j is considered to be the same thing as an edge from node j to node i.

Tanner graph of the Hamming code

- The parity-check matrix of the Hamming code $\mathbb{B}(7, 4, 3)$

$$\mathbf{H} = \begin{pmatrix} 1 & 0 & 0 & 1 & 1 & 0 & 1 \\ 0 & 1 & 0 & 1 & 0 & 1 & 1 \\ 0 & 0 & 1 & 0 & 1 & 1 & 1 \end{pmatrix}$$

can be considered as adjacency matrix of the following Tanner graph

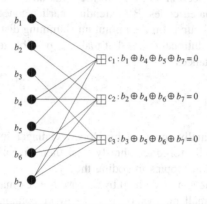

Figure 4.2: Tanner graph of the Hamming code

and b_7. Note that edges only connect two nodes not residing in the same class. A vector $\mathbf{b} = (b_1, \ldots, b_n)$ is a code word if and only if all parity-check equations are satisfied, i.e. for all check nodes the sum of the neighbouring positions among the message nodes is zero. Hence, the graph defines a linear code of block length n. The dimension is at least $k = n - r$. It might be larger than $n - r$, because some of the check equations could be linearly dependent.

Actually, the $r \times n$ parity-check matrix can be considered as the adjacency matrix of the Tanner graph.[3] The entry h_{ji} of the parity-check matrix $\mathbf{H}$ is 1 if and only if the jth check node is connected to the ith message node. Consequently, the jth row of the parity-check matrix determines the connections of the check node c_j, i.e. c_j is connected to all message nodes corresponding to 1s in the jth row. We call those nodes the neighbourhood of c_j. The neighbourhood of c_j is represented by the set $\mathbb{P}_j = \{i : h_{ji} = 1\}$. Similarly, the 1s in the ith column of the parity-check determine the connections of the message node

---

[3]In general, the adjacency matrix $\mathbf{A}$ for a finite graph with N nodes is an $N \times N$ matrix where the entry a_{ij} is the number of edges connecting node i and node j. In the special case of a bipartite Tanner graph, there exist no edges between check nodes and message nodes. Therefore, we do not require a square $(n + r) \times (n + r)$ adjacency matrix. The parity-check matrix is sufficient. For sparse graphs an adjacency list is often the preferred representation because it uses less space.

TURBO CODES

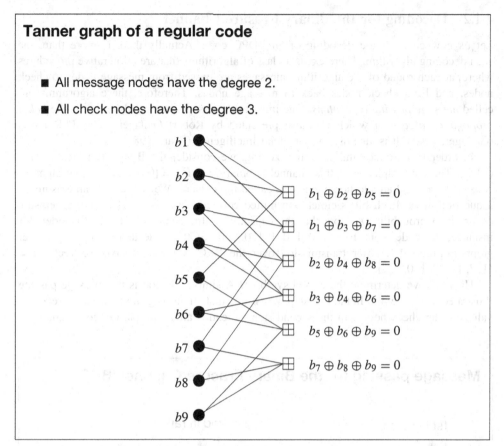

Figure 4.3: Tanner graph of a regular code

b_i. We call all check nodes that are connected to b_i the neighbourhood of b_i and denote it by the set $\mathbb{M}_i = \{j : h_{ji} = 1\}$. For instance, we have $\mathbb{P}_1 = \{1, 4, 5, 7\}$ and $\mathbb{M}_7 = \{1, 2, 3\}$ for the Tanner graph in Figure 4.2.

The Tanner graph in Figure 4.2 defines an *irregular code*, because the different message nodes have different degrees (different numbers of connected edges).[4] A graph where all message nodes have the same degree and all check nodes have the same degree results in a *regular code*. An example of a regular code is given in Figure 4.3. The LDPC codes as invented by Gallager were regular codes. Gallager defined the code with the parity-check matrix so that every column contains a small fixed number d_m of 1s and each row contains a small fixed number d_c of 1s. This is equivalent to defining a Tanner graph with message node degree d_m and check node degree d_c.

---

[4] In graph theory, the degree of a node is the number of edges incident to the node.

4.1.2 Decoding for the Binary Erasure Channel

Let us now consider the decoding of an LDPC code. Actually there is more than one such decoding algorithm. There exists a class of algorithms that are all iterative procedures where, at each round of the algorithm, messages are passed from message nodes to check nodes, and from check nodes back to message nodes. Therefore, these algorithms are called *message-passing algorithms*. One important message-passing algorithm is the *belief propagation algorithm* which was also presented by Robert Gallager in his PhD thesis (Gallager, 1963). It is also used in Artificial Intelligence (Pearl, 1988).

In order to introduce this message passing, we consider the Binary Erasure Channel (BEC). The input alphabet of this channel is binary, i.e. $\mathbb{F}_2 = \{0, 1\}$. The output alphabet consists of $\mathbb{F}_2$ and an additional element, called the erasure. We will denote an erasure by a question mark. Each bit is either transmitted correctly or it is erased where an erasure occurs with probability ε. Note that the capacity of this channel is $1 - \varepsilon$. Consider, for instance, the code word $\mathbf{b} = (1, 0, 1, 0, 1, 1, 0, 0, 0)$ of the code defined by the Tanner graph in Figure 4.3. After transmission over the BEC we may receive the vector $\mathbf{r} = (1, ?, 1, ?, ?, 1, 0, 0, ?)$.

How can we determine the erased symbols? A simple method is the message passing illustrated in Figure 4.4. In the first step we assume that all message nodes send the received values to the check nodes. In the second step we can evaluate all parity-check equations.

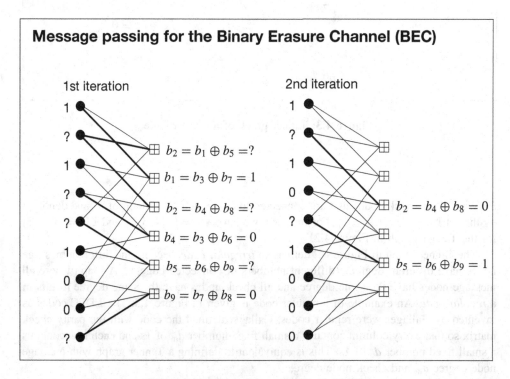

Figure 4.4: Message passing for the BEC

TURBO CODES

If a single symbol in one of the equations is unknown, e.g. b_4 in $b_3 \oplus b_4 \oplus b_6$, then the parity-check equation determines the value of the erased symbol. In our example, $b_4 = b_3 \oplus b_6 = 0$. However, if more than one erasure occurs within a parity-check equation, we cannot directly infer the corresponding values. In this case we assume that evaluation of the parity-check equation results in an erasure, e.g. $b_2 = b_1 \oplus b_5 = ?$. All results from the parity-check equations are then forwarded to the message nodes. Basically, every check node will send three messages. In Figure 4.4 we have only stated one equation per check node. The connection for this message is highlighted. With this first iteration of the message-passing algorithm we could determine the bits $b_4 = 0$ and $b_9 = 0$. Now, we run the same procedure for a second iteration, where the message nodes b_4 and b_9 send the corrected values. In Figure 4.4 we have highlighted the two important parity-check equations and connections that are necessary to determine the remaining erased bits $b_2 = 0$ and $b_5 = 1$.

4.1.3 Log-Likelihood Algebra

To discuss the general belief propagation algorithm, we require the notion of a *log-likelihood ratio*. The *log-likelihood algebra*, as introduced in this section, was developed by Hagenauer (Hagenauer et al., 1996). Let x be a binary random variable and let $\Pr\{x = x'\}$ denote the probability that the random variable x takes on the value x'. The log-likelihood ratio of x is defined as

$$L(x) = \ln \frac{\Pr\{x = 0\}}{\Pr\{x = 1\}},$$

where the logarithm is the natural logarithm. The log-likelihood ratio $L(x)$ is also called the *L-value* of the binary random variable x. From the *L*-value we can calculate the probabilities

$$\Pr\{x = 0\} = \frac{1}{1 + e^{-L(x)}} \text{ and } \Pr\{x = 1\} = \frac{1}{1 + e^{L(x)}}.$$

If the binary random variable x is conditioned on another random variable y, we obtain the conditional *L*-value

$$L(x|y) = \ln \frac{\Pr\{x = 0|y\}}{\Pr\{x = 1|y\}} = \ln \frac{\Pr\{y|x = 0\}}{\Pr\{y|x = 1\}} + \ln \frac{\Pr\{x = 0\}}{\Pr\{x = 1\}} = L(y|x) + L(x).$$

Consider, for example, a binary symbol transmitted over a Gaussian channel. With binary phase shift keying we will map the code bit $b_i = 0$ to the symbol -1 and $b_i = 1$ to $+1$. Let r_i be the received symbol. For the Gaussian channel with variance σ and signal-to-noise ratio $\frac{E_s}{N_0} = \frac{1}{2\sigma^2}$, we have the probability density function

$$p(r_i|b_i) = \frac{1}{\sqrt{2\pi}\sigma} \exp\left(\frac{-(r_i - (1 - 2b_i))^2}{2\sigma^2}\right).$$

Thus, for the conditional *L*-values we obtain

$$L(r_i|b_i) = \ln \frac{\Pr\{r_i|b_i = 0\}}{\Pr\{r_i|b_i = 1\}}$$

$$= \ln \frac{p(r_i|b_i = 0)}{p(r_i|b_i = 1)}$$

$$= \ln \frac{\exp\left(-\frac{E_s}{N_0}(r_i - 1)^2\right)}{\exp\left(-\frac{E_s}{N_0}(r_i + 1)^2\right)}$$

$$= -\frac{E_s}{N_0}\left((r_i - 1)^2 - (r_i + 1)^2\right)$$

$$= 4\frac{E_s}{N_0}r_i.$$

As $L(r_i|b_i)$ only depends on the received value r_i and the signal-to-noise ratio, we will usually use the shorter notation $L(r_i)$ for $L(r_i|b_i)$. The a-posteriori L-value $L(b_i|r_i)$ is therefore

$$L(b_i|r_i) = L(r_i) + L(b_i) = 4\frac{E_s}{N_0}r_i + L(b_i).$$

The basic properties of log-likelihood ratios are summarised in Figure 4.5. Note that the hard decision of the received symbol can be based on this L-value, i.e.

$$\hat{b}_i = \begin{cases} 0 & \text{if } L(b_i|r_i) > 0, \text{i.e. } \Pr\{b_i = 0|r_i\} > \Pr\{b_i = 1|r_i\} \\ 1 & \text{if } L(b_i|r_i) < 0, \text{i.e. } \Pr\{b_i = 0|r_i\} < \Pr\{b_i = 1|r_i\} \end{cases}.$$

Furthermore, note that the magnitude $|L(b_i|r_i)|$ is the reliability of this decision. To see this, assume that $L(b_i|r_i) > 0$. Then, the above decision rule yields an error if the transmitted bit was actually $b_i = 1$. This happens with probability

$$\Pr\{b_i = 1\} = \frac{1}{1 + e^{L(b_i|r_i)}}, \quad L(b_i|r_i) > 0.$$

Now, assume that $L(b_i|r_i) < 0$. A decision error occurs if actually $b_i = 0$ was transmitted. This event has the probability

$$\Pr\{b_i = 0\} = \frac{1}{1 + e^{-L(b_i|r_i)}} = \frac{1}{1 + e^{|L(b_i|r_i)|}}, \quad L(b_i|r_i) < 0.$$

Hence, the probability of a decision error is

$$\Pr\{b_i \neq \hat{b}_i\} = \frac{1}{1 + e^{|L(b_i|r_i)|}}$$

and for the probability of a correct decision we obtain

$$\Pr\{b_i = \hat{b}_i\} = \frac{e^{|L(b_i|r_i)|}}{1 + e^{|L(b_i|r_i)|}}.$$

Up to now, we have only considered decisions based on a single observation. In the following we deal with several observations. The resulting rules are useful for decoding. If the binary random variable x is conditioned on two statistically independent random variables y_1 and y_2, then we have

$$L(x|y_1, y_2) = \ln \frac{\Pr\{x = 0|y_1, y_2\}}{\Pr\{x = 1|y_1, y_2\}} = \ln \frac{\Pr\{y_1|x = 0\}}{\Pr\{y_1|x = 1\}} + \ln \frac{\Pr\{y_2|x = 0\}}{\Pr\{y_2|x = 1\}} + \ln \frac{\Pr\{x = 0\}}{\Pr\{x = 1\}}$$

$$= L(y_1|x) + L(y_2|x) + L(x).$$

TURBO CODES

Log-likelihood ratios

- The log-likelihood ratio of the binary random variable x is defined as

$$L(x) = \ln \frac{\Pr\{x = 0\}}{\Pr\{x = 1\}} \qquad (4.1)$$

- From the L-value we can calculate the probabilities

$$\Pr\{x = 0\} = \frac{1}{1 + e^{-L(x)}} \text{ and } \Pr\{x = 1\} = \frac{1}{1 + e^{L(x)}} \qquad (4.2)$$

- If the binary random variable x is conditioned on another random variable y, or on two statistically independent random variables y_1 and y_2, we obtain the conditional L-values

$$L(x|y) = L(y|x) + L(x) \text{ or } L(x|y_1, y_2) = L(y_1|x) + L(y_2|x) + L(x) \qquad (4.3)$$

- For the Gaussian channel with binary phase shift keying and signal-to-noise ratio $\frac{E_s}{N_0}$ we have the a-posteriori L-value

$$L(b_i|r_i) = L(r_i|b_i) + L(b_i) = 4\frac{E_s}{N_0}r_i + L(b_i) \qquad (4.4)$$

where b_i is the transmitted bit and r_i is the received symbol.

Figure 4.5: Log-likelihood ratios

This rule is useful whenever we have independent observations of a random variable, for example for decoding a repetition code. Consider, for instance, the code $\mathbb{B} = \{(0,0),(1,1)\}$. We assume that the information bit u is equally likely to be 0 or 1. Hence, for a memoryless symmetrical channel we can simply sum over the received L-values to obtain $L(u|\mathbf{r}) = L(r_1) + L(r_2)$ with the received vector $\mathbf{r} = (r_1, r_2)$.

Consider now two statistically independent random variables x_1 and x_2. Let $\oplus$ denote the addition modulo 2. Then, $x_1 \oplus x_2$ is also a binary random variable with the L-value $L(x_1 \oplus x_2)$. This L-value can be calculated from the values $L(x_1)$ and $L(x_2)$

$$L(x_1 \oplus x_2) = \ln \frac{\Pr\{x_1 = 0\}\Pr\{x_2 = 0\} + \Pr\{x_1 = 1\}\Pr\{x_2 = 1\}}{\Pr\{x_1 = 1\}\Pr\{x_2 = 0\} + \Pr\{x_1 = 0\}\Pr\{x_2 = 1\}}.$$

Using $\Pr\{x_1 \oplus x_2 = 0\} = \Pr\{x_1 = 0\}\Pr\{x_2 = 0\} + (1 - \Pr\{x_1 = 0\})(1 - \Pr\{x_2 = 0\})$ and Equation (4.2), we obtain

$$\Pr\{x_1 \oplus x_2 = 0\} = \frac{1 + e^{L(x_1)}e^{L(x_2)}}{(1 + e^{L(x_1)})(1 + e^{L(x_2)})}.$$

Similarly, we have

$$\Pr\{x_1 \oplus x_2 = 1\} = \frac{e^{L(x_1)} + e^{L(x_2)}}{(1 + e^{L(x_1)})(1 + e^{L(x_2)})}$$

which yields

$$L(x_1 \oplus x_2) = \ln \frac{1 + e^{L(x_1)} e^{L(x_2)}}{e^{L(x_1)} + e^{L(x_2)}}.$$

This operation is called the *boxplus operation*, because the symbol $\boxplus$ is usually used for notation, i.e.

$$L(x_1 \oplus x_2) = L(x_1) \boxplus L(x_2) = \ln \frac{1 + e^{L(x_1)} e^{L(x_2)}}{e^{L(x_1)} + e^{L(x_2)}}.$$

Later on we will see that the boxplus operation is a significant, sometimes dominant portion of the overall decoder complexity with iterative decoding. However, a fixed-point Digital Signal Processor (DSP) implementation of this operation is rather difficult. Therefore, in practice the boxplus operation is often approximated. The computationally simplest estimate is the so-called *max-log approximation*

$$L(x_1) \boxplus L(x_2) \approx \operatorname{sign}(L(x_1) \cdot L(x_2)) \cdot \min\{|L(x_1)|, |L(x_2)|\}.$$

The name expresses the similarity to the max-log approximation introduced in Section 3.5.2. Both approximations are derived from the *Jacobian logarithm*.

Besides a low computational complexity, this approximation has another advantage, i.e. the estimated L-values can be arbitrarily scaled, because constant factors can be cancelled. Therefore, an exact knowledge of the signal-to-noise ratio is not required. The max-log approximation is illustrated in Figure 4.6 for a fixed value of $L(x_2) = 2.5$. We observe that maximum deviation from the exact solution occurs for $||L(x_1)| - |L(x_2)|| = 0$.

We now use the boxplus operation to decode a single parity-check code $\mathbb{B}(3, 2, 2)$ after transmission over the Additive White Gaussian Noise (AWGN) channel with a signal-to-noise ratio of 3 dB ($\sigma = 0.5$). Usually, we assume that the information symbols are 0 or 1 with a probability of 0.5. Hence, all a-priori L-values $L(b_i)$ are zero. Assume that the code word $\mathbf{b} = (0, 1, 1)$ was transmitted and the received word is $\mathbf{r} = (0.71, 0.09, -1.07)$. To obtain the corresponding channel L-values, we have to multiply $\mathbf{r}$ by $4\frac{E_s}{N_0} = \frac{2}{\sigma^2} = 8$. Hence, we have $L(r_0) = 5.6$, $L(r_1) = 0.7$ and $L(r_2) = -8.5$. In order to decode the code, we would like to calculate the a-posteriori L-values $L(b_i|\mathbf{r})$. Consider the decoding of the first code bit b_0 which is equal to the first information bit u_0. The hard decision for the information bit $\hat{u}_0$ should be equal to the result of the modulo addition $\hat{b}_1 \oplus \hat{b}_2$. The log-likelihood ratio of the corresponding received symbols is $L(r_1) \boxplus L(r_2)$. Using the max-log approximation, this can approximately be done by

$$\begin{aligned} L_e(u_0) &= L(r_1) \boxplus L(r_2) \\ &\approx \operatorname{sign}(L(r_1) \cdot L(r_2)) \cdot \min\{|L(r_1)|, |L(r_2)|\} \\ &\approx \operatorname{sign}(0.7 \cdot (-8.5)) \cdot \min\{|0.7|, |-8.5|\} \\ &\approx -0.7. \end{aligned}$$

TURBO CODES

Boxplus operation

- For x_1 and x_2, two statistically independent binary random variables, $x_1 \oplus x_2$ is also a binary random variable. The L-value $L(x_1 \oplus x_2)$ of this random variable is calculated with the *boxplus operation*

$$L(x_1 \oplus x_2) = L(x_1) \boxplus L(x_2) = \ln \frac{1 + e^{L(x_1)} e^{L(x_2)}}{e^{L(x_1)} + e^{L(x_2)}} \qquad (4.5)$$

- This operation can be approximated by

$$L(x_1) \boxplus L(x_2) \approx \text{sign}(L(x_1) \cdot L(x_2)) \cdot \min\{|L(x_1)|, |L(x_2)|\} \qquad (4.6)$$

as illustrated in the following figure for $L(x_2) = 2.5$

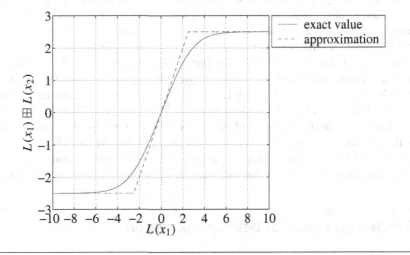

Figure 4.6: Illustration of the boxplus operation and its approximation. Reprinted with permission from © 2001 IEEE.

The value $L_e(u_0)$ is called an *extrinsic log-likelihood ratio*. It can be considered as the information that results from the code constraints. Note that this extrinsic information is statistically independent of the received value r_0. Therefore, we can simply add $L_e(u_0)$ and $L(u_0|r_0) = L(r_0)$ to obtain the a-posteriori L-value

$$L(u_0|\mathbf{r}) = L(r_0) + L_e(u_0) \approx 4.9.$$

For the two other bits we calculate the extrinsic L-values $L_e(u_1) = -5.6$ and $L_e(b_2) = 0.7$, as well as the a-posteriori L-values $L(u_1|\mathbf{r}) = -4.9$ and $L(b_2|\mathbf{r}) = -7.7$. The hard decision results in $\hat{\mathbf{b}} = (0, 1, 1)$.

4.1.4 Belief Propagation

The general belief propagation algorithm is also a message-passing algorithm similar to the one discussed in Section 4.1.2. The difference lies in the messages that are passed between nodes. The messages passed along the edges in the belief propagation algorithm are log-likelihood values. Each round of the algorithm consists of two steps. In the first half-iteration, a message is sent from each message node to all neighbouring check nodes. In the second half-iteration, each check node sends a message to the neighbouring message nodes. Let $L_l[b_i \to c_j]$ denote the message from a message node b_i to a check node c_j in the lth iteration. This message is computed on the basis of the observed channel value r_i and some of the messages received from the neighbouring check nodes except c_j according to Equation (4.7) in Figure 4.7. It is an important aspect of belief propagation that the message sent from a message node b_i to a check node c_j must not take into account the message sent in the previous round from c_j to b_i. Therefore, this message is explicitly excluded in the update Equation (4.7). Remember that $\mathbb{M}_i$ denotes the neighbourhood of the node b_i.

The message $L_l[c_j \to b_i]$ from the check node c_j to the message node b_i is an extrinsic log-likelihood value based on the parity-check equation and the incoming messages from all neighbouring message nodes except b_i. The update rule is given in Equation (4.8). The symbol $\sum\boxplus$ denotes the sum with respect to the boxplus operation.

Consider now the code defined by the Tanner graph in Figure 4.3. We consider transmission over a Gaussian channel with binary phase shift keying. Suppose the transmitted code word is $\mathbf{b} = (+1, -1, -1, -1, -1, +1, -1, +1, -1)$. For this particular code word we may obtain the following channel L-values $L(\mathbf{r}) = 4\frac{E_s}{N_0} \cdot \mathbf{r} = (5.6, -10.2, 0.7, 0.5, -7.5, 12.2, -8.5, 6.9, -7.7)$. In the first step of the belief propagation algorithm the message nodes pass these received values to the neighbouring check nodes. At each check node we calculate a message for the message nodes. This message takes the code constraints into account. For the first check node this is the parity-check equation $b_1 \oplus b_2 \oplus b_5 = 0$. Based on this

---

Update equations for belief propagation

- Messages from message nodes to check nodes

$$L_l[b_i \to c_j] = \begin{cases} L(r_i) & \text{if } l = 1 \\ L(r_i) + \sum_{j' \in \mathbb{M}_i \setminus \{j\}} L_{l-1}[c_{j'} \to b_i] & \text{if } l > 1 \end{cases} \quad (4.7)$$

- Messages from check nodes to message nodes

$$L_l[c_j \to b_i] = \sum_{i' \in \mathbb{P}_j \setminus \{i\}} \boxplus L_{l-1}[b_{i'} \to c_j] \quad (4.8)$$

Figure 4.7: Update equations for the message passing with belief propagation

TURBO CODES

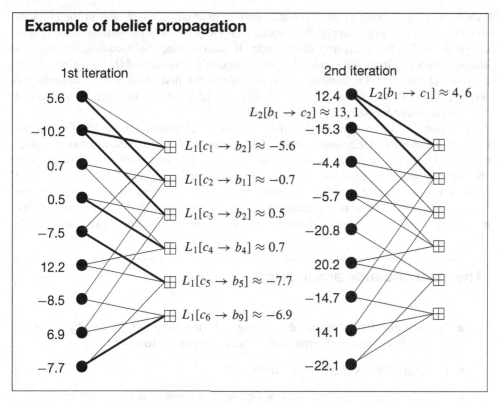

Figure 4.8: Example of belief propagation

equation, we calculate three different extrinsic messages for the three neighbouring message nodes

$$L_e(b_1) = L(b_2 \oplus b_5) = L(b_2) \boxplus L(b_5) \approx 7.5,$$
$$L_e(b_2) = L(b_1 \oplus b_5) = L(b_1) \boxplus L(b_5) \approx -5.6,$$
$$L_e(b_5) = L(b_1 \oplus b_2) = L(b_1) \boxplus L(b_2) \approx -5.6.$$

Similarly, we can calculate the messages for the remaining check nodes. Some of these messages are provided in Figure 4.8, where the corresponding connections from check node to message node are highlighted.

The extrinsic L-values from the check nodes are based on observations that are statistically independent from the received value. Therefore, we can simply add the received channel L-value and the extrinsic messages for each message node. For instance, for the node b_1 we have $L(b_1) = L(r_1) + L(b_2 \oplus b_5) + L(b_3 \oplus b_7) \approx 12.4$. If we do this for all message nodes we obtain the values given in Figure 4.8. These are the L-values for the message nodes after the first iteration.

As mentioned above, it is an important aspect of belief propagation that we only pass extrinsic information between nodes. In particular, the message that is sent from a message node b_i to a check node c_j must not take into account the message sent in the previous round from c_j to b_i. For example, the L-value $L(b_1) \approx 12.4$ contains the message $L_e(b_1) = L(b_2 \oplus b_5) = 7.5$ from the first check node. If we continue the decoding, we have to subtract this value from the current L-value to obtain the message $L(b_1) - L(b_2 \oplus b_5) = L(r_1) + L(b_3 \oplus b_7) \approx 4.6$ from the message node to the first check node. Similarly, we obtain $L(b_1) - L(b_3 \oplus b_7) = L(r_1) + L(b_2 \oplus b_5) \approx 13.1$ for the message to the second check node. These messages are indicated in Figure 4.8.

The reason for these particular calculations is the *independence assumption*. For the first iteration, the addition at the message node of the received L-value and the extrinsic L-values from check nodes is justified, because these values resulted from statistically independent observations. What happens with further iterations? This depends on the neighbourhood of the message nodes. Consider, for example, the graph in Figure 4.9 which illustrates the neighbourhood of the first message node in our example. This graph represents the message flow during the iterative decoding process. In the first iteration we receive extrinsic

The independence assumption

- The independence assumption is correct for l iterations of the algorithm if the neighbourhood of a message node is a tree up to depth l.
- The graph below is a tree up to depth $l = 1$.
- At depth $l = 2$ there are several nodes that represent the same code bit. This causes cycles as indicated for the node b_4.
- The shortest possible cycle has a length of 4.

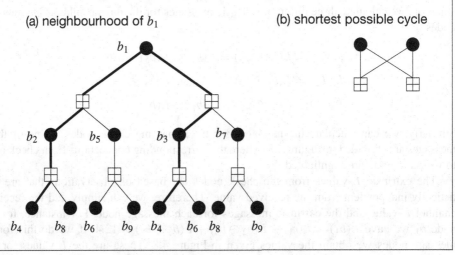

Figure 4.9: The independence assumption

TURBO CODES

information from all nodes of depth $l = 1$. In the second iteration we receive messages from the nodes of depth $l = 2$, and so on.

The messages for further rounds of the algorithm are only statistically independent if the graph is a so-called tree.[5] In particular, if all message nodes are unique. In our simple example this holds only for $l = 1$. At depth $l = 2$ there are several nodes that represent the same code bit. This causes cycles[6] as indicated for the node b_4. However, for longer codes we can apply significantly more independent iterations. In general, if the neighbourhood of all message nodes is a tree up to depth l, then the incoming messages are statistically independent and the update equation correctly calculates the corresponding log-likelihood based on the observations.

In graph theory, we call the length of the shortest cycle the *girth* of a graph. If the graph has no cycles, its girth is defined to be infinite. According to figure (b) in Figure 4.9, the girth is at least 4. In our example graph the girth is 8. The girth determines the maximum number of independent iterations of belief propagation. Moreover, short cycles in the Tanner graph of a code may lead to a small minimum Hamming distance. Therefore, procedures to construct Tanner graphs maximising the girth were proposed. However, in practice it is not clear how the girth actually determines the decoding performance. Most methods to analyse the performance of belief propagation are based on the independence assumption, and little is known about decoding for graphs with cycles.

4.2 A First Encounter with Code Concatenation

In this section we introduce the notion of code concatenation, i.e. the concept of constructing long powerful codes from short component codes. The first published concatenated codes were the *product codes* introduced by Elias (Elias, 1954). This construction yields a block code of length $n \times n$ based on component codes of length n. We will consider product codes to introduce the basic idea of code concatenation and to discuss the connection to Tanner graphs and LDPC code. Later, in Section 4.6, we will consider another kind of product code. These so-called woven convolutional codes were first introduced by Höst, Johannesson and Zyablov (Höst *et al.*, 1997) and are product codes based on convolutional component codes.

4.2.1 Product Codes

Code concatenation is a method for constructing good codes by combining several simple codes. Consider, for instance, a simple parity-check code of rate $R = 2/3$. For each 2-nit information word, a parity bit is attached so that the 3-bit code word has an even weight, i.e. an even number of 1s. This simple code can only detect a single error. For the Binary Symmetric Channel (BSC) it cannot be used for error-correction. However, we can construct a single error-correcting code combining several simple parity-check codes. For this construction, we assume that the information word is a block of 2×2 bits, e.g.

$$\mathbf{u} = \begin{pmatrix} 0 & 1 \\ 1 & 0 \end{pmatrix}.$$

---

[5]A tree is a graph in which any two nodes are connected by exactly one path, where a path in a graph is a sequence of edges such that from each of its nodes there is an edge to the next node in the sequence.

[6]A cycle is a path where the start edge and the end edge are the same.

We encode each row with a single parity-check code and obtain

$$\mathbb{B}^o = \begin{pmatrix} 0 & 1 & 1 \\ 1 & 0 & 1 \end{pmatrix}.$$

Next, we encode column-wise, i.e. encode each column of $\mathbb{B}^o$ with a single parity-check code and have the overall code word

$$\mathbf{b} = \begin{pmatrix} 0 & 1 & 1 \\ 1 & 0 & 1 \\ 1 & 1 & 0 \end{pmatrix}.$$

The overall code is called a *product code*, because the code parameters are obtained from the products of the corresponding parameters of the component codes. For instance, let k^i and k^o denote the dimension of the inner and outer code respectively. Then, the overall dimension is $k = k^i k^o$. Similarly, we obtain the overall code length $n = n^i n^o$, where n^i and n^o are the lengths of the inner code and outer code respectively. In our example, we have $k = 2 \cdot 2 = 4$ and $n = 3 \cdot 3 = 9$. Moreover the overall code can correct any single transmission error. By decoding the parity-check codes, we can detect both the row and the column where the error occurred. Consider, for instance, the received word

$$\mathbf{r} = \begin{pmatrix} 0 & 1 & 1 \\ 0 & 0 & 1 \\ 1 & 1 & 0 \end{pmatrix}.$$

We observe that the second row and the first column have odd weights. Hence, we can correct the transmission error at the intersection of this row and column.

The construction of product codes is not limited to parity-check codes as component codes. Basically, we could construct product codes over an arbitrary finite field. However, we will restrict ourselves to the binary case. In general, we organise the $k = k^o k^i$ information bits in k^o columns and k^i rows. We apply first row-wise outer encoding. The code bits of the outer code words are then encoded column-wise. It is easy to see that the overall code is linear if the constituent codes are linear. Let $\mathbf{G}^o$ and $\mathbf{G}^i$ be the generator matrices of the outer and the inner code respectively. Using the Kronecker product, we can describe the encoding of the k^i outer codes by $\mathbf{G}^o \otimes \mathbf{I}_{k^i}$, where $\mathbf{I}_{k^i}$ is a $k^i \times k^i$ identity matrix. Similarly, the encoding of the k^o inner codes can be described by $\mathbf{I}_{k^o} \otimes \mathbf{G}^i$ with the $k^o \times k^o$ identity matrix $\mathbf{I}_{k^o}$. The overall generator matrix is then

$$\mathbf{G} = (\mathbf{G}^o \otimes \mathbf{I}_{k^i})(\mathbf{I}_{k^o} \otimes \mathbf{G}^i).$$

The linearity simplifies the analysis of the minimum Hamming distance of a product code. Remember that the minimum Hamming distance of a linear code is equal to the minimum weight of a non-zero code word. The minimum weight of a non-zero code word of a product code can easily be estimated. Let d^o and d^i denote the minimum Hamming distance of the outer and inner code. First, consider the encoding of the outer codes. Note that any non-zero information word leads to at least one non-zero outer code word. This code word has at least weight d^o. Hence, there are at least d^o non-zero columns. Every non-zero column results in a non-zero code word of the inner code with a weight of at least d^i. Therefore, a non-zero code word of the product code consists of at least d^o non-zero columns each

TURBO CODES

Product codes

- A binary product code $\mathbb{B}(n = n^o n^i, k = k^o k^i, d = d^o d^i)$ is a concatenation of k^i binary outer codes $\mathbb{B}^o(n^o, k^o, d^o)$ and k^o binary inner codes $\mathbb{B}^i(n^i, k^i, d^i)$.

- To encode a product code, we organise the $k = k^o k^i$ information bits in k^o columns and k^i rows. We apply first row-wise outer encoding. The code bits of the outer code words are then encoded column-wise. This encoding is illustrated in the following figure.

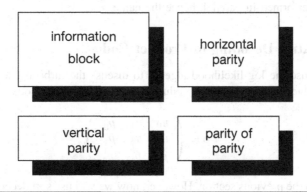

Figure 4.10: Encoding of product codes

of at least d^i non-zero symbols. We conclude that the minimum Hamming distance of the overall code

$$d \geq d^o d^i.$$

We can also use product codes to discuss the concept of parallel concatenation, i.e. of turbo codes. We will consider possibly the most simple turbo code constructed from systematically encoded single parity-check codes $\mathbb{B}(3, 2, 2)$. Take the encoding of the product code as illustrated in Figure 4.10. We organise the information bits in k^o columns and k^i rows and apply first row-wise outer and then column-wise inner encoding. This results in the parity bits of the outer code (horizontal parity) and the parity bits of the inner code, which can be divided into the parity bits corresponding to information symbols, and parity of outer parity bits. Using two systematically encoded single parity-check codes $\mathbb{B}(3, 2, 2)$ as above, this results in code words of the form

$$\mathbf{b} = \begin{pmatrix} u_0 & u_1 & p_0^- \\ u_2 & u_3 & p_1^- \\ p_0^| & p_1^| & p_2^| \end{pmatrix},$$

where p_i^- denotes a horizontal parity bit and $p_i^|$ denotes a vertical parity bit.

A product code is a serial concatenation as we first encode the information with the outer code(s) and apply the outer code bits as information to the inner encoder(s). With the parallel concatenated construction, we omit the parity bits of parity. That is, we simply encode the information bits twice, once in the horizontal and once in the vertical direction. For our example, we obtain

$$\mathbf{b} = \begin{pmatrix} u_0 & u_1 & p_0^- \\ u_2 & u_3 & p_1^- \\ p_0^| & p_1^| & \end{pmatrix}.$$

Note that the encoding in the horizontal and vertical directions is independent and can therefore be performed in parallel, hence the name.

4.2.2 Iterative Decoding of Product Codes

We will now use the log-likelihood algebra to discuss the turbo decoding algorithm. We will use this algorithm to decode a product code $\mathbb{B}(9, 4, 4)$ with code words

$$\mathbf{b} = \begin{pmatrix} u_0 & u_1 & p_0^- \\ u_2 & u_3 & p_1^- \\ p_0^| & p_1^| & p_2^| \end{pmatrix},$$

as described in the previous section. However, now we will use soft decoding, i.e. soft channel values. For this particular product code, we could use the belief propagation algorithm as discussed in Section 4.1.4. In fact, the Tanner graph given in Figure 4.3 on page 167 is a representation of the code $\mathbb{B}(9, 4, 4)$. To see this, note that the first check node in the Tanner graph represents the parity equation of the first row of the product code. Similarly, c_4 and c_6 represent the second and third row. The column constraints are represented by the check nodes c_2, c_3 and c_5.

In practice, a so-called *turbo decoding algorithm* is used to decode concatenated codes. This is also an iterative message-passing algorithm similar to belief propagation. The iterative decoding is based on decoders for the component codes that use reliability information at the input and provide symbol-by-symbol reliability information at the output. In the case of a simple single parity-check code, this is already performed by executing a single boxplus operation. When we use more complex component codes, we apply in each iteration a so-called soft-in/soft-out (SISO) decoding algorithm. Such an algorithm is the Bahl, Cocke, Jelinek, Raviv (BCJR) algorithm discussed in Section 3.5. Commonly, all symbol reliabilities are represented by log-likelihood ratios (L-values). In general, a SISO decoder, such as an implementation of the BCJR algorithm, may expect channel L-values and a-priori information as an input and may produce reliabilities for the estimated information symbols $L(u_i|\mathbf{r})$, the extrinsic L-value for code $L_e(b_j|\mathbf{r})$ and information bits $L_e(u_i|\mathbf{r})$ (see Figure 4.11). The turbo decoding algorithm determines only the exchange of symbol reliabilities between the different component decoders. In the following we discuss this turbo decoding, considering the parallel concatenated code with single parity-check component codes. The same concept of message passing applies to more complex codes, as we will see later on.

TURBO CODES

Soft-in/soft-out decoder

- In general, a soft-in/soft-out (SISO) decoder, such as an implementation of the BCJR algorithm, may expect channel L-values and a-priori information as an input and may produce reliabilities for the estimated information symbols $L(u_i|\mathbf{r})$, the extrinsic L-value for code $L_e(b_j|\mathbf{r})$ and information bits $L_e(u_i|\mathbf{r})$.

Figure 4.11: Soft-in/soft-out decoder

Suppose that we encode the information block $\mathbf{u} = (0, 1, 1, 1)$ and use binary phase shift keying with symbols from $\{+1, -1\}$. The transmitted code word is

$$\mathbf{b} = \begin{pmatrix} +1 & -1 & -1 \\ -1 & -1 & +1 \\ -1 & +1 & \end{pmatrix}.$$

For this particular code word we may obtain the following channel L-values after transmission over a Gaussian channel

$$4\frac{E_s}{N_0} \cdot \mathbf{r} = \begin{pmatrix} 5.6 & -10.2 & -7.5 \\ 0.7 & 0.5 & 12.2 \\ -8.5 & 6.9 & \end{pmatrix}.$$

Again, we assume that no a-priori information is available. Hence, we have $L_a(u_i) = 0$ for all information bits. Let us start with decoding the first row. On account of the parity-check equation we have $L_e^-(u_0) = L(u_1) \boxplus L(p_0^-) \approx 7.5$ and $L_e^-(u_1) = L(u_0) \boxplus L(p_0^-) \approx -5.6$. Similarly, we can evaluate the remaining extrinsic values of $L_e^-(\mathbf{u})$ and obtain

$$L_e^-(\mathbf{u}) = \begin{pmatrix} 7.5 & -5.6 \\ 0.5 & 0.7 \end{pmatrix}.$$

Now, we use this extrinsic information as a-priori knowledge for the column-wise decoding, i.e. we assume $L_a(\mathbf{u}) = L_e^-(\mathbf{u})$. As we use systematic encoded component codes and the log-likelihood values are statistically independent, we can simply add up the channel and extrinsic values for the information bits

$$L_{ch} \cdot \mathbf{r} + L_e^-(\mathbf{u}) \cdot \mathbf{r} = \begin{pmatrix} 13.1 & -15.8 & -7.5 \\ 1.2 & 1.2 & 12.2 \\ -8.5 & 6.9 & \end{pmatrix}.$$

These values are the basis for the second decoding step, i.e. column-wise decoding. For the first column we have $L_e^|(u_0) = L(u_2) \boxplus L(p_0^|) \approx -1.2$ and $L_e(^|u_2) = L(u_0) \boxplus L(p_0^|) \approx -8.5$. Hence, we obtain

$$L_e^|(\mathbf{u}) = \begin{pmatrix} -1.2 & -1.2 \\ -8.5 & -6.9 \end{pmatrix}.$$

If we wish to continue the iterative decoding, we use $L_e^|(\hat{\mathbf{u}})$ as a-priori information for a new iteration. Hence, we calculate

$$L_{ch} \cdot \mathbf{r} + L_e^|(\mathbf{u}) \cdot \mathbf{r} = \begin{pmatrix} 4.4 & -11.4 & -7.5 \\ -7.8 & -6.4 & 12.2 \\ -8.5 & 6.9 & \end{pmatrix}$$

as input for the next iteration of row-wise decoding.

When we stop the decoding, we use $L(\hat{\mathbf{u}}) = L_{ch} \cdot \mathbf{r} + L_e^|(\mathbf{u}) + L_e^-(\mathbf{u})$ as output of the iterative decoder. After the first iteration we have

$$L(\hat{\mathbf{u}}) = \begin{pmatrix} 11.9 & -17 \\ -7.3 & -5.7 \end{pmatrix}.$$

4.3 Concatenated Convolutional Codes

To start with concatenated convolutional codes, it is convenient to consider their encoding scheme. We introduce three different types of encoder in this section: the original turbo encoder, i.e. a parallel concatenation, and serially and partially concatenated codes.

4.3.1 Parallel Concatenation

As mentioned above, turbo codes result from a parallel concatenation of a number of systematic encoders linked by interleavers. The general encoding scheme for turbo codes is depicted in Figure 4.12. All encoders are systematic. The information bits are interleaved after encoding the first component code, but the information bits are only transmitted once. With two component codes, the code word of the resulting code has the structure

$$\mathbf{b} = (\mathbf{u}, \mathbf{u}\mathbf{A}_1, \pi(\mathbf{u})\mathbf{A}_2),$$

where $\pi(\cdot)$ denotes a permutation of the information bits and $\mathbf{A}_j$ denotes a submatrix of the systematic generator matrix $\mathbf{G}_j = (\mathbf{I}_k \mathbf{A}_j)$.

As we see, the parallel concatenation scheme can easily be generalised to more than two component codes. If not mentioned otherwise, we will only consider turbo codes with two equal component codes in the following discussion. The rate of such a code is

$$R = \frac{R_c}{2 - R_c},$$

where R_c is the rate of the component codes.

Although the component codes are usually convolutional codes, the resulting turbo code is a block code. This follows from the block-wise encoding scheme and the fact that convolutional component codes have to be terminated.

TURBO CODES

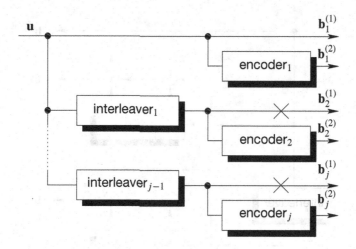

Encoder scheme for turbo codes

- Most commonly, turbo codes are constructed from two equal component codes of rate R_c.
- In this case, the rate of the overall code is

$$R = \frac{R_c}{2 - R_c} \tag{4.9}$$

Figure 4.12: Encoder scheme for turbo codes

4.3.2 The UMTS Turbo Code

As an example of a turbo code we will consider the code defined in the Universal Mobile Telecommunications System (UMTS) standard. The corresponding encoder is shown in Figure 4.13. The code is constructed from two parallel concatenated convolutional codes of memory $m = 3$. Both encoders are recursive and have the generator matrix $\mathbf{G}(D) = \left(1, \frac{1+D+D^3}{1+D^2+D^3}\right)$. In octal notation, the feedforward generator is $(15)_8$ and the feedback generator is $(13)_8$. The information is encoded by the first encoder in its original order. The second encoder is applied after the information sequence is interleaved. The information is only transmitted once, therefore the systematic output of the second decoder is omitted in the figure. Hence, the overall code rate is $R = 1/3$.

According to the UMTS standard, the length K of the information sequence will be in the range $40 \leq K \leq 5114$. After the information sequence is encoded, the encoders are forced back to the all-zero state. However, unlike the conventional code termination, as discussed in Section 3.1.4, the recursive encoders cannot be terminated with m zero bits. The termination sequence for a recursive encoder depends on the encoder state. Because the

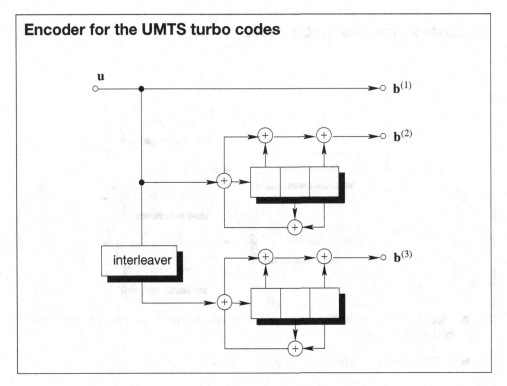

Figure 4.13: Encoder for the UMTS turbo codes

states of the two encoders will usually be different after the information has been encoded, the tails for each encoder must be determined separately. An example of this termination is given in Figure 3.8 on page 106. The tail bits are then transmitted at the end of the encoded code sequence. Hence, the actual code rate is slightly smaller than 1/3.

4.3.3 Serial Concatenation

At high signal-to-noise ratio (SNR), turbo codes typically reveal an error floor, which means that the slope of the bit error rate curve declines with increasing SNR (see, for example, Figure 4.18 on page 189). In order to improve the performance of parallel concatenated codes for higher SNR, Benedetto and Montorsi applied iterative decoding to serially concatenated convolutional codes (SCCC) with interleaving (Benedetto and Montorsi, 1998). The corresponding encoders (see Figure 4.14) consist of the cascade of an outer encoder, an interleaver permuting the outer code word and an inner encoder whose input words are the permuted outer code bits.

The information sequence **u** is encoded by a rate $R^o = k^o/n^o$ outer encoder. The outer code sequence **b**^{o} is interleaved and then fed into the inner encoder. The inner code has rate $R^i = k^i/n^i$. Hence, the overall rate is

$$R = R^i R^o.$$

TURBO CODES

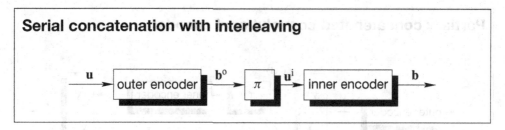

Figure 4.14: Serial concatenation with interleaving

4.3.4 Partial Concatenation

We will consider the partially concatenated convolutional encoder as depicted in Figure 4.15. Such an encoder consists of one outer and one inner convolutional encoder. In between there are a partitioning device and an interleaver, denoted by **P** and π respectively. The information sequence **u** is encoded by a rate $R^o = k^o/n^o$ outer encoder. The outer code sequence $\mathbf{b}^o$ is partitioned into two so-called partial code sequences $\mathbf{b}^{o,(1)}$ and $\mathbf{b}^{o,(2)}$. The partial code sequence $\mathbf{b}^{o,(1)}$ is interleaved and then fed into the inner encoder. The other symbols of the outer code sequence ($\mathbf{b}^{o,(2)}$, dashed lines in Figure 4.15) are not encoded by the inner encoder. This sequence, together with the inner code sequence $\mathbf{b}^i$, constitutes the overall code sequence **b**.

Similarly to the puncturing of convolutional codes, we describe the partitioning of the outer code sequence by means of a partitioning matrix **P**. Consider a rate $R^o = k^o/n^o$ outer convolutional code. **P** is a $t_p \times n^o$ matrix with matrix elements $p_{s,j} \in \{0, 1\}$, where $t_p \geq 1$ is an integer determining the partitioning period. A matrix element $p_{s,j} = 1$ means that the corresponding code bit will be mapped to the partial code sequence $\mathbf{b}^{o,(1)}$, while a code bit corresponding to $p_{s,j} = 0$ will appear in the partial code sequence $\mathbf{b}^{o,(2)}$. We define the *partial rate* R_p as the fraction of outer code bits that will be encoded by the inner encoder. With $\sum_{s,j} p_{s,j}$, the number of 1s in the partitioning matrix, and with the total number of elements $n^o t_p$ in **P**, we have

$$R_p = \frac{\sum_{s,j} p_{s,j}}{n^o t_p}.$$

Finally, we obtain the rate of the overall code

$$R = \frac{R^o R^i}{R_p + R^i(1 - R_p)},$$

where R^i is the inner code rate.

Clearly, if $R_p = 1$, this construction results in a serially concatenated convolutional code (Benedetto and Montorsi, 1998). If we choose $R_p = R^o$, and use systematic outer and inner encoding, such that we encode the information sequence with the inner encoder, we obtain a parallel (turbo) encoder. Besides these two *classical* cases, partitioning provides a new degree of freedom for code design.

Throughout the following sections we give examples where the properties of different concatenated convolutional codes are compared. Therefore, we will refer to the codes

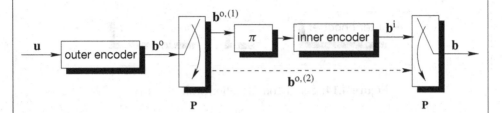

Partially concatenated convolutional encoder

- The partial rate R_p is the fraction of outer code bits that will be encoded by the inner encoder

$$R_p = \frac{\sum_{s,j} p_{s,j}}{n^o t_p} \qquad (4.10)$$

- The rate of the overall code is

$$R = \frac{R^o R^i}{R_p + R^i(1 - R_p)} \qquad (4.11)$$

where R^i and R^o are the rate of the inner code and outer code respectively.

- $R_p = 1$ results in a serially concatenated convolutional code.
- $R_p = R^o$ and systematic outer and inner encoding result in a parallel concatenated convolutional code.

Figure 4.15: Partially concatenated convolutional encoder

constructed in Figure 4.16. Those codes are all of overall rate $R = 1/3$, but with different partial rates. There are *classical* parallel $R_p = 1/2$ and serially concatenated convolutional codes $R_p = 1$, as well as partially concatenated convolutional codes.

4.3.5 Turbo Decoding

All turbo-like codes can be decoded iteratively with a message-passing algorithm similar to belief propagation. The iterative decoding is based on symbol-by-symbol soft-in/soft-out (SISO) decoding of the component codes, i.e. algorithms that use reliability information at the input and provide reliability information at the output, like the BCJR algorithm which evaluates symbol-by-symbol a-posteriori probabilities. For the following discussion, all symbol reliabilities are represented by log-likelihood ratios (L-values).

We consider only the decoder structure for partially concatenated codes as given in Figure 4.17. Decoders of parallel and serial constructions can be derived as special cases from this structure. The inner component decoder requires channel L-values as well as

TURBO CODES

> **Example of concatenated convolutional codes**
>
> - We construct different concatenated codes of overall rate $R = 1/3$. For the inner encoding we always employ rate $R^i = 1/2$ codes with generator matrix $\mathbf{G}^i(D) = (1, \frac{1+D^2}{1+D+D^2})$.
> - For outer encoding we employ the mother code with generator matrix $\mathbf{G}(D) = (1+D+D^2, 1+D^2)$, but we apply puncturing and partitioning to obtain the desired overall rate.
> - We consider:
>   - Two parallel concatenated codes: both with partial rate $R_p = R^o = 1/2$, but with different partitioning schemes ($\mathbf{P} = (1, 0)$, $\mathbf{P} = (0, 1)$).
>   - A serially concatenated code with $R_p = 1$, $R^o = 2/3$.
>   - Two partially concatenated constructions with $R^o = 3/5$, $R_p = 4/5$ and $\mathbf{P} = (0, 1, 1, 1, 1)$ and with $R^o = 4/7$, $R_p = 5/7$ and $\mathbf{P} = (1, 1, 0, 1, 1, 0, 1)$. Note that here the partitioning matrices are obtained by optimising the partial distances (see Section 4.6.2).

Figure 4.16: Example of concatenated convolutional codes

a-priori information for the inner information symbols as an input. The output of the inner decoder consists of channel and extrinsic information for the inner information symbols. The outer decoder expects channel L-values and provides extrinsic L-values $L_e(\mathbf{b}^o)$ for the outer code symbols and estimates $\hat{\mathbf{u}}$ for the information symbols. Initially, the received sequence $\mathbf{r}$ is split up into a sequence $\mathbf{r}^{(1)}$ which is fed into the inner decoder and a sequence $\mathbf{r}^{(2)}$ which is not decoded by the inner decoder. Moreover, the a-priori values for the inner decoder $L_a(\mathbf{u}^i)$ are set to zero.

For each iteration we use the following procedure. We first perform inner decoding. The output symbols $L(\mathbf{u}^i)$ of the inner decoder are de-interleaved (π^{-1}). This de-interleaved sequence and the received symbols $\mathbf{r}^{(2)}$ are multiplexed according to the partitioning scheme $\mathbf{P}$. The obtained sequence $L(\mathbf{r}^o)$ is regarded as the channel values of the outer code, i.e. it is decoded by the outer decoder. The outer decoder provides extrinsic L-values for the outer code bits $L_e(\mathbf{b}^o)$ and an estimated information sequence $\hat{\mathbf{u}}$. For the next round of iterative decoding, the extrinsic output $L_e(\mathbf{b}^o)$ is partitioned according to $\mathbf{P}$ into $L_e(\mathbf{b}^{o,(1)})$ and $L_e(\mathbf{b}^{o,(2)})$, whereas the sequence $L_e(\mathbf{b}^{o,(2)})$ is not regarded any more, because $\mathbf{b}^{o,(2)}$ was not encoded by the inner encoder. $L_e(\mathbf{b}^{o,(1)})$ is interleaved again such that the resulting sequence $L_a(\mathbf{u}^i)$ can be used as a-priori information for the next iteration of inner decoding.

Finally, we give some simulation results for the AWGN channel with binary phase shift keying. All results are obtained for ten iterations of iterative decoding, where we utilise the

Decoding structure

For each iteration we use the following procedure:

- Perform inner decoding. The output symbols $L(\mathbf{u}^i)$ of the inner decoder are de-interleaved (π^{-1}), and the received symbols $\mathbf{r}^{(2)}$ are multiplexed according to the partitioning scheme $\mathbf{P}$.

- The obtained sequence $L(\mathbf{b}^o)$ is decoded by the outer decoder. The extrinsic output $L_e(\mathbf{b}_o)$ is partitioned according to $\mathbf{P}$ into $L_e(\mathbf{b}^{o,(1)})$ and $L_e(\mathbf{b}^{o,(2)})$, where $L_e(\mathbf{b}^{o,(2)})$ is discarded.

- $L_e(\mathbf{b}^{o,(1)})$ is interleaved again such that the resulting sequence $L_a(\mathbf{b}^i)$ can be used as a priori information for the next iteration of inner decoding.

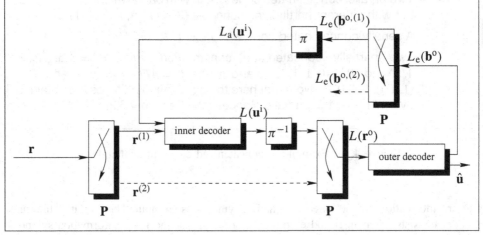

Figure 4.17: Decoding structure

max-log version of the BCJR algorithm as discussed in Section 3.5.2. Simulation results are depicted in Figure 4.18, where the figures presents the Word Error Rate (WER) for codes of length $n = 600$ or $n = 3000$ respectively. The component codes are as given in Figure 4.16. The two partially concatenated codes outperform the serial concatenated code for the whole considered region of WER. At high SNR we notice the typical error floor of the turbo code ($\mathbf{P} = (1, 0)$). Here, the partially concatenated codes achieve significantly better performance. Simulation results regarding the bit error rate are similar.

4.4 EXIT Charts

For the analysis of the iterative decoding algorithm we utilise the EXtrinsic Information Transfer (EXIT) characteristics as proposed by ten Brink (ten Brink, 2000; see also ten Brink, 2001). The basic idea of the EXIT chart method is to predict the behaviour of

TURBO CODES

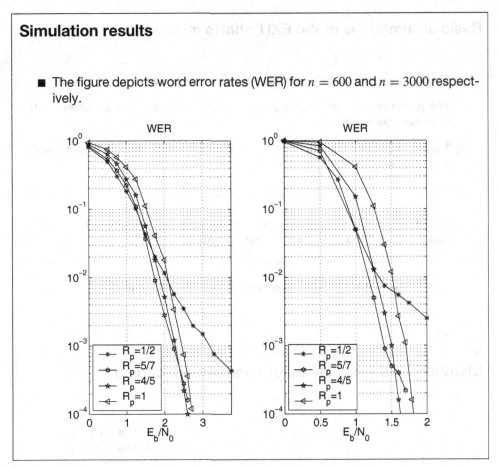

Figure 4.18: Simulation results

the iterative decoder by looking at the input/output relations of the individual constituent decoders.

4.4.1 Calculating an EXIT Chart

We assume that a-priori information can be modelled by jointly independent Gaussian random variables. The decoder can then be characterised by its transfer function where the mutual information at the output is measured depending on the a-priori information and the channel input. All considerations are based on the assumptions in Figure 4.19. These assumptions are motivated by simulation results, which show that with large interleavers the a priori values remain fairly uncorrelated from the channel values. Moreover, the histograms of the extrinsic output values are Gaussian-like. This is indicated in Figure 4.20 which depicts the distribution of the L-values at the input and output of the Soft-Input Soft-Output (SISO) decoder. For this figure we have simulated a transmitted

Basic assumptions of the EXIT charts method

(1) The a-priori values are independent of the respective channel values.

(2) The probability density function $p_e(\cdot)$ of the extrinsic output values is the Gaussian density function.

(3) The probability density function $p_e(\cdot)$ fulfils the symmetry condition (Divsalar and McEliece, 1998)

$$p_e(\xi|\mathcal{X}=x) = p_e(-\xi|\mathcal{X}=x) \cdot e^{x\xi} \qquad (4.12)$$

where the binary random variable $\mathcal{X}$ denotes a transmitted symbol (either outer code symbol or inner information symbol) with $x \in \{\pm 1\}$.

Figure 4.19: Basic assumptions of the EXIT charts method

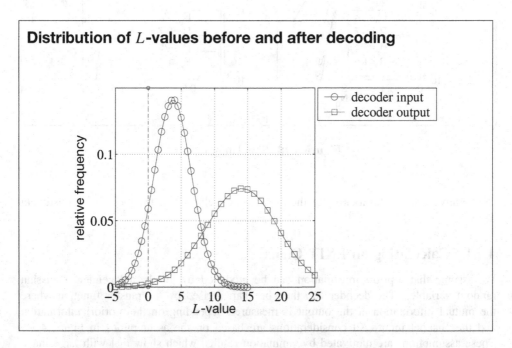

Figure 4.20: Distribution of the L-values at the decoder input and output

TURBO CODES

all-zero code sequence of the (7 5) convolutional code over a Gaussian channel with binary phase shift keying. The Gaussian distribution of the input values is a consequence of the channel model. The distribution of the output L-values of the SISO decoder is also Gaussian-like. Note that the area under the curve on the right of the axis $L = 0$ determines the bit error rate. Obviously, this area is smaller after decoding, which indicates a smaller bit error rate.

4.4.2 Interpretation

Let L_a denote the a-priori L-value corresponding to the transmitted symbol $\mathcal{X}$. Based on the assumptions in Figure 4.19, the a-priori input of the SISO decoders is modelled as independent random variables

$$L_a = \mu_a \cdot \mathcal{X} + \mathcal{N}_a,$$

where μ_a is a real constant and $\mathcal{N}_a$ is a zero-mean Gaussian distributed random variable with variance σ_a^2. Thus, the conditional probability density function of the a-priori L-values is

$$p_a(\xi|\mathcal{X}=x) = \frac{1}{\sqrt{2\pi}\sigma_a} e^{-\frac{(\xi-\mu_a x)^2}{2\sigma_a^2}}.$$

Moreover, the symmetry condition (4.12) implies

$$\mu_a = \frac{\sigma_a^2}{2}.$$

The mutual information $I_a = I(\mathcal{X}; L_a)$ between the transmitted symbol $\mathcal{X}$ and the corresponding L-value L_a is employed as a measure of the information content of the a-priori knowledge

$$I_a = \frac{1}{2} \sum_{x \in \{-1,1\}} \int_{-\infty}^{\infty} p_a(\xi|X=x)$$

$$\times \log_2 \frac{2 \cdot p_a(\xi|X=x)}{p_a(\xi|X=-1) + p_a(\xi|X=1)} d\xi.$$

With Equation (4.13) and $\mu_a = \frac{\sigma_a^2}{2}$ we obtain

$$I_a = 1 - \frac{1}{\sqrt{2\pi}\sigma_a} \int_{-\infty}^{\infty} e^{-\frac{(\xi-\frac{\sigma_a^2}{2}x)^2}{2\sigma_a^2}} \log_2(1 + e^{-\xi}) d\xi.$$

Note that I_a is only a function of σ_a. Similarly, for the extrinsic L-values we obtain

$$I_e = I(X; L_e) = \frac{1}{2} \sum_{x \in \{-1,1\}} \int_{-\infty}^{\infty} p_e(\xi|X=x)$$

$$\times \log_2 \frac{2 \cdot p_e(\xi|X=x)}{p_e(\xi|X=-1) + p_e(\xi|X=1)} d\xi.$$

Calculating an EXIT chart

- The a-priori input of the SISO decoders is modelled as independent Gaussian random variables L_a with variance σ_a^2. The conditional probability density function of the a-priori L-values is

$$p_a(\xi|\mathcal{X}=x) = \frac{1}{\sqrt{2\pi}\,\sigma_a} e^{-\frac{(\xi-\frac{\sigma_a^2}{2}x)^2}{2\sigma_a^2}} \qquad (4.13)$$

- The mutual information $I_a = I(\mathcal{X}; L_a)$ between the transmitted symbol $\mathcal{X}$ and the corresponding L-value L_a is employed as a measure of the information content of the a-priori knowledge. I_a is only a function of σ_a

$$I_a = 1 - \frac{1}{\sqrt{2\pi}\,\sigma_a} \int_{-\infty}^{\infty} e^{-\frac{(\xi-\frac{\sigma_a^2}{2}x)^2}{2\sigma_a^2}} \log_2(1+e^{-\xi})d\xi \qquad (4.14)$$

- The mutual information $I_e = I(\mathcal{X}; L_e)$ between the transmitted symbol $\mathcal{X}$ and the corresponding extrinsic L-value L_e is employed as a measure of the information content of the code correlation

$$I_e = I(X; L_e) = \frac{1}{2} \sum_{x \in \{-1,1\}} \int_{-\infty}^{\infty} p_e(\xi|X=x)$$

$$\times \log_2 \frac{2 \cdot p_e(\xi|X=x)}{p_e(\xi|X=-1)+p_e(\xi|X=1)} d\xi \qquad (4.15)$$

where $p_e(\cdot)$ is estimated by means of Monte Carlo simulations.

Figure 4.21: Calculating an EXIT chart

Note that it is not possible to express $I_a = I(\mathcal{X}; L_a)$ and $I_e = I(\mathcal{X}; L_e)$ in closed form. However, both quantities can be evaluated numerically. In order to evaluate Equation (4.15), we estimate $p_e(\cdot)$ by means of Monte Carlo simulations. For this purpose, we generate independent random variables L_a according to Equation (4.13) and apply them as a-priori input to the SISO decoder. We obtain a similar distribution $p_e(\cdot)$ to Figure 4.20. With this distribution we can calculate $I_e = I(\mathcal{X}; L_e)$ numerically according to Figure 4.21. $I_a = I(\mathcal{X}; L_a)$ is also evaluated numerically according to Equation (4.14).

In order to analyse the iterative decoding algorithm, the resulting extrinsic information transfer (EXIT) characteristics $I_e = T(I_a, E_b/N_0)$ of the inner and outer decoder are plotted into a single diagram, where for the outer decoder the axes of the transfer characteristic are swapped. An example of an EXIT chart is given in Figure 4.22. This EXIT chart can be

TURBO CODES

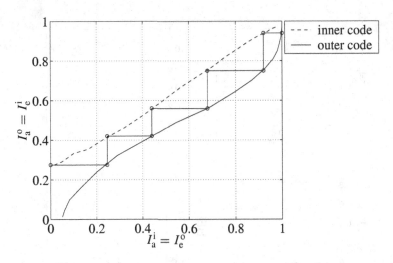

Interpretation of an EXIT chart

- The resulting extrinsic information transfer (EXIT) characteristics $I_e = T(I_a, E_b/N_0)$ of the inner and outer decoder are plotted into a single diagram, where for one decoder the axes of the transfer characteristic is swapped.

- During the iterative decoding procedure the extrinsic output values of one decoder become the a-priori values of the other decoder. This is indicated by the line between the transfer functions.

- Each line indicates a single decoding step of the iterative decoding procedure.

Figure 4.22: Interpretation of an EXIT chart

interpreted as follows. During the iterative decoding procedure, the extrinsic output values of one decoder become the a-priori values of the other decoder. This is indicated by the line between the transfer functions, where each line indicates a single decoding step of one of the constituent decoders (a half-iteration of the iterative decoding procedure).

EXIT charts provide a visualisation of the exchange of extrinsic information between the two component decoders. Moreover, for very large interleaver sizes, EXIT charts make it possible to predict the convergence of the iterative decoding procedure according to Figure 4.23. Convergence is only possible if the transfer characteristics do not intersect. In the case of convergence, the average number of required decoding steps can be estimated. Furthermore, the EXIT chart method make it possible to predict the so-called *waterfall region* of the bit error rate performance. This means that we observe

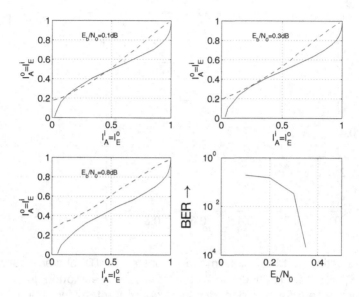

Convergence of iterative decoding

These EXIT charts can be interpreted as follows:

- **Pinch-off region:** the region of low signal-to-noise ratios where the bit error rate is hardly improved with iterative decoding.

- **Bottleneck region:** here the transfer characteristics leave a narrow tunnel. During the iterative decoding the convergence towards low bit error rates is slow, but possible.

- **Wide-open region:** region of fast convergence.

Figure 4.23: EXIT charts and the simulated bit error rate (BER) for a serially concatenated code with overall rate $R = 1/3$

an E_b/N_0-decoding threshold that corresponds to the region of the EXIT charts where the transfer characteristics leave a narrow tunnel. For this signal-to-noise ratio, the iterative decoding algorithm converges towards low bit error rates with a large number of iterations. Therefore, this region of signal-to-noise ratios is also called the *bottleneck region*. The region of low signal-to-noise ratios, the *pinch-off region*, is characterised by an intersection of the transfer characteristics of the two decoders. Here, the bit error rate is hardly improved with iterative decoding. Finally, we have the region of fast convergence, where the signal-to-noise ratio is significantly larger than the decoding threshold. In

TURBO CODES

the so-called *wide-open region* the decoding procedure converges with a small number of iterations.

For instance, we construct a serially concatenated code. For the inner encoding we employ a rate $R^i = 1/2$ code with generator matrix $\mathbf{G}^i(D) = (1, \frac{1+D^2}{1+D+D^2})$. For the outer encoding we employ the same code, but we apply puncturing and partitioning to obtain the outer rate $R^o = 2/3$ and the desired overall rate $R = 1/3$. For this example, the EXIT charts for different values of E_b/N_0 are depicted in Figure 4.23. The bottleneck region corresponds to a signal-to-noise ratio $E_b/N_0 \approx 0.3$ dB. For the code of rate $R = 1/3$ and length $n = 100\,000$, the required signal-to-noise ratio in order to obtain a bit error rate of 10^{-4} is less than 0.1 dB away from the predicted threshold value, where the iterative decoding procedure was run for 30 iterations.

Now, consider our running example with the codes from Figure 4.16. The corresponding EXIT charts for a signal-to-noise ratio $E_b/N_0 = 0.1$ dB can be found in Figure 4.24. The solid line is the transfer characteristic of the inner code. The two other charts correspond to the parallel concatenation ($R_p = 1/2$) with the partitioning matrix $\mathbf{P} = (1, 0)$ and the partially concatenated code with $R_p = 5/7$. For the parallel concatenation the EXIT chart leaves a narrow tunnel. Thus, the iterative decoding should converge towards low bit error rates. Similarly, with $R_p = 5/7$ the convergence of the iterative decoding is

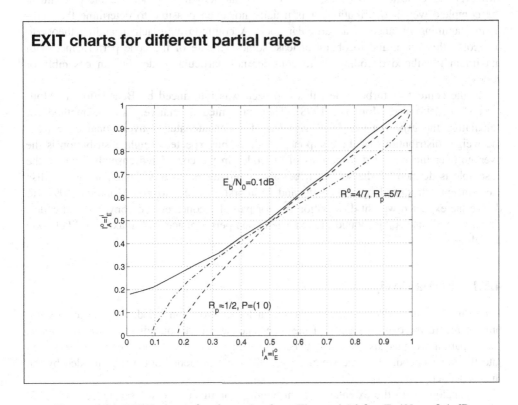

Figure 4.24: EXIT charts for the codes from Figure 4.16 for $E_b/N_0 = 0.1$ dB

slow, but possible. In comparison with $\mathbf{P} = (1, 0)$, we observe a wider tunnel for the $R_p = 5/7$ construction. Therefore, the iterative decoding should converge with a smaller number of iterations. Note that the choice of the partitioning pattern is also important. The parallel construction with $\mathbf{P} = (0, 1)$ gets stuck at values > 0.5. Likewise, the EXIT charts for the serially concatenated code as well as for $R_p = 4/5$ get instantly stuck, so no bit error rate reduction is expected at this signal-to-noise ratio. These curves are omitted in Figure 4.24.

For large interleaver sizes the EXIT chart method allows accurate prediction of the waterfall region. For interleavers of short or moderate length bounding techniques that include the interleaving depth are more appropriate, particularly if we also consider the region of fast convergence, i.e. if we wish to determine possible error floors.

4.5 Weight Distribution

We have seen in Section 3.3.6 that the performance of a code with maximum likelihood decoding is determined by the weight distribution, of the code. If we know the weight distribution, we can bound the word error rate using, for example, the union bound. However, for concatenated codes, even of moderate length, it is not feasible to evaluate the complete weight distribution. It is usually not even possible to determine the minimum Hamming distance of a particular code. A common approach in coding theory to overcome this issue, and to obtain at least some estimate of the code performance with maximum likelihood decoding, is to consider not particular codes but an ensemble of codes.

In the context of turbo codes, this approach was introduced by Benedetto and Montorsi (Benedetto and Montorsi, 1998). They presented a relatively simple method for calculating the *expected weight distribution* of a concatenated convolutional code from the weight distributions of the component codes. The expected weight distribution is the average over the weight distributions of all codes in the considered ensemble, where the ensemble is defined by the set of all possible interleavers. In this section we will utilise the concept introduced by Benedetto and Montorsi. (Benedetto and Montorsi, 1998) to derive the expected weight distribution $\overline{A}_w$ for partially concatenated convolutional codes. Then, we will use $\overline{A}_w$ to bound the expected code performance with maximum likelihood decoding.

4.5.1 Partial Weights

For further consideration of partially concatenated convolutional codes it is necessary not only to regard the overall weight of the outer encoder output but also the *partial weights*. These partial weights distinguish between the weight w_1 of the code sequence being fed into the inner encoder and the weight w_2 of the code sequence not being encoded by the inner encoder.

By analogy with the extended weight enumerator function discussed in Section 3.3.3, we introduce the labels W_1 and W_2 instead of W so that we can determine the partial weight enumerator function $A_{\text{PEF}}(W_1, W_2)$ as described in Section 3.3.4.

TURBO CODES

For further calculations, assuming $\mathbf{b}_1$ is fed into the inner encoder and $\mathbf{b}_2$ is not encoded, the matrix $\mathbf{A}$ from the above examples is modified as

$$\tilde{\mathbf{A}} = \begin{pmatrix} 1 & 0 & W_1 W_2 & 0 \\ W_1 W_2 & 0 & 1 & 0 \\ 0 & W_1 & 0 & W_2 \\ 0 & W_2 & 0 & W_1 \end{pmatrix}.$$

The modification of $\mathbf{A}$ can be done by introducing labels $V_1, \ldots, V_n$ which represent the output weight of each output of the encoder so that $\mathbf{A}$ contains polynomials $\mathbf{A}_{ij}(V_1, \ldots, V_n)$. Then the polynomials $\tilde{\mathbf{A}}_{ij}(W_1, W_2)$ can be easily obtained by replacing V_k ($k = 1, \ldots, c$) with W_1 or W_2 according to the partitioning scheme. In this context, $\mathbf{A}_{ij}$ marks the element in row i and column j of matrix $\mathbf{A}$.

4.5.2 Expected Weight Distribution

Now we derive the expected weight distribution $\overline{A}_w$ for partially concatenated convolutional codes, where $\overline{A}_w$ denotes the expected number of code words with weight w. We restrict ourselves to randomly chosen permutations of length N and terminated convolutional component codes. First, we evaluate the expected number of code words $E\{A^i_{w_2,\tilde{w}}\}$ with weight $\tilde{w}$ at the output of the inner encoder conditioned on the weight w_2 of the partial outer code sequence $\mathbf{v}^{o,(2)}$. Let $A^i_{w_1,\tilde{w}}$ denote the number of code words generated by the inner encoder with input weight w_1 and output weight $\tilde{w}$ (Benedetto and Montorsi, 1998). Let $A^o_{w_1,w_2}$ denote the number of code words with partial weights w_1 and w_2 corresponding to the outer encoder and its partitioning. We have

$$E\{A^i_{w_2,\tilde{w}}\} = \sum_{w_1=0}^{N} \frac{A^o_{w_1,w_2} \cdot A^i_{w_1,\tilde{w}}}{\binom{N}{w_1}}.$$

This formula becomes plausible if we note that the term $A^i_{w_1,\tilde{w}}/\binom{N}{w_1}$ is the probability that a random input word $\mathbf{u}^i$ to the inner encoder of weight w_1 will produce an output word $\mathbf{b}^i$ of weight $\tilde{w}$. With $w = \tilde{w} + w_2$ we obtain

$$\overline{A}_w = E\{A_w\} = \sum_{w_2 \leq w} E\{A^i_{w_2,(w-w_2)}\}.$$

This is summarized in Figure 4.25. formula (4.16) can be used for bounding the average maximum likelihood performance of the code ensemble given by all possible permutations. Let $\overline{P}_W$ denote the expected word error rate and n the overall code length. Using the standard union bound for the additive white Gaussian noise (AWGN) channel with binary phase shift keying, we have

$$\overline{P}_W \leq \sum_{w=1}^{n} \overline{A}_w e^{-w \cdot R \cdot E_b/N_0}.$$

We consider the codes as given in Figure 4.16. However, for inner encoding we use the inner rate $R^i = 1/2$ code with generator matrix $\mathbf{G}^i(D) = (1, \frac{1}{1+D})$. With code length $n = 300$

Expected weight distribution

- The expected weight distribution $\overline{A}_w$ for a partially concatenated convolutional code is

$$\overline{A}_w = E\{A_w\} = \sum_{w_2 \leq w} E\left\{A^i_{w_2,(w-w_2)}\right\} \tag{4.16}$$

where $E\{A^i_{w_2,\tilde{w}}\}$ is the expected number of code words with weight $\tilde{w}$ at the output of the inner encoder conditioned on the weight w_2 of the partial outer code sequence.

- Using the standard union bound for the AWGN channel with Binary Phase Shift Keying (BPSK), we can bound the expected word error rate

$$\overline{P}_W \leq \sum_{w=1}^{n} \overline{A}_w e^{-w \cdot R \cdot E_b/N_0} \tag{4.17}$$

Figure 4.25: Expected weight distribution

we obtain similar overall rates $R \approx 1/3$. The results for bounding the word error rate by the union bound in inequality (4.17) are depicted in Figure 4.26. $R_p = 1$ corresponds to a serially concatenated convolutional code, and $R_p = 1/2$ to a turbo code. For a partially concatenated code with $R_p = 4/5$ we use a rate $R^o = 3/5$ punctured outer code. With the partitioning period $t_p = 1$ there exist five different partitioning schemes. The results for the schemes with best and worst performance are also given in Figure 4.26. The performance difference between $\mathbf{P} = (1, 1, 1, 1, 0)$ and $\mathbf{P} = (0, 1, 1, 1, 1)$ indicates that the particular choice of the partitioning scheme is important. In this example the partially concatenated code with $\mathbf{P} = (0, 1, 1, 1, 1)$ outperforms the serially and parallel concatenated constructions.

4.6 Woven Convolutional Codes

In this section we concentrate on the minimum Hamming distance of the constructed concatenated code. With respect to the minimum Hamming distance of the concatenated code, especially for codes of short lengths, the choice of the particular interleaver is very important. With turbo-like codes, the use of designed interleavers is motivated by the asymptotic coding gain, which is the gain in terms of transmit power that can be achieved with coding compared with the uncoded case for very low residual error rates. For unquantised channels the asymptotic coding gain is (Clark and Cain, 1988)

$$G_a = 10 \log_{10}(R \cdot d) \text{dB},$$

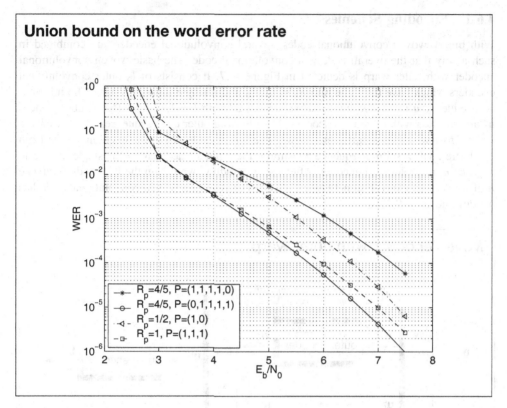

Figure 4.26: Union bound on the word error rate for rate $R = 1/3$ codes (code length $n = 300$) with different partial rates (cf. Figure 4.16)

where R is the rate and d is the minimum Hamming distance of the code. This formula implies that for fixed rates the codes should be constructed with minimum Hamming distances as large as possible, in order to ensure efficient performance for high signal-to-noise ratios.

We start our discussion by introducing the class of *woven convolutional codes*. Woven code constructions yield a larger designed minimum Hamming distance than ordinary serial or parallel constructions. In the original proposal (Höst *et al.*, 1997), two types of woven convolutional code are distinguished: those with outer warp and those with inner warp. In this section we consider encoding schemes that are variations of woven codes with outer warp. We propose methods for evaluating the distance characteristics of the considered codes on the basis of the active distances of the component codes. With this analytical bounding technique, we derive lower bounds on the minimum (or free) distance of the concatenated code. These considerations also lead to design criteria for interleavers.

Note that some of the figures and results of this section are reprinted, with permission, from Freudenberger *et al.* (2001), © 2001 IEEE.

4.6.1 Encoding Schemes

With binary woven convolutional codes, several convolutional encoders are combined in such a way that the overall code is a convolutional code. The basic woven convolutional encoder with outer warp is depicted in Figure 4.27. It consists of l_o outer convolutional encoders which have the same rate $R_o = k_o/n_o$ and a single inner encoder.[7] The information sequence $\mathbf{u}$ is divided into l_o subsequences $\mathbf{u}_l^o$ with $l = 1, \ldots, l_o$. These subsequences $\mathbf{u}_l^o$ are encoded with the outer encoders. The resulting outer code sequences $\mathbf{b}_1^o, \ldots, \mathbf{b}_{l_o}^o$ are written row-wise into a buffer of l_o rows. The binary code bits are read column-wise from this buffer. The resulting sequence constitutes the input sequence $\mathbf{u}^i$ of the single inner rate $R_i = k_i/n_i$ convolutional encoder. After inner encoding, we obtain the final code sequence $\mathbf{b}$ of the Woven Convolutional Code (WCC). The resulting woven convolutional code has overall rate $R = R_i R_o$.

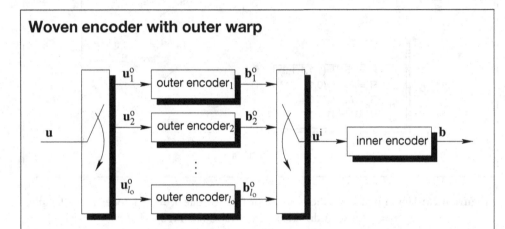

Woven encoder with outer warp

- A woven convolutional encoder with outer warp consists of l_o outer convolutional encoders that have the same rate $R_o = k_o/n_o$.

- The information sequence $\mathbf{u}$ is divided into l_o subsequences $\mathbf{u}_l^o$ with $l = 1, \ldots, l_o$.

- The outer code sequences $\mathbf{b}_1^o, \ldots, \mathbf{b}_{l_o}^o$ are written row-wise into a buffer of l_o rows. The binary code bits are read column-wise and the resulting sequence constitutes the input sequence $\mathbf{u}^i$ of the single inner rate $R_i = k_i/n_i$ convolutional encoder.

- The resulting woven convolutional code has the overall rate

$$R = R_i R_o.$$

Figure 4.27: Woven encoder with outer warp

---

[7] In contrast, a woven encoder with inner warp has a single outer and several inner encoders.

TURBO CODES

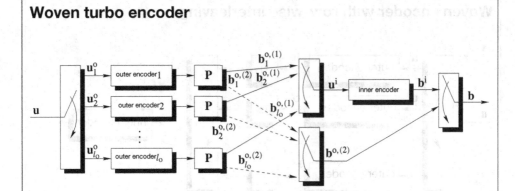

Woven turbo encoder

- The overall woven turbo code has the rate

$$R = \frac{R_o R_i}{R_p + R_i(1 - R_p)},$$

where R_p is the fraction of outer code bits that will be encoded by the inner encoder.

- The partitioning of the outer code sequences into two partial code sequences $\mathbf{b}_l^{o,(1)}$ and $\mathbf{b}_l^{o,(2)}$ is described by means of a partitioning matrix $\mathbf{P}$.

Figure 4.28: Woven turbo encoder

The concept of partial concatenation as discussed in Section 4.3.4 was first introduced by Freudenberger *et al.* (Freudenberger *et al.*, 2000a, 2001) in connection with *woven turbo codes*. Woven turbo codes belong to the general class of woven convolutional codes. Figure 4.28 presents the encoder of a Woven Turbo Code (WTC). Like an ordinary woven encoder, a woven turbo encoder consists also of l_o outer convolutional encoders and one inner convolutional encoder. The information sequence $\mathbf{u}$ is subdivided into l_o sequences which are the input sequences to the l_o rate $R_o = k_o/n_o$ outer encoders.

Parts of the symbols of the outer code sequences ($\mathbf{b}_l^{o,(1)}$), which are located in the same bit positions, are multiplexed to the sequence $\mathbf{u}^i$. The sequence $\mathbf{u}^i$ is the input sequence of the inner encoder. The other symbols of the outer code sequences ($\mathbf{b}_l^{o,(2)}$, dashed lines in Figure 4.28) are not encoded by the inner encoder. These sequences $\mathbf{b}_l^{o,(2)}$ form the sequence $\mathbf{b}^{o,(2)}$ which, together with the inner code sequence $\mathbf{b}^i$, constitutes the overall code sequence $\mathbf{b}$.

As with partially concatenated codes, the partitioning of the outer code sequences into two partial code sequences $\mathbf{b}_l^{o,(1)}$ and $\mathbf{b}_l^{o,(2)}$ is described by means of a partitioning matrix $\mathbf{P}$. Furthermore, the overall code rate is

$$R = \frac{R_o R_i}{R_p + R_i(1 - R_p)},$$

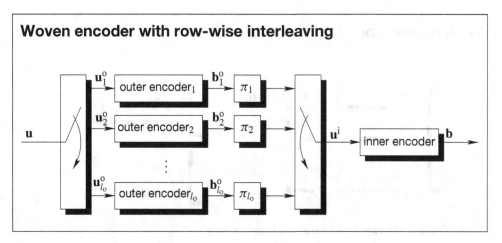

Figure 4.29: Woven encoder with row-wise interleaving

where R_p is the fraction of outer code bits that will be encoded by the inner encoder as defined in Section 4.3.4.

Up to now, we have only considered woven constructions without interleavers. These constructions lead to overall convolutional codes. Usually, when we employ interleaving, we use block-wise interleavers and terminated convolutional codes so that the resulting concatenated code is actually a block code. This is possibly also the case with woven codes for most applications.

However, there are a number of interesting aspects of the construction without interleaving, e.g. a sliding window decoding algorithm for WCC can be introduced (Jordan *et al.*, 2004a). This algorithm requires no code termination and might be interesting for applications where low-latency decoding is required. In the following we will mainly consider constructions with interleaving, in particular row-wise interleaving as indicated in Figure 4.29. Here, each outer code sequence $\mathbf{b}_l^o$ is interleaved by arbitrary and independent interleavers. The interleaved sequences are fed into the inner encoder. Of course, the same interleaver concept can also be applied to woven turbo codes.

4.6.2 Distance Properties of Woven Codes

Now, we investigate the distance properties of the considered codes on the basis of the active distances of the component codes. With this analytical bounding technique, we derive lower bounds on the minimum (or free) distance of the concatenated code.

Below, we define the generating tuples of an input sequence $\mathbf{u}(D)$ of a convolutional encoder. Then, we show that each generating tuple generates at least d_g non-zero bits in the encoded sequence $\mathbf{b}(D)$, where d_g is some weight in the region $\beta^b \leq d_g \leq d_{\text{free}}$.

Consider a convolutional encoder and its active burst distance (cf. Section 3.3.2 on page 122). We call a segment of a convolutional code sequence *burst* if it corresponds to an encoder state sequence starting in the all-zero state and ending in the all-zero states and having no consecutive all-zero states in between which correspond to an all-zero input

Generating tuples

- Let d_g be an integer with $\beta^b \leq d_g \leq d_{\text{free}}$. We define the generating length for d_g as

$$j_g = \left\lceil \frac{2d_g - \beta^b}{\alpha} \right\rceil \qquad (4.18)$$

i.e. j_g is the minimum j for which the lower bound on the active burst distance satisfies $\alpha j + \beta^b \geq 2d_g$.

- Let t_1 be the time index of the first non-zero tuple $\mathbf{u}_t = (u_t^{(1)}, u_t^{(2)}, \ldots, u_t^{(k)})$ of the sequence $\mathbf{u}$. Let t_2 be the time index of the first non-zero tuple with $t_2 \geq t_1 + j_g$, and so on. We call the information tuples $\mathbf{u}_{t_1}, \mathbf{u}_{t_2}, \ldots$ *generating tuples*.

- Let $\mathbf{u}$ be the input sequence of a convolutional encoder with N_g generating tuples with generating length j_g. Then the weight of the corresponding code sequence $\mathbf{b}$ satisfies

$$\text{wt}(\mathbf{b}) \geq N_g d_g \qquad (4.19)$$

Figure 4.30: Definition of generating tuples

tuple of the encoder. Note that a burst of length $j+1$ has at least weight $a^b(j)$, where length is defined in n-tuples and the corresponding number of bits is equal to $n(j+1)$.

For an encoder characterised by its active burst distance, we will now bound the weight of the generated code sequence given the weight of the corresponding information sequence. Of course, this weight of the code sequence will depend on the distribution of the 1s in the input sequence. In order to consider this distribution, we introduce the notion of *generating tuples* as defined in Figure 4.30. Let d_g be an integer satisfying $\beta^b \leq d_g \leq d_{\text{free}}$. Remember that d_{free} is the free distance of the code and β^b is a constant in the lower bound of the active burst distance defined in Equation (3.11). We define the *generating length* for d_g as

$$j_g = \left\lceil \frac{2d_g - \beta^b}{\alpha} \right\rceil,$$

where α is the slope of the lower bound on the active distances. The generating length j_g is the minimum length j for which the lower bound on the active burst distance satisfies $\alpha j + \beta^b \geq 2d_g$, i.e. j_g is the length of a burst that guarantees that the burst has at least weight $2d_g$.

Now consider an arbitrary information sequence $\mathbf{u}$. We call the first non-zero k-tuple $\mathbf{u}_t = (u_t^{(1)}, u_t^{(2)}, \ldots, u_t^{(k)})$ a generating tuple, because it will generate a weight of at least d_{free} in the code sequence $\mathbf{b}$. But what happens if there are more non-zero input bits? Now the definition of the generating tuples comes in handy. Let t_1 be the time index of the first generating tuple, i.e. of the first non-zero tuple $\mathbf{u}_t = (u_t^{(1)}, u_t^{(2)}, \ldots, u_t^{(k)})$ in the sequence $\mathbf{u}$. Moreover, let t_2 be the time index of the first non-zero tuple with $t_2 \geq t_1 + j_g$, and so on. We call the information tuples $\mathbf{u}_{t_1}, \mathbf{u}_{t_2}, \ldots$ generating tuples. If the number of generating tuples is N_g, then the weight of the code sequence is at least $N_g \cdot d_g$.

Why is this true? Consider an encoder with generating length j_g according to Equation (4.18). The weight of a burst that is started by a generating tuple of the encoder will be at least d_{free} if the next generating tuple enters the encoder in a new burst, and at least $2d_g$ if the next generating tuple enters the encoder inside the burst. This approach can be generalised. Let N_i denote the number of generating tuples corresponding to the ith burst. For $N_i = 1$ the weight of the ith burst is greater or equal to d_{free}, which is greater or equal to d_g. The length of the ith burst is at least $(N_i - 1)j_g + 1$ for $N_i > 1$ and we obtain

$$\text{wt}(burst_i) \geq \tilde{a}^b \left((N_i - 1)j_g\right)$$
$$\geq \alpha(N_i - 1)j_g + \beta^b.$$

With Equation (4.18) we have $\alpha j_g \geq 2d_g - \beta^b$ and it follows that

$$\text{wt}(burst_i) \geq (N_i - 1)(2d_g - \beta^b) + \beta^b$$
$$\geq N_i d_g + (N_i - 2)(d_g - \beta^b).$$

Taking into account that $d_g \geq \beta^b$, i.e. $d_g - \beta^b \geq 0$, we obtain

$$\text{wt}(burst_i) \geq N_i d_g \; \forall \; N_i \geq 1.$$

Finally, with $N_g = \sum_i N_i$ we obtain

$$\text{wt}(\mathbf{b}) \geq \sum_i \text{wt}(burst_i) \geq \sum_i N_i d_g = N_g d_g.$$

We will now use this result to bound the free distance of a WCC. Consider the encoder of a woven convolutional code with outer warp as depicted in Figure 4.29. Owing to the linearity of the considered codes, the free distance of the woven convolutional code is given by the minimal weight of all possible inner code sequences, except the all-zero sequence.

If one of the outer code sequences $\mathbf{b}_l^o$ is non-zero, then there exist at least d_{free}^o non-zero bits in the inner information sequence. Can we guarantee that there are at least d_{free}^o generating tuples? In fact we can if we choose a large enough l_o. Let d_g be equal to the free distance d_{free}^i of the inner code. We define the *effective length* of a convolutional encoder as

$$l_{\text{eff}} = k \left\lceil \frac{2d_{\text{free}} - \beta^b}{\alpha} \right\rceil.$$

Let l_{eff}^i be the effective length of the inner encoder. If $l_o \geq l_{\text{eff}}^i$ holds and one of the outer code sequences $\mathbf{b}_l^o$ is non-zero, then there exist at least d_{free}^o generating tuples in the inner information sequence that generate a weight greater or equal to d_{free}^i. Consequently, it follows from inequality (4.19) that $d_{\text{free}}^{\text{WCC}} \geq d_{\text{free}}^o d_{\text{free}}^i$. This result is summarised in Figure 4.31.

TURBO CODES

Free distance of woven convolutional codes

- We define the *effective length* of a convolutional encoder as

$$l_{\text{eff}} = k \left\lceil \frac{2d_{\text{free}} - \beta^b}{\alpha} \right\rceil \quad (4.20)$$

- Let l_{eff}^i be the effective length of the inner encoder. The free distance of the WCC with $l_o \geq l_{\text{eff}}^i$ outer convolutional encoders satisfies the inequality

$$d_{\text{free}}^{\text{WCC}} \geq d_{\text{free}}^o d_{\text{free}}^i \quad (4.21)$$

Figure 4.31: Free distance of woven convolutional codes

For instance, let us construct a woven encoder employing l_o outer codes. For outer codes we use the punctured convolutional code defined in Section 3.1.5 with the generator matrix

$$\mathbf{G}^o(D) = \begin{pmatrix} 1+D & 1+D & 1 \\ D & 0 & 1+D \end{pmatrix}.$$

This code has free distance $d_{\text{free}}^o = 3$. The generator matrix of the inner convolutional code is $\mathbf{G}^i(D) = (1, \frac{1+D^2}{1+D+D^2})$, i.e. $d_{\text{free}}^i = 5$. The lower bound on the active burst distance of the inner encoder is given by $\tilde{a}^{b,i}(j) = 0.5j + 4$ and we have $l_{\text{eff}}^i = 12$. Then, with $l_o = l_{\text{eff}}^i = 12$, we obtain the free distance $d_{\text{free}}^{\text{WCC}} = 15$. The rate of the overall code is $R = R_i R_o = \frac{1}{3}$.

4.6.3 Woven Turbo Codes

We will now analyse the free distance of a woven turbo code. In Section 4.6.2 we have used the concept of generating tuples and the fact that any non-zero outer code sequence in a woven convolutional encoder has at least weight d_{free}^o.

With woven turbo codes, however, the number of non-zero bits encoded by the inner encoder will be smaller than d_{free}^o. We will now define a distance measure that enables us to estimate the free distance in the case of partitioned outer code sequences. We call this distance measure the *partial distance*.

The partitioning matrix $\mathbf{P}$ determines how a code sequence $\mathbf{b}$ is partitioned into the two sequences $\mathbf{b}^{(1)}$ and $\mathbf{b}^{(2)}$. Note that, using the notion of a burst, we could define the free distance as the minimum weight of a burst. Analogously, we define the partial distance as the minimum weight of partial code sequence $\mathbf{b}_{[0,j]}^{(2)}$. However, we condition this value on the weight of the partial code sequence $\mathbf{b}_{[0,j]}^{(1)}$. The formal definition is given in Figure 4.32.

Partial distance and free distance of a WTC

- Let $\mathbf{b}_{[0,j]}$ denote a burst of length $j+1$.
- Considering all possible bursts, we define the partial distance with respect to a partitioning matrix $\mathbf{P}$ as the minimum weight of the partial code sequence $\mathbf{b}^{(2)}_{[0,j]}$, where we fix the weight of the partial code sequence $\mathbf{b}^{(1)}_{[0,j]}$

$$d_p(w) = \min_{\mathbf{b}_{[0,j]}, \text{wt}(\mathbf{b}^{(1)}_{[0,j]}) = w} \left\{ \text{wt}\left(\mathbf{b}^{(2)}_{[0,j]}\right) \right\} \qquad (4.22)$$

- The free distance of the WTC with $l_o \geq l^i_{\text{eff}}$ outer codes satisfies

$$d^{\text{WTC}}_{\text{free}} \geq \min_w \left\{ w \cdot d^i_{\text{free}} + d^o_p(w) \right\} \qquad (4.23)$$

Figure 4.32: Definition of the partial distance and lower bound on the free distance of a woven turbo code (WTC)

For instance, the rate $R = 1/2$ convolutional code with generator matrix $\mathbf{G}(D) = (1 + D + D^2, 1 + D^2)$ has free distance $d_{\text{free}} = 5$. With partitioning matrix

$$\mathbf{P} = \begin{pmatrix} 1 \\ 0 \end{pmatrix}$$

we obtain the partial distances $d_p(w) = 4, 2, 2, 2, 2, \ldots$ with $w = 2, 3, 4, \ldots$. Note that $d_p(w = 1)$ does not exist. With partitioning matrix

$$\mathbf{P}' = \begin{pmatrix} 0 \\ 1 \end{pmatrix}$$

we obtain the partial distances $d'_p(w) = 3, 2, 2, 2, 2, \ldots$ with $w = 2, 4, 6, \ldots$. These partial distances exist only for even values of w.

Using the partial distance, we can now bound the free distance of a WTC. Owing to the linearity of the considered codes, the free distance of the woven turbo code is given by the minimal weight of all possible overall code sequences $\mathbf{b}$, except the all-zero sequence. If $l_o \geq l^i_{\text{eff}}$ holds and one of the outer partial code sequences $\mathbf{b}^{o,(1)}_l$ has weight w, then there exist at least w generating tuples in the inner information sequence.

Thus, with inequality (4.19) we have $\text{wt}(\mathbf{b}) \geq w \cdot d^i_{\text{free}}$. However, the weight of the corresponding partial code sequence $\mathbf{b}^{o,(2)}_l$ is at least $d^o_p(w)$, where $d^o_p(w)$ denotes the partial distance of the outer code. Hence, we have $\text{wt}(\mathbf{b}) \geq w \cdot d^i_{\text{free}} + d^o_p(w)$. Consequently, the free distance of the WTC with $l_o \geq l^i_{\text{eff}}$ outer codes satisfies

$$d^{\text{WTC}}_{\text{free}} \geq \min_w \left\{ w \cdot d^i_{\text{free}} + d^o_p(w) \right\},$$

TURBO CODES

where d_{free}^i is the free distance of the inner code and l_{eff}^i denotes the effective length of the inner encoder according to Equation (4.20).

For instance, we construct a WTC with equal inner and outer rate $R^i = R^o = 1/2$ codes with generator matrices $\mathbf{G}^i(D) = \mathbf{G}^o(D) = (1, \frac{1+D^2}{1+D+D^2})$ similar to the examples given in Figure 4.16. After outer encoding, we apply the partitioning matrix $\mathbf{P}$ as given in the last example. Then, the overall rate is $R^{\text{WTC}} = 1/3$. The lower bound on the active burst distance of the inner encoder is given by $\tilde{a}^{b,i}(j) = 0.5j + 4$ and we have $l_{\text{eff}}^i = 12$.

Thus, with $l_o \geq 12$ outer codes we obtain an overall code with free distance $d_{\text{free}}^{\text{WTC}} \geq 14$ according to inequality (4.23). Note that partitioning of the outer code sequences according to $\mathbf{P}'$ would lead to $d_{\text{free}}^{\text{WTC}} \geq 13$. For an ordinary rate $R = 1/3$ turbo code with the same inner and outer generator matrices as given above, and with a randomly chosen block interleaver, we can only guarantee a minimum Hamming distance $d \geq 7$.

We should note that the results from inequalities (4.21) and (4.23) also hold when we employ row-wise interleaving and terminate the component codes. In this case, the overall code is a block code and the bounds become bounds for the overall minimum Hamming distance.

How close are those bounds? Up to now, no tight upper bounds on the minimum or free distance have been known. However, we can estimate the actual minimum Hamming distance of a particular code construction by searching for low-weight code words. For this purpose we randomly generate information sequences of weight 1, 2, and 3. Then we encode these sequences and calculate the weight of the corresponding code words. It is obvious that the actual minimum distance can be upper bounded by the lowest weight found.

In Figure 4.33 we depict the results of such a simulation. This histogram shows the distribution of the lowest weights, where we have investigated 1000 different random interleavers per construction. All constructions are rate $R = 1/3$ codes with memory $m = 2$ component codes, as in the examples above. The number of information symbols was $K = 100$. Note that for almost 90 percent of the considered turbo codes (TC) we get $d^{\text{TC}} \leq 10$, and we even found one example where the lowest weight was equal to the lower bound 7.

However, with woven turbo codes with row-wise interleaving the lowest weight found was 15, and therefore $d^{\text{WTC}} \leq 15$. For the serial concatenated code (SCC) with random interleaving and for the WCC with row-wise interleaving, the smallest lowest weights were $d^{\text{SCC}} \leq 7$ and $d^{\text{WCC}} \leq 17$ respectively. For $K = 1000$ we have investigated 100 different interleavers per construction and obtained $d^{\text{WTC}} \leq 18$, $d^{\text{TC}} \leq 10$, $d^{\text{WCC}} \leq 17$ and $d^{\text{SCC}} \leq 9$. Thus, for these examples we observe only small differences between analytical lower bounds and the upper bounds obtained by simulation.

In Figure 4.34 we give some simulation results for the AWGN channel with binary phase shift keying. We employ the iterative decoding procedure discussed in Section 4.3.5. For the decoding of the component codes we use the suboptimum symbol-by-symbol a-posteriori probability algorithm from Section 3.5.2. All results are obtained for ten iterations of iterative decoding. We compare the performance of a rate $R = 1/3$ turbo code and woven turbo code as discussed above. Both codes have dimension $K = 1000$.

Because of the row-wise interleaving, the WTC construction possesses less randomness than a turbo code. Nevertheless, this row-wise interleaving guarantees a higher minimum Hamming distance. We observe from Figure 4.34 that both constructions have the same

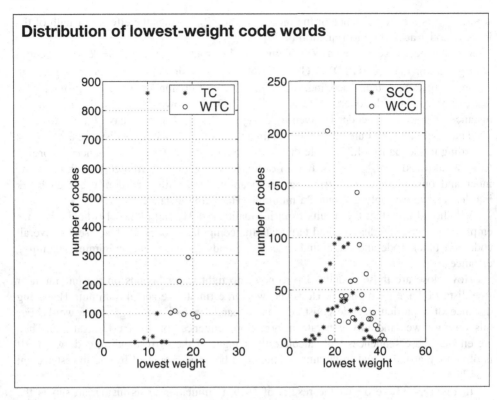

Figure 4.33: Distribution of lowest-weight code words for different rate $R = 1/3$ code constructions

performance at low and moderate Bit Error Rate (BER). Owing to the higher minimum distance, the performance of the WTC becomes better at high SNR. Furthermore, the WTC outperforms the Turbo Code (TC) for the whole considered region of WER and achieves significantly better performance at high SNR.

4.6.4 Interleaver Design

Now we apply the concept of generating tuples to serially concatenated codes as introduced in Section 4.3.3. A serially concatenated encoder consists of a cascade of an outer encoder, an interleaver and an inner encoder, where we have assumed that the interleaver is randomly chosen. The aim of this section is an interleaver design for this serial construction that guarantees a minimum Hamming distance similar to that of woven codes.

Let us first consider a random interleaver. The minimum Hamming distance of the concatenated code is given by the minimal weight of all possible inner code sequences, except the all-zero sequence. A non-zero outer code sequence has at least weight d_{free}^o. Owing to the random structure of the interleaver, those d_{free}^o non-zero bits may occur in direct sequence after interleaving and therefore may be encoded as a burst of length $\lceil d_{\text{free}}^o / k^i \rceil$.

TURBO CODES

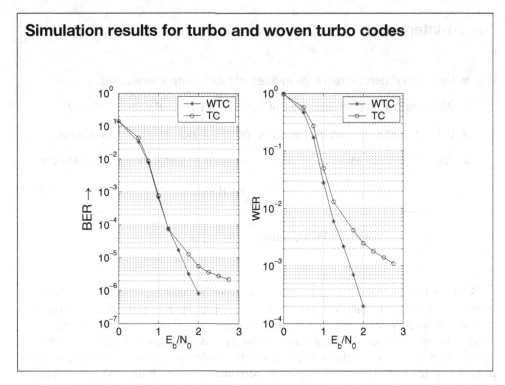

Figure 4.34: Simulation results for rate $R = 1/3$ turbo and woven turbo codes for the AWGN channel

Such a burst has at least weight $a^{b,i}(\lceil d^o_{\text{free}}/k^i \rceil - 1)$, where $a^{b,i}(\cdot)$ is the active burst distance of the inner code. However, the weight of the inner code sequence cannot be less then d^i_{free}. We conclude that the minimum Hamming distance of the Serially Concatenated Code (SCC) with a randomly chosen interleaver satisfies the inequality

$$d^{\text{SCC}} \geq \max \left\{ d^i_{\text{free}}, a^{b,i} \left(\left\lceil \frac{d^o_{\text{free}}}{k^i} \right\rceil - 1 \right) \right\}.$$

Of course, we have considered a worst-case scenario, and other interleavers may lead to a much higher minimum Hamming distance. For the rate $R = 1/3$ SCC with memory $m = 2$ component codes as discussed in the previous section, this results in the estimate $d^{\text{SCC}} \geq 7$. However, we have seen in Figure 4.33 that it is actually not too difficult to find interleavers that lead to such a low minimum Hamming distance.

Nevertheless, the above worst-case considerations lead to a criterion for *good* interleavers with respect to the overall minimum distance. Obviously, the bits of an outer code sequence should be separated after interleaving. Formally, we can specify this characteristic of the interleaver as in the definition of Figure 4.35. This criterion guarantees that an (l_1, l_2)-interleaver separates bits that are in close proximity before the interleaving by at least l_2 positions after the interleaving.

> **(l_1, l_2)-interleaver**
>
> - Let t and t' denote two time indices of bits before interleaving.
> - We consider all possible pairs of indices $(t, t'), t \neq t'$ with $|t - t'| < l_1$.
> - Let $\pi(t)$ and $\pi(t')$ denote the corresponding indices after interleaving.
> - We call an interleaver (l_1, l_2)-*interleaver* if for all such pairs (t, t') it satisfies
> $$|\pi(t) - \pi(t')| \geq l_2 \qquad (4.24)$$

Figure 4.35: Definition of (l_1, l_2)-interleaving

Such an (l_1, l_2)-interleaver can be realised by an ordinary block interleaver. A block interleaver is a two-dimensional array where we write the bits into the interleaver in a row-wise manner, whereas we read the bits column-wise. However, such a simple block interleaving usually leads to a significant performance degradation with iterative decoding. To introduce some more randomness in the interleaver design, we can permute the symbols in each row randomly, and with a different permutation for each row. When we read the bits from the interleaver, we still read them column-wise, but we first read the bits from the even-numbered rows, then from the odd-numbered rows. This still achieves an (l_1, l_2)-structure. Such an interleaver is called an *odd/even interleaver* and was first introduced for turbo codes. The turbo coding scheme in the UMTS standard (see Section 4.3.2) uses a similar approach.

A more sophisticated approach uses search algorithms for pseudorandom permutations, guaranteeing that the conditions in Figure 4.35 are fulfilled (Hübner and Jordan, 2006; Hübner et al., 2004). We will not discuss the particular construction of an (l_1, l_2)-interleaver, but only investigate whether such an interleaver ensures a minimum Hamming distance that is at least the product of the free distances of the component codes. This result was first published by Freudenberger et al. (Freudenberger et al. 2000b; see also Freudenberger et al. 2001).

But how do we choose the parameters l_1 and l_2 in order to achieve a larger minimum Hamming distance? In the case of l_2 the answer follows from the investigations in the previous sections. In order to bound the minimum Hamming distance of the concatenated code, we are again looking for the minimum weight sequence among all possible inner code sequences. If the outer code sequence has only weight d_{free}^o, those successive bits in the outer code sequence should be sufficiently interleaved to belong to independent generating tuples. Hence, we require $l_2 = l_{\text{eff}}^i$, where l_{eff}^i is the effective length of the inner encoder.

The parameter l_1 has to consider the distribution of the code bits in the outer code sequence. Again, we can use the active distances to determine this parameter. In Figure 4.36, we define the *minimum length* of a convolutional encoder on the basis of the active column distance and the active reverse column distance (see Section 3.3.2). Both distance measures

TURBO CODES

> **Minimum length**
>
> - Consider a rate $R = k/n$ convolutional encoder and its active distances.
> - Let j_c denote the minimum j for which $a^c(j) \geq d_{\text{free}}$ holds
>
> $$j_c = \operatorname*{argmin}_{j}\{a^c(j) \geq d_{\text{free}}\} \qquad (4.25)$$
>
> - Let j_{rc} denote the minimum j for which $a^{\text{rc}}(j) \geq d_{\text{free}}$ holds
>
> $$j_{\text{rc}} = \operatorname*{argmin}_{j}\{a^{\text{rc}}(j) \geq d_{\text{free}}\} \qquad (4.26)$$
>
> - We define the *minimum length* as
>
> $$l_{\min} = \min\{n(j_c + 1), n(j_{\text{rc}} + 1)\} \qquad (4.27)$$

Figure 4.36: Minimum length

consider the weight growth of a burst with increasing segment length. The column distance considers the weight growth in the direction from the start to the end of the burst, while the reverse column distance regards the opposite direction.

The *minimum length* is an estimate of the positions of the 1s in a code sequence. A burst may have d_{free} or more non-zero bits. However, when we consider a span of $l_{\min}$ positions at the start or end of the burst, this span includes at least d_{free} 1s.

Let us summarise these two results. The definition of the minimum length ensures that we have to consider at most $l_{\min}^o$ code bits of a burst to obtain an outer code segment with at least d_{free}^o non-zero bits. The definition of the effective length guarantees that those bits in the outer code sequence are sufficiently interleaved to belong to independent generating tuples.

Consequently, using an (l_1, l_2)-block interleaver with $l_1 \geq l_{\min}^o$ and $l_2 \geq l_{\text{eff}}^i$, there exist at least d_{free}^o generating tuples in each non-zero input sequence to the inner encoder. With the results from Figure 4.30, it follows that the minimum Hamming distance of the SCC with an (l_1, l_2)-block interleaver with $l_1 \geq l_{\min}^o$ and $l_2 \geq l_{\text{eff}}^i$ satisfies the inequality

$$d^{\text{SCC}} \geq d_{\text{free}}^o d_{\text{free}}^i.$$

It should be clear from this discussion that the concept of (l_1, l_2)-interleaving can also be applied to partially concatenated codes as introduced in Section 4.3.4. In this case we obtain the same bound as for woven turbo codes in Figure 4.32.

Furthermore, we should note that the concept of designing interleavers on the basis of the active distances is not limited to the product distance. Freudenberger *et al.* presented an interleaver design for woven codes with row-wise interleaving that resulted in a minimum

Hamming distance of about twice the product of the distances of the component codes (Freudenberger *et al.*, 2001). A similar result was obtained by Hübner and Richter, but with a design that enabled much smaller interleaver sizes (Hübner and Richter, 2006).

4.7 Summary

There exist numerous possible code constructions for turbo-like concatenated convolutional codes. Most of these constructions can be classified into the two major classes of parallel (Berrou *et al.*, 1993) and serially (Benedetto and Montorsi, 1998) concatenated codes. However, other classes such as multiple concatenations (Divsalar and Pollara, 1995) and hybrid constructions (Divsalar and McEliece, 1998) are known, i.e. combinations of parallel and serial concatenations.

Therefore, we have concentrated our discussion on some, as we think, interesting code classes. In Section 4.3.4 we introduced the concept of partial concatenation (Freudenberger *et al.*, 2004). Partially concatenated convolutional codes are based on the idea of partitioning the code sequences of the outer codes in a concatenated coding system. Partially concatenated convolutional codes provide a general framework to investigate concatenated convolutional codes. For example, parallel and serially concatenated convolutional codes can be regarded as special cases of this construction.

The concept of partial concatenation was first introduced (Freudenberger *et al.*, 2001) in connection with woven turbo codes which belong to the general class of woven convolutional codes. The woven code construction was first introduced and investigated by Höst, Johannesson and Zyablov (Höst *et al.*, 1997). A series of papers on the asymptotic behaviour of WCC show their distance properties (Zyablov *et al.*, 1999a) and error-correcting capabilities (Zyablov *et al.*, 1999b, 2001). The characteristics of woven codes were further investigated (Freudenberger *et al.*, 2001; Höst, 1999; Höst *et al.*, 2002, 1998; Jordan *et al.*, 2004a).

In the context of turbo codes, the idea of looking at code ensembles rather than individual codes was introduced by Benedetto and Montorsi. Methods for estimating the weight distribution of turbo codes and serial concatenations with randomly chosen interleavers were presented (Benedetto and Montorsi, 1996; Perez *et al.*, 1996; Benedetto and Montorsi, 1998). These weight distributions can be used for bounding the average maximum likelihood performance of the considered code ensemble. Based on this technique, Benedetto and Montorsi derived design rules for turbo codes as well as for serially concatenated codes.

The analysis of the iterative decoding algorithm is the key to understanding the remarkably good performance of LDPC and turbo-like codes. The first analysis for a special type of belief propagation was made by Luby *et al.* (Luby *et al.*, 1998). This analysis was applied to hard decision decoding of LDPC codes (Luby *et al.*, 2001) and generalised to belief propagation over a large class of channels (Richardson and Urbanke, 2001).

The analysis for turbo-like codes was pioneered by ten Brink (ten Brink, 2000). In Section 4.4 we considered the analysis of the iterative decoding algorithm for concatenated convolutional codes. Therefore, we utilised the extrinsic information transfer characteristics as proposed by ten Brink. The idea is to predict the behaviour of the iterative decoder by looking at the input/output relations of the individual constituent decoders.

In Section 4.6 the minimum Hamming distance was the important criterion for the code search and code construction. We derived lower bounds on the minimum Hamming distance for different concatenated code constructions. Restricting the interleaver to the class of (l_1, l_2)-permutations made it possible to improve the lower bounds.

Naturally, the reader will be interested in design guidelines to construct *good* concatenated convolutional codes. However, giving design guidelines remains a subtle task, because none of the three considered methods for analysing the code properties gives a complete picture of the code performance with iterative decoding.

A possible approach could be based on the results of Section 4.5, where we considered the expected weight distribution. However, bounds on performance based on the expected weight distribution usually assume maximum likelihood decoding.

EXIT charts are a suitable method for investigating the convergence behaviour of the iterative decoding for long codes. For codes of moderate length it becomes important to take the interleaving depth into account.

With respect to the minimum Hamming distance of the concatenated code, especially for codes of short length, the choice of the particular interleaver is very important. The use of designed interleavers is motivated by the asymptotic coding gain. In order to ensure efficient performance for high signal-to-noise ratios, the concatenated code should have a large minimum Hamming distance, which motivates the use of designed interleavers.

Considering the convergence behaviour of the iterative decoding or the overall distance spectrum, a connection with the active distances or partial distances is not obvious. Anyhow, we would like to note that, in all presented examples, optimising the slope of the active distances or optimising the partial distances led to the best results. Tables of convolutional codes and of punctured convolutional codes with good active distances can be found elsewhere (Jordan *et al.*, 2004b).

Certainly, it would be desirable to predict the absolute code performance with suboptimum iterative decoding. First results on the finite-length analysis of iterative coding systems are available (Di *et al.*, 2002). However, these results are at present limited to the class of low-density parity-check codes and to the binary erasure channel.

5

Space–Time Codes

5.1 Introduction

Today's communication systems approach more and more the capacity limits predicted by Shannon. Sophisticated coding and signal processing techniques exploit a physical radio link very efficiently. Basically, the capacity is limited by the bandwidth restriction. In the course of the deployment of new services and applications, especially multimedia services, the demand of high data rates is currently increasing and will continue in the near future. Since bandwidth is an expensive resource, network providers are looking for efficient opportunities to increase the system capacity without requiring a larger portion of the spectrum. In this context, multiple antennas represent an efficient opportunity to increase the spectral efficiency of wireless communication systems by exploiting the resource space. They became popular a decade ago as a result of fundamental work (Alamouti, 1998; Foschini, 1996; Foschini and Gans, 1998; Kühn and Kammeyer, 2004; Seshadri and Winters, 1994; Seshadri et al., 1997; Tarokh et al., 1998; Telatar, 1995; Wittneben, 1991; Wolniansky et al., 1998; and many others). After initial research activities, their large potential has been widely recognised, so that they have been incorporated into several standards. As an example, very simple structures can already been found in Release 99 of Universal Mobile Telecommunications System (UMTS) systems (Holma and Toskala, 2004). More sophisticated methods are in discussion for further evolutions (3GPP, 2007; Hanzo et al., 2002).

Basically, two different categories of how to use multiple antennas can be distinguished. In the first category, the link reliability is improved by increasing the instantaneous signal-to-noise ratio (SNR) or by reducing the variations in the fading SNR. This leads to lower outage probabilities. The first goal can be accomplished by beamforming, i.e. the main lobe of the antenna pattern steers in distinct directions. A reduced SNR variance is obtained by increasing the system's diversity degree, e.g. by space–time coding concepts. Moreover, spatially separable interferers can be suppressed with multiple antennas, resulting in a signal-to-interference-plus-noise-ratio (SINR).

In the second category of multiple-input multiple-output (MIMO) techniques, the data rate is multiplied by transmitting several independent data streams simultaneously, one

---

Coding Theory – Algorithms, Architectures, and Applications André Neubauer, Jürgen Freudenberger, Volker Kühn
© 2007 John Wiley & Sons, Ltd

over each antenna. If all data streams belong to a single user, this approach is denoted as Space Division Multiplexing (SDM), otherwise it is termed Space Division Multiple Access (SDMA). It will be shown in Section 5.3 that the potential capacity gain of multiple-antenna systems is much larger than the gain obtained by simply increasing the transmit power. Certainly, diversity and multiplexing techniques do not exclude each other but can be combined. An interesting example are multistratum codes (Böhnke et al., 2004a,b,c).

This chapter introduces two MIMO strategies, orthogonal space–time block codes and spatial multiplexing, and is therefore not comprehensive. Well-known coding techniques such as space–time trellis codes (Bäro et al., 2000a,b; Naguib et al., 1997, 1998; Seshadri et al., 1997; Tarokh et al., 1997, 1998) and non-orthogonal space–time block codes (Bossert et al., 2000, 2002; Gabidulin et al., 2000; Lusina et al., 2001, 2003, 2002) are not considered here. Moreover, detection strategies such as the sphere detector (Agrell et al., 2002; Fincke and Pohst, 1985; Schnoor and Euchner, 1994) achieving maximum likelihood performance or lattice-reduction-based approaches (Kühn, 2006; Windpassinger and Fischer, 2003a,b; Wübben, 2006; Wübben et al., 2004a,b) will not be presented. A comprehensive overview of space–time coding is available elsewhere (Liew and Hanzo, 2002).

The first four chapters introduced coding and decoding techniques of error-correcting codes. In order to focus on the main topic, they simplified the communication system very much and hid all system components that were not actually needed for the coding itself in a channel model with appropriate statistics. This is not possible for multiple-antenna techniques because they directly influence the channel model. Although we try to keep the system as simple as possible, some more components are required.

Therefore, this introduction contains two sections describing briefly digital modulation schemes and the principle of diversity. Section 5.2 extends the scalar channel used in previous chapters to a MIMO channel with multiple inputs and outputs. Some information about standardisation issues for MIMO channels are also presented. Section 5.3 derives different performance measures for MIMO transmission strategies. In Section 5.4, orthogonal space–time block codes are introduced. Owing to their simplicity, they have already found their way into existing mobile radio standards. Section 5.5 explains spatial multiplexing, and Section 5.6 gives a short overview of currently discussed MIMO techniques for UMTS.

5.1.1 Digital Modulation Schemes

Before we start to describe MIMO channels and specific space–time coding techniques, we will review linear modulation schemes and the principle of diversity. We will abandon a detailed analysis and refer instead to the rich literature (Benedetto and Biglieri, 1999; Kammeyer, 2004; Proakis, 2001; Schulze and Lüders, 2005; Sklar, 2003). According to Figure 5.1, the modulator located at the transmitter maps an m-bit tuple

$$\mathbf{b}[\ell] = \big(b_1[\ell], \ldots, b_m[\ell]\big)^{\mathrm{T}} = \big(\tilde{b}[i], \ldots, \tilde{b}[i+m-1]\big)^{\mathrm{T}} \quad \text{with} \quad \ell = \left\lfloor \frac{i}{m} \right\rfloor$$

onto one of $M = 2^m$ possible symbols $S_\mu \in \mathbb{S}$, where $M = |\mathbb{S}|$ denotes the size of the alphabet $\mathbb{S}$. The average symbol energy is given in Equation (5.1), where the last equality holds if all symbols and, hence, all bit tuples are equally likely.

The mapping $\mathcal{M}$ of $\mathbf{b}[\ell]$ onto $s[\ell] = \mathcal{M}(\mathbf{b}[\ell])$ can be performed in different ways. While the symbol error rate only depends on the geometrical arrangement of the symbols

SPACE–TIME CODES

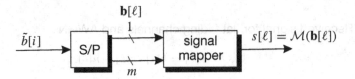

- m-bit tuple $\mathbf{b}[\ell]$ obtained from serial-to-parallel conversion.
- m-bit tuple is mapped onto one of $M = 2^m$ symbols $S_\mu \in \$$.
- Mapping strategies:
  - Gray mapping: neighbouring symbols differ only in a single bit
  - Natural mapping: counting symbols counterclockwise
  - Anti-Gray mapping: neighbouring symbols differ in many bits
- Average symbol energy for equiprobable symbols S_μ

$$E_s = T_s \cdot \mathrm{E}\{|S_\mu|^2\} = T_s \cdot \sum_{\mu=0}^{M-1} \Pr\{S_\mu\} \cdot |S_\mu|^2 \stackrel{\text{i.i.d.}}{=} \frac{T_s}{M} \cdot \sum_{\mu=0}^{M-1} |S_\mu|^2 \qquad (5.1)$$

Figure 5.1: Principles of linear digital modulation

as well as the signal-to-noise ratio, the bit error rate is also affected by the specific mapping of the m-tuples onto the symbols S_μ. In uncoded systems, Gray mapping delivers the lowest bit error probability because neighbouring symbols differ only in one bit, leading to single-bit errors when adjacent symbols are mixed up. In the context of concatenated systems with turbo detection, different strategies such as anti-Gray mapping should be preferred because they ensure a better convergence of the iterative detection scheme (Sezgin et al., 2003).

At the receiver, the demodulator has to deliver an estimate for each bit in $\mathbf{b}[\ell]$ on the basis of the received symbol $r[\ell] = h[\ell]s[\ell] + n[\ell]$. The factor $h[\ell]$ represents the complex-valued channel coefficient which is assumed to be perfectly known or estimated at the receiver. A look at Figure 5.2 shows two different approaches. The ML symbol detector chooses the hypothesis $\hat{s}[\ell]$ that minimizes the smallest squared Euclidean distance between $h[\ell]\hat{s}[\ell]$ and the received symbol $r[\ell]$. It therefore performs a hard decision, and the bits $b_\mu[\ell]$ are obtained by the inverse mapping procedure $\hat{\mathbf{b}}[\ell] = \mathcal{M}^{-1}(\hat{s}[\ell])$. In practical implementations, thresholds are defined between adjacent symbols and a quantisation with respect to these thresholds delivers the estimates $\hat{s}[\ell]$.

Especially in concatenated systems, hard decisions are not desired because subsequent decoding stages may require reliability information, e.g. log-likelihood values, on the bit level. In these cases, the best way is to apply the bit-by-bit Maximum A-Posteriori (MAP) detector which delivers an LLR for each bit $b_\mu[\ell]$ in $\mathbf{b}[\ell]$ according to Equation (5.3).

Digital demodulation

- Received signal for flat fading channels and AWGN

$$r[\ell] = h[\ell] \cdot s[\ell] + n[\ell].$$

- Maximum Likelihood (ML) symbol detector decides in favour of the symbol S_μ closest to $r[\ell]$

$$\hat{s}[\ell] = \underset{\tilde{s}}{\operatorname{argmin}} \left| r[\ell] - h[\ell] \cdot \tilde{s} \right|^2 \quad \text{and} \quad \hat{\mathbf{b}}[\ell] = \mathcal{M}^{-1}(\hat{s}[\ell]) \qquad (5.2)$$

- A-Posteriori Probability (APP) bit detector delivers a Log-Likelihood Ratio (LLR) for each bit in $\mathbf{b}[\ell]$

$$L(b_\mu[\ell]) = \ln \frac{\Pr\{b_\mu = 0 \mid r[\ell]\}}{\Pr\{b_\mu = 1 \mid r[\ell]\}} = \ln \frac{\sum_{\tilde{s} \in \mathbb{S}, b_\mu=0} p(r[\ell] \mid \tilde{s}) \cdot \Pr\{\tilde{s}\}}{\sum_{\tilde{s} \in \mathbb{S}, b_\mu=1} p(r[\ell] \mid \tilde{s}) \cdot \Pr\{\tilde{s}\}} \qquad (5.3)$$

Figure 5.2: Principles of digital demodulation

The signal alphabet $\mathbb{S} = \{S_0, \ldots, S_{M-1}\}$ strongly depends on the type of modulation. Generally, amplitude, phase and frequency modulation are distinguished. Since we are confining ourselves in this chapter to linear schemes, only the first two techniques will be considered.

Amplitude Shift Keying (ASK)

ASK is a modulation scheme that maps the information onto the amplitude of real-valued symbols. Two examples, a binary signalling scheme (2-ASK) and 4-ASK, are depicted in the upper row of Figure 5.3. Generally, the amplitudes are chosen from a signal alphabet

$$\mathbb{S} = \left\{ (2\mu + 1 - M) \cdot e \mid 0 \leq \mu < M \right\}$$

consisting of M odd multiples of a constant e. The parameter e is adjusted such that the energy constraint in Equation (5.1) is fulfilled. For equally likely symbols, the result in Equation (5.4) is obtained, leading to the minimum normalised squared Euclidean distance given in Equation (5.5).

Quadrature Amplitude Modulation (QAM)

QAM is deployed in many modern communication systems such as Wireless Local Area Network (WLAN) systems (ETSI, 2001; Hanzo et al., 2000) and the High-Speed Downlink Packet Access (HSDPA) in UMTS (Holma and Toskala, 2004). It differs from the

SPACE-TIME CODES

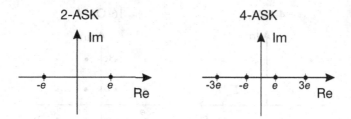

- Energy normalisation

$$\frac{T_s}{M} \sum_{\mu=0}^{M-1} [(2\mu + 1 - M)e]^2 \stackrel{!}{=} E_s \quad \Rightarrow \quad e = \sqrt{\frac{3}{M^2 - 1} \cdot \frac{E_s}{T_s}} \quad (5.4)$$

- Normalised minimum squared Euclidean distance

$$\Delta_0^2 = \frac{(2e)^2}{E_s/T_s} = \frac{12}{M^2 - 1} \quad (5.5)$$

Figure 5.3: ASK modulation: symbol alphabets for 2-ASK ($e = \sqrt{E_s/T_s}$) and 4-ASK ($e = \sqrt{E_s/T_s/5}$) and main modulation parameters

real-valued ASK in that the symbols' imaginary parts are also used for data transmission. Figure 5.4 shows examples of 4-QAM and 16-QAM. Real and imaginary parts of $s[\ell]$ can be chosen independently from each other so that the number of bits per symbol, and hence the spectral efficiency, is doubled compared with ASK schemes. Furthermore, both real and imaginary parts contribute equally to the total symbol energy, resulting in Equation (5.6). Accordingly, the minimum squared Euclidean distance is given in Equation (5.7).

Phase Shift Keying (PSK)

The Phase Shift Keying arranges all symbols on a circle with radius $\sqrt{E_s/T_s}$, resulting in identical symbol energies. This is an important property, especially in mobile radio communications because the mobile unit has a limited battery lifetime and an energy-efficient power amplifier is essential. They are generally designed for a certain operating point so that an energy-efficient transmission is obtained if the transmitted signal has a constant complex envelope. Within the Global System for Mobile communications (GSM) extension Enhanced Data rates for GSM Evolution (EDGE) (Olofsson and Furuskär, 1998; Schramm et al., 1998), 8-PSK is used in contrast to Gaussian Minimum Key Shifting (GMSK) as in standard GSM systems. This enlarges the data rate

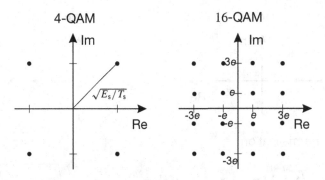

Figure 5.4: QAM modulation: symbol alphabets for 4-QAM ($e = 1$) and 16-QAM ($e = \sqrt{E_s/T_s/10}$) and main modulation parameters

significantly since 3 bits are transmitted per symbol compared with only a single bit for GMSK.

The bits of the m-tuples $\mathbf{b}[\ell]$ determine the symbols' phases which are generally multiples of $2\pi/M$. Alternatively, an offset of π/M can be chosen, as shown in Figure 5.5 for Quaternary Phase Shift Keying (QPSK), 8-PSK and 16-PSK. Binary Phase Shift Keying for $M = 2$ and QPSK for $M = 4$ represent special cases because they are identical to 2-ASK and 4-QAM respectively. For $M > 4$, real and imaginary parts are not independent from each other and have to be detected simultaneously. The normalised minimum squared Euclidean distance is given in Equation (5.8).

Error Rate Performance

Unfortunately, the exact symbol error probability cannot be expressed in closed form for all considered modulation schemes. However, a common tight approximation exists. Assuming that most error events mix up adjacent symbols, the average error probability is dominated by this event. For the Additive White Gaussian Noise (AWGN) channel, we obtain the

SPACE–TIME CODES

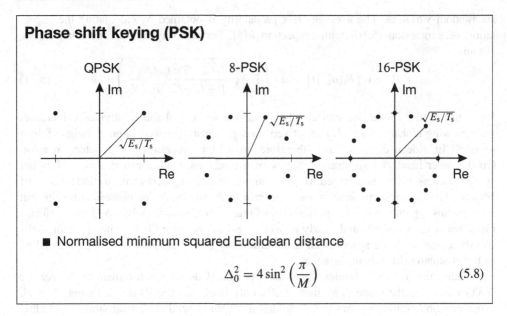

Figure 5.5: PSK modulation: symbol alphabets and main modulation parameters

error probability approximation (Kühn, 2006)

$$P_s \approx a \cdot \mathrm{erfc}\left(\sqrt{\left(\frac{\Delta_0}{2}\right)^2 \frac{E_s}{N_0}}\right) \tag{5.9}$$

In expression (5.9) the parameter a depends on the modulation scheme and amounts to

$$a = \begin{cases} (M-1)/M & \text{for ASK} \\ 2 \cdot (\sqrt{M}-1)/\sqrt{M} & \text{for QAM} \\ 1 & \text{for PSK} \end{cases}.$$

The well-known result for Binary Phase Shift Keying (BPSK)

$$P_b^{\mathrm{BPSK}} = \frac{1}{2} \cdot \mathrm{erfc}\left(\sqrt{\frac{E_s}{N_0}}\right)$$

is obtained by setting $M = 2$ for ASK.

For flat fading channels, the instantaneous signal-to-noise ratio depends on the actual channel coefficient $h[\ell]$. Hence, the SNR, and, consequently, the error probability

$$P_s(h[\ell]) \approx a \cdot \mathrm{erfc}\left(\sqrt{\left(\frac{\Delta_0}{2}\right)^2 |h[\ell]|^2 \frac{E_s}{N_0}}\right) \tag{5.10}$$

are random variables. The average error probability is obtained by calculating the expectation of expression (5.10) with respect to $h[\ell]$. For the Rayleigh fading channel, we obtain

$$P_s = \mathrm{E}\{P_s(h[\ell])\} \approx a \cdot \left(1 - \sqrt{\frac{(\Delta_0/2)^2 E_s/N_0}{1+(\Delta_0/2)^2 E_s/N_0}}\right) \quad (5.11)$$

From the results above, we can conclude that the error probability obviously increases with growing alphabet size M. For a fixed average transmit power, large alphabets $ lead to small Euclidean distances and, therefore, to a high probability of a detection error. On the other hand, the spectral efficiency of a modulation scheme measured in bits per symbol increases if $|$|$ is enlarged. Hence, communication engineers have to find a trade-off between high spectral efficiencies and low error probabilities. Some numerical results that confirm this argumentation are presented in Figure 5.6. Astonishingly, 16-QAM performs much better than 16-PSK and nearly as well as 8-PSK because QAM schemes exploit the two-dimensional signal space much more efficiently than their PSK counterparts, leading to larger squared Euclidean distances.

While the error probabilities show an exponential decay when transmitting over the AWGN channel, the slope is asymptotically only linear for the Rayleigh fading channel. From this observation we can conclude that a reliable uncoded transmission over fading channels requires a much higher transmit power in order to obtain the same error probability as for the white Gaussian noise channel. Instead of increasing the transmit power, appropriate coding represents a powerful alternative.

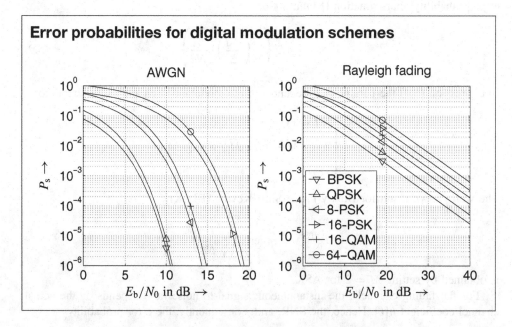

Figure 5.6: Error rate performance for digital modulation schemes:
(a) AWGN channel, (b) flat Rayleigh fading channel

SPACE–TIME CODES

5.1.2 Diversity

The detrimental effect of fading on the error rate performance can be overcome if sources of diversity are available. Having diversity means that a symbol $s[\ell]$ is transmitted over different independent propagation paths. The risk that all available paths experience simultaneously a deep fade is much smaller than the probability that a single channel has a small gain. Therefore, variations in the signal-to-noise ratio owing to fading can be significantly reduced, leading to much smaller error probabilities. If, however, correlations between the propagation paths exist, the potential diversity gain is smaller.

Diversity can originate from different sources which are listed in Figure 5.7. The general principle will be explained for the simple example of receive diversity. More sophisticated techniques are described later in subsequent sections. Let us assume for the moment that a symbol $x[\ell]$ arrives at the receiver via D independent parallel propagation paths with channel coefficients $h_\mu[\ell]$, $1 \leq \mu \leq D$, as depicted in Figure 5.8. These paths may stem from

---

Sources of diversity

- **Frequency diversity**:
  Frequency-selective channels provide frequency diversity. The signal is transmitted over different propagation paths that differ in strength and delay. They are often modelled time varying and statistically independent (uncorrelated scattering assumption). With appropriate receiver structures such as linear FIR filters, the Viterbi equaliser or the Rake receiver for Code Division Multiple Access (CDMA) systems, frequency diversity can be exploited.

- **Time diversity**:
  The application of Forward Error Correction (FEC) coding yields time diversity if the channel varies significantly during one code word or coded frame. In this case, the decoder performs a kind of averaging over good and bad channel states.

- **Space diversity**:
  In this chapter, systems using multiple antennas at transmitter or receiver are deployed to use spatial diversity. If the antenna elements are more than several wavelengths apart from each other, the channels can be assumed independent and diversity is obtained.

- **Polarisation diversity**:
  If antennas support different polarisations, this can be used for polarisation diversity.

Figure 5.7: List of different diversity sources

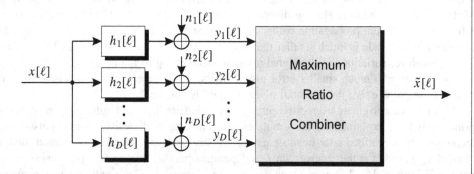

Figure 5.8: Illustration of receive diversity and optimum Maximum Ratio Combining (MRC). Reproduced by permission of John Wiley & Sons, Ltd

the deployment of D receive antennas collecting the emitted signal at different locations. The received samples

$$y_\mu[\ell] = h_\mu[\ell] \cdot x[\ell] + n_\mu[\ell], \quad 1 \leq \mu \leq D \tag{5.15}$$

are disturbed by independent noise contributions $n_\mu[\ell]$ resulting in instantaneous signal-to-noise ratios

$$\gamma_\mu[\ell] = |h_\mu[\ell]|^2 \cdot \frac{E_s}{N_0}$$

with expectations $\bar{\gamma}_\mu = \sigma_\mu^2 E_s/N_0$. The samples have to be appropriately combined. We will confine ourselves in this chapter to the optimal Maximum Ratio Combining (MRC) technique which maximises the signal-to-noise ratio at the combiner's output. A description of further approaches such as equal gain combining, square-law combining and selection combining can be found elsewhere (Kühn, 2006; Simon and Alouini, 2000).

SPACE–TIME CODES

Using vector notations, the received vector $\mathbf{y}[\ell] = (y_1[\ell], \ldots, y_D[\ell])^T$ has the form

$$\mathbf{y}[\ell] = \mathbf{h}[\ell] \cdot x[\ell] + \mathbf{n}[\ell] \tag{5.16}$$

writing the coefficients of the combiner into a vector $\mathbf{w}$ leads to Equation (5.12) and the corresponding SNR in Equation (5.13). Since the product $\mathbf{w}^H \mathbf{h}$ describes the projection of $\mathbf{w}$ onto $\mathbf{h}$, the SNR in Equation (5.13) becomes largest if $\mathbf{w}$ and $\mathbf{h}$ are parallel, e.g. if $\mathbf{w} = \mathbf{h}$ holds. In this case, the received samples $y_\mu[\ell]$ are weighted by the corresponding complex conjugate channel coefficient $h_\mu^*[\ell]$ and combined to the signal $\tilde{x}[\ell]$. This procedure is called MRC and maximises the signal-to-noise ratio (SNR). For independent channel coefficients and noise samples, it amounts to

$$\gamma[\ell] = \|\mathbf{h}[\ell]\|^2 \cdot \frac{E_s}{N_0} = \sum_{\mu=1}^{D} |h_\mu[\ell]|^2 \cdot \frac{E_s}{N_0}.$$

Essentially, the maximum ratio combiner can be interpreted as a matched filter that also maximises the SNR at its output.

If the channel coefficients in all paths are identically Rayleigh distributed with average power $\sigma_\mathcal{H}^2 = 1$, the sum of their squared magnitudes is chi-squared distributed with $2D$ degrees of freedom (Bronstein et al., 2000; Simon and Alouini, 2000)

$$p_{\sum |h_\mu|^2}(\xi) = \frac{\xi^{D-1}}{(D-1)!} \cdot e^{-\xi} \tag{5.17}$$

The achievable gain due to the use of multiple antennas at the receiver is twofold. First, the mean of the new random variable $\sum_\mu |h_\mu|^2$ equals D. Hence, the average SNR is increased by a factor D and amounts to

$$\bar{\gamma} = E\{\gamma[\ell]\} = D \cdot \frac{E_s}{N_0}.$$

This enhancement is termed *array gain* and originates from the fact that the D-fold signal energy is collected by the receive antennas. The second effect is called *diversity gain* and can be illuminated best by normalising $\gamma[\ell]$ to unit mean. Using the relation

$$y = a \cdot x \quad \Rightarrow \quad p_y(y) = \frac{1}{|a|} \cdot p_x\left(\frac{y}{a}\right),$$

we can transform the Probability Density Function (PDF) in Equation (5.18) with $a = E_s/N_0/D$ into that of the normalised SNR in Equation (5.17). Numerical results are shown in Figure 5.9. With growing diversity degree D, the instantaneous signal-to-noise ratio concentrates more and more on E_s/N_0 and the SNR variations become smaller. Since very low signal-to-noise ratios occur less frequently, the average error probability will decrease. For $D \to \infty$, the SNR does not vary any more and the AWGN channel without fading is obtained.

Ergodic Error Probability for MRC

Regarding the average error probability, the solution in expression (5.10) can be reused. First, the squared magnitude $|h[\ell]|^2$ has to be replaced by the sum $\sum_\mu |h_\mu[\ell]|^2$. Second,

SNR distribution for diversity reception

- Probability density function of SNR normalised to unit mean

$$p(\xi) = \left[\frac{D}{E_s/N_0}\right]^D \cdot \frac{\xi^{D-1}}{(D-1)!} \cdot e^{-\xi D/(E_s/N_0)} \qquad (5.18)$$

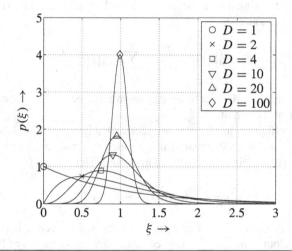

Figure 5.9: Probability density function of SNR for D-fold diversity, Rayleigh distributed coefficients and Maximum Ratio Combining (MRC). Reproduced by permission of John Wiley & Sons, Ltd

the expectation with respect to the chi-square distribution with $2D$ degrees of freedom has to be determined. For BPSK and i.i.d. diversity branches, an exact solution is given in Equation (5.21) (Proakis, 2001). The parameter $\bar{\gamma}$ denotes the average SNR in each branch. A well-known approximation for $\bar{\gamma} \gg 1$ provides a better illustration of the result. In this case, $(1 + \alpha)/2 \approx 1$ holds and the sum in Equation (5.21) becomes

$$\sum_{\ell=0}^{D-1} \binom{D-1+\ell}{\ell} = \binom{2D-1}{D} \qquad (5.19)$$

Furthermore, the Taylor series yields $(1 - \alpha)/2 \approx 1/(4\bar{\gamma})$. With these results, the symbol error rate can be approximated for large signal-to-noise ratios by

$$P_s \approx \left(\frac{1}{4\bar{\gamma}}\right)^D \cdot \binom{2D-1}{D} = (4E_s/N_0)^{-D} \cdot \binom{2D-1}{D} \qquad (5.20)$$

Obviously, P_s is proportional to the Dth power of the reciprocal signal-to-noise ratio. Since error rate curves are scaled logarithmically, their slope will be dominated by the diversity

SPACE–TIME CODES

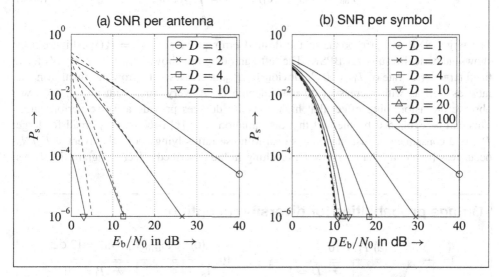

Figure 5.10: Error probability for D-fold diversity, BPSK, Rayleigh distributed coefficients and MRC: **(a)** SNR measured per receive antenna (dashed lines = approximation from expression (5.20)), **(b)** SNR per combined symbol (bold dashed line = AWGN reference)

degree D at high SNRs. This is confirmed by the results depicted in Figure 5.10. In diagram (a) the symbol error probability is plotted versus the signal-to-noise ratio E_s/N_0 per receive antenna. Hence, we observe the array gain as well as the diversity gain. The array gain amounts to 3 dB if D is doubled, e.g. from $D=1$ to $D=2$. The additional gap is caused by the diversity gain. It can be illuminated by the curves' slopes at high SNRs, as indicated by the dashed lines.

In diagram (b), the same error probabilities are plotted versus the SNR after the combiner. Therefore, the array gain is invisible and only the diversity gain can be observed. The gains are largest at high signal-to-noise ratios and if the diversity degree increases from a low level. Asymptotically for $D \to \infty$, the AWGN performance without any fading is reached.

Outage probability

For slowly fading channels, the average error probability is often of minor interest. Instead, operators are particularly interested in the probability that the system will not be able to guarantee a specified target error rate. This probability is called the *outage probability* P_{out}. Since the instantaneous error probability $P_s(\gamma[\ell])$ depends directly on the actual signal-to-noise ratio $\gamma[\ell]$, the probability of an outage event when maximum ratio combining i.i.d. diversity paths is obtained by integrating the distribution in Equation (5.17)

$$P_{\text{out}} = \Pr\{P_s(\gamma[\ell]) > P_t\} = \int_0^{\gamma_t} p(\xi)\,d\xi \qquad (5.22)$$

The target SNR γ_t corresponds to the desired error probability $P_t = P_s(\gamma_t)$. Figure 5.11 shows numerical results for BPSK. The left-hand diagram shows P_{out} versus E_b/N_0 for a fixed target error rate of $P_t = 10^{-3}$. Obviously, P_{out} decreases with growing signal-to-noise ratio as well as with increasing diversity degree D. At very small values of E_s/N_0, we observe the astonishing effect that high diversity degrees provide a worse performance. This can be explained by the fact that the variations in $\gamma[\ell]$ become very small for large D. As a consequence, instantaneous signal-to-noise ratios lying above the average E_s/N_0 occur less frequently than for low D, resulting in the described effect. Diagram (b) shows

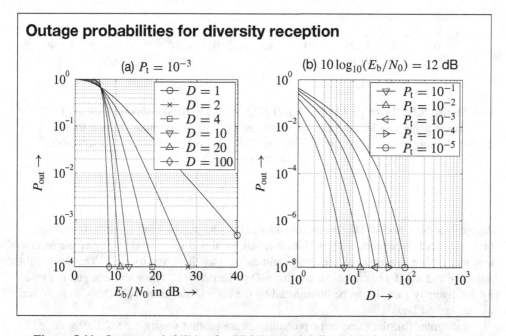

Figure 5.11: Outage probabilities for BPSK and i.i.d. Rayleigh fading channels: (a) a target error rate of $P_t = 10^{-3}$ and (b) $10\log_{10}(E_b/N_0) = 12$ dB. Reproduced by permission of John Wiley & Sons, Ltd

SPACE–TIME CODES

the results for a fixed value $10\log_{10}(E_b/N_0) = 12$ dB. Obviously, diversity can reduce the outage probability remarkably.

5.2 Spatial Channels

5.2.1 Basic Description

Contrary to scalar channels with a single input and a single output introduced in Chapter 1, spatial channels resulting from the deployment of multiple transmit and receive antennas are vector channels with an additional degree of freedom, namely the spatial dimension. A general scenario is depicted in Figure 5.12. It illustrates a two-dimensional view that considers only the azimuth but not the elevation. For the scope of this chapter, this two-dimensional approach is sufficient and will be used in the subsequent description. An extension to a truly three-dimensional spatial model is straightforward and differs only by an additional parameter, the elevation angle.

As shown in Figure 5.12, the azimuth angle θ_R denotes the DoA of an impinging waveform with respect to the orientation of the antenna array. Equivalently, θ_T represents

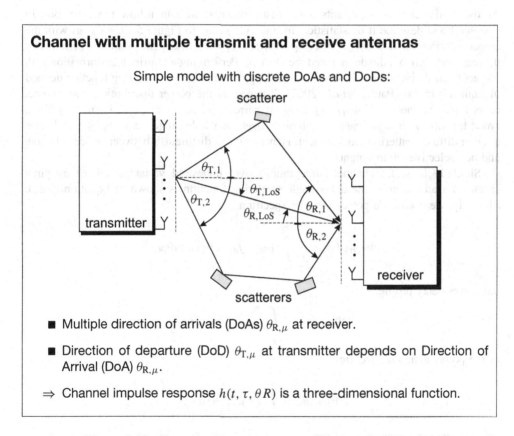

Figure 5.12: Channel with multiple transmit and receive antennas

the DoD of a waveform leaving the transmitting antenna array. Obviously, both angles depend on the orientation of the antenna arrays at transmitter and receiver as well as on the location of the scatterers. For the purpose of this book it is sufficient to presuppose a one-to-one correspondence between θ_R and θ_T. In this case, the DoD is a function of the DoA and the channel impulse response has to be parameterised by only one additional parameter, the azimuth angle θ_R. Hence, the generally time-varying channel impulse response $h(t, \tau)$ known from Chapter 1 is extended by a third parameter, the direction of arrival θ_R.[1] Therefore, the augmented channel impulse response $h(t, \tau, \theta_R)$ bears information about the angular power distribution.

Principally, Line of Sight (LoS) and non-line of sight (NLoS) scenarios are distinguished. In Figure 5.12, an LoS path with azimuth angles $\theta_{T,LoS}$ and $\theta_{R,LoS}$ exists. Those paths are mainly modelled by a Ricean fading process and occur for rural outdoor areas. NLoS scenarios typically occur in indoor environments and urban areas with rich scattering, and the corresponding channel coefficients are generally modelled as Rayleigh fading processes.

Statistical Characterisation

As the spatial channel represents a stochastic process, we can follow the derivation in Chapter 1 and describe it by statistical means. According to Figure 5.13, we start with the autocorrelation function $\phi_{\mathcal{HH}}(\Delta t, \tau, \theta_R)$ of $h(t, \tau, \theta_R)$ which depends on three parameters, the temporal shift Δt, the delay τ and the DoA θ_R. Performing a Fourier transformation with respect to Δt delivers the three-dimensional Doppler delay-angle scattering function defined in Equation (5.24) (Paulraj et al., 2003). It describes the power distribution with respect to its three parameters. If $\Phi_{\mathcal{HH}}(f_d, \tau, \theta_R)$ is narrow, the angular spread is small, while a broad function with significant contributions over the whole range $-\pi < \theta_R \leq \pi$ indicates a rather diffuse scattering environment. Hence, we can distinguish between space-selective and non-selective environments.

Similarly to scalar channels, integrations over undesired variables deliver marginal spectra. As an example, the delay-angle scattering function is shown in Equation (5.25). Similarly, we obtain the power Doppler spectrum

$$\Phi_{\mathcal{HH}}(f_d) = \int_{-\pi}^{\pi} \int_{0}^{\infty} \Phi_{\mathcal{HH}}(f_d, \tau, \theta_R) \, d\tau \, d\theta_R \, ,$$

the power delay profile

$$\Phi_{\mathcal{HH}}(\tau) = \int_{-\pi}^{\pi} \Phi_{\mathcal{HH}}(\tau, \theta_R) \, d\theta_R$$

or the power azimuth spectrum

$$\Phi_{\mathcal{HH}}(\theta) = \int_{0}^{\infty} \Phi_{\mathcal{HH}}(\tau, \theta) \, d\tau \, .$$

---

[1] In order to simplify notation, we will use in this subsection a channel representation assuming that all parameters are continuously distributed.

Statistical description of spatial channels

- Extended channel impulse response $h(t, \tau, \theta_R)$.
- Autocorrelation function of $h(t, \tau, \theta_R)$

$$\phi_{\mathcal{HH}}(\Delta t, \tau, \theta_R) = \int_{-\infty}^{\infty} h(t, \tau, \theta_R) \cdot h^*(t + \Delta t, \tau, \theta_R) \, dt \quad (5.23)$$

- Doppler delay-angle scattering function

$$\Phi_{\mathcal{HH}}(f_d, \tau, \theta_R) = \int_{-\infty}^{\infty} \phi_{\mathcal{HH}}(\xi, \tau, \theta_R) \cdot e^{-j2\pi f_d \xi} \, d\xi \quad (5.24)$$

- Delay-angle scattering function

$$\Phi_{\mathcal{HH}}(\tau, \theta_R) = \int_{-f_{d\,\text{max}}}^{f_{d\,\text{max}}} \Phi_{\mathcal{HH}}(f_d, \tau, \theta_R) df_d \quad (5.25)$$

Figure 5.13: Statistical description of spatial channels

Typical Scenarios

Looking for typical propagation conditions, some classifications can be made. First, outdoor and indoor environments have to be distinguished. While indoor channels are often affected by rich scattering environments with low mobility, the outdoor case shows a larger variety of different propagation conditions. Here, macro-, micro- and picocells in cellular networks show a significantly different behaviour. Besides different cell sizes, these cases also differ as regards the number and spatial distribution of scatterers, mobile users and the position of the base station or access point. At this point, we will not give a comprehensive overview of all possible scenarios but confine ourselves to the description of an exemplary environment and some settings used for standardisation.

A typical scenario of a macrocell environment is depicted in Figure 5.14. Since the uplink is being considered, θ_R denotes the DoA at the base station while θ_T denotes the DoD at the mobile unit. The latter is often assumed to be uniformly surrounded by local scatterers, resulting in a diffuse field of impinging waveforms arriving from nearly all directions. Hence, the angular spread $\Delta\theta_T$ that defines the azimuthal range within which signal paths depart amounts to $\Delta\theta_T = 2\pi$. By contrast, the base station is generally elevated above rooftops. Scatterers are not as close to the base station as for the mobile device and are likely to occur in clusters located at discrete azimuth angles θ_R. Depending on the size and the structure of a cluster, a certain angular spread $\Delta\theta_R \ll 2\pi$ is associated with it. Moreover, the propagation delay τ_i corresponding to

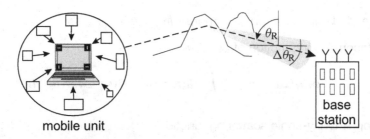

Figure 5.14: Spatial channel properties in a typical macrocell environment

cluster i depends on the length of the path from transmitter over the scatterer towards the receiver.

In order to evaluate competing architectures, multiple access strategies, etc., of future telecommunication systems, their standardisation requires the definition of certain test cases. These cases include a set of channel profiles under which the techniques under consideration have to be evaluated. Figure 5.15 shows a set of channel profiles defined by the third-Generation Partnership Project 3GPP (3GPP, 2003). The Third-Generation Partnership Project (3GPP) is a collaboration agreement between the following telecommunications standards bodies and supports the standardisation process by elaborating agreements on technical proposals:

- ARIB: Association of Radio Industries and Business,
  http://www.arib.or.jp/english/index.html
- CCSA: China Communications Standards Association,
  http://www.ccsa.org.cn/english
- ETSI: European Telecommunications Standards Institute,
  http://www.etsi.org
- ATIS: Alliance for Telecommunications Industry Solutions,
  http://www.atis.org
- TTA: Telecommunications Technology Association,
  http://www.tta.or.kr/e_index.htm

SPACE-TIME CODES

- TTC: Telecommunications Technology Committee,
  http://www.ttc.or.jp/e/index.html

In Figure 5.15, Δd represents the antenna spacing in multiples of the wavelength λ. For the mobile unit, two antennas are generally assumed, while the base station can deploy more antennas. Regarding the mobile, NLoS and LoS scenarios are distinguished for the modified pedestrian A power delay profile. With a line-of-sight (LoS) component, the scattered paths are uniformly distributed in the range $[-\pi, \pi]$ and have a large angular spread. Only the LoS component arrives and departs from a preferred direction. In the absence of an LoS component and for all other power delay profiles, the power azimuth spectrum is assumed to be Laplacian distributed

$$p_\Theta(\theta) = K \cdot e^{-\frac{\sqrt{2}|\theta - \bar{\Theta}|}{\sigma_\Theta}} \cdot G(\theta)$$

with mean $\bar{\Theta}$ and root mean square σ_Θ. The array gain $G(\theta)$ certainly depends on the angle θ, and K is a normalisation constant ensuring that the integral over $p_\Theta(\theta)$ equals unity. Here, the base station array, generally mounted above rooftops, is characterised by small angular spreads and distinct DoDs and DoAs. In the next subsection, it will be shown how to classify and model spatial channels.

3GPP SCM link level parameters (3GPP, 2003)

Model		Case 1	Case 2	Case 3
power delay profile		mod. pedestrian A	vehicular A	pedestrian B
Mobile station	Δd	0.5λ	0.5λ	0.5λ
	θ_R, θ_T	1) LoS on: LoS path 22.5°, rest uniform 2) LoS off: 67.5°, (Laplacian)	67.5°	Odd paths: 22.5°, Even paths: −67.5°
	σ_{θ_R}	1) LoS on: uniform for NLoS paths 2) LoS off: 35°	35° (Laplacian) or uniform	35° (Laplacian)
Base station	Δd	Uniform linear array with 0.5λ, 4λ or 10λ		
	θ_T	50°		
	θ_R	20°		
	$\sigma_{\theta_{T,R}}$	2° for DoD, 5° for DoA		

Figure 5.15: 3GPP SCM link level parameters (3GPP, 2003)

5.2.2 Spatial Channel Models

General Classification

In Chapter 1, channels with a single input and a single output have been introduced. For a frequency-selective channel with L_h taps, their output was described by

$$r[k] = \sum_{\kappa=0}^{L_h-1} h[k,\kappa] \cdot x[k-\kappa] + n[k].$$

In this chapter, we extend the scenario to multiple-input and multiple-output (MIMO) systems as depicted in Figure 5.16. The MIMO channel has N_T inputs represented by the signal vector

$$\mathbf{x}[k] = \begin{pmatrix} x_1[k] & \cdots & x_{N_T}[k] \end{pmatrix}^T$$

and N_R outputs denoted by

$$\mathbf{r}[k] = \begin{pmatrix} r_1[k] & \cdots & r_{N_R}[k] \end{pmatrix}^T.$$

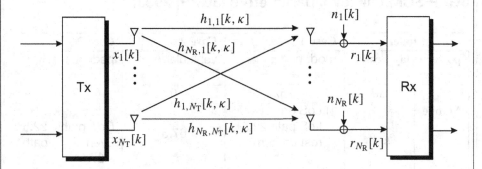

General structure of a frequency-selective MIMO channel

- Received signal at antenna μ

$$r_\mu[k] = \sum_{\nu=1}^{N_T} \sum_{\kappa=0}^{L_h-1} h_{\mu,\nu}[k,\kappa] \cdot x_\nu[k-\kappa] + n_\mu[k] \qquad (5.26)$$

- Entire received signal vector

$$\mathbf{r}[k] = \sum_{\kappa=0}^{L_h-1} \mathbf{H}[k,\kappa] \cdot \mathbf{x}[k-\kappa] + \mathbf{n}[k] = \mathbf{H}[k] \cdot \mathbf{x}_{L_h}[k] + \mathbf{n}[k] \qquad (5.27)$$

Figure 5.16: General structure of a frequency-selective MIMO channel

SPACE–TIME CODES

Each pair (ν, μ) of transmit and receive antennas is connected by a generally frequency-selective single-input single-output channel $h_{\mu,\nu}[k, \kappa]$, the properties of which have already been described in Chapter 1. At each receive antenna, the N_T transmit signals and additive noise superpose, so that the νth output at time instant k can be expressed as in Equation (5.26) in Figure 5.16. The parameter L_h represents the largest number of taps among all contributing channels.

A more compact description is obtained by using vector notations. According to Equation (5.27), the output vector

$$\mathbf{r}[k] = \begin{pmatrix} r_1[k] & \cdots & r_{N_R}[k] \end{pmatrix}^T$$

can be described by the convolution of a sequence of channel matrices

$$\mathbf{H}[k, \kappa] = \begin{pmatrix} h_{1,1}[k, \kappa] & \cdots & h_{1,N_T}[k, \kappa] \\ \vdots & \ddots & \vdots \\ h_{N_R,1}[k, \kappa] & \cdots & h_{N_R,N_T}[k, \kappa] \end{pmatrix}$$

with $0 \leq \kappa < L_h$ and the input vector $\mathbf{x}[k]$ plus the N_R dimensional noise vector $\mathbf{n}[k]$. Each row of the channel matrix $\mathbf{H}[k, \kappa]$ contains the coefficients corresponding to a specific receive antenna, and each column comprises the coefficients of a specific transmit antenna, all for a certain delay κ at time instant k. Arranging all matrices $\mathbf{H}[k, \kappa]$ side by side to an overall channel matrix

$$\mathbf{H}[k] = \begin{pmatrix} \mathbf{H}[k, 0] & \cdots & \mathbf{H}[k, L_h - 1] \end{pmatrix}$$

and stacking all input vectors $\mathbf{x}[k - \kappa]$ on top of each other

$$\mathbf{x}_{L_h}[k] = \begin{pmatrix} \mathbf{x}[k]^T & \cdots & \mathbf{x}[k - L_h - 1]^T \end{pmatrix}^T$$

leads to the expression on the right-handside of Equation (5.27) in Figure 5.16.

The general MIMO scenario of Figure 5.16 contains some special cases. For the frequency-non-selective MIMO case, $\mathbf{H}[k, \kappa] = \mathbf{0}_{N_R \times N_T}$ holds for $\kappa > 0$ and the model simplifies to

$$\mathbf{r}[k] = \mathbf{H}[k] \cdot \mathbf{x}[k] + \mathbf{n}[k]$$

with $\mathbf{H}[k] = \mathbf{H}[k, 0]$. Many space–time coding concepts have been originally designed especially for this flat fading case where the channel does not provide frequency diversity. In Orthogonal Frequency Division Multiplexing (OFDM) systems with multiple transmit and receive antennas, the data symbols are spread over different subcarriers each experiencing a flat fading MIMO channel. Two further cases depicted in Figure 5.17 are obtained if the transmitter or the receiver deploys only a single antenna. The channel matrix of these Single-input Multiple-Output (SIMO) systems reduces in this case to a column vector

$$\mathbf{h}[k, \kappa] = \begin{pmatrix} h_{1,1}[k, \kappa] & \cdots & h_{N_R,1}[k, \kappa] \end{pmatrix}^T .$$

By Contrast, the Multiple-Input Single-Output (MISO) system has multiple inputs but only a single receive antenna and is specified by the row vector

$$\underline{\mathbf{h}}[k, \kappa] = \begin{pmatrix} h_{1,1}[k, \kappa] & \cdots & h_{N_R,1}[k, \kappa] \end{pmatrix} .$$

In order to distinguish between column and row vectors, the latter is underlined. Appropriate transmission strategies for MIMO, SIMO and MISO scenarios are discussed in later sections.

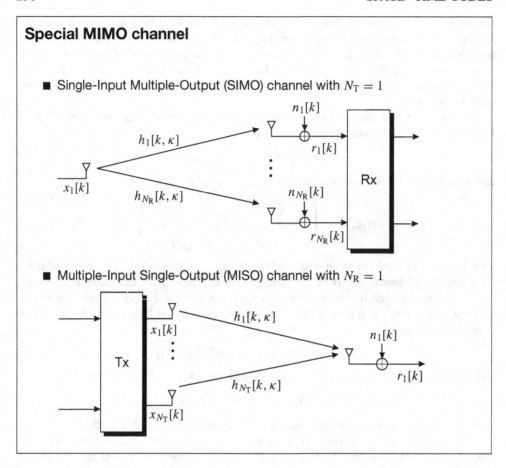

Figure 5.17: Special MIMO channel

Modelling Spatial Channels

In many scientific investigations, the elements in **H** are assumed to be independent and identically Rayleigh distributed (i.i.d.). This case assumes a rich scattering environment without an LoS connection between transmitter and receiver. While this special situation represents an interesting theoretical point of view and will also be used in this chapter for benchmarks, real-world scenarios often look different. In many situations, the channel coefficients in **H** are highly correlated. This correlation may be caused by a small distance between neighbouring antenna elements, which is desired for beamforming purposes. For statistically independent coefficients, this distance has to be much larger than half the wavelength. A second reason for correlations is the existence of only a few dominant scatterers, leading to preferred directions of departure and arrival and small angular spreads.

SPACE–TIME CODES

Modelling spatial channels

- General construction of MIMO channel matrix

$$\mathbf{H}[k,\kappa] = \sum_{\nu} \sum_{\xi=0}^{\kappa_{\max}} h[k,\xi,\theta_{R,\nu}] \cdot \mathbf{a}[k,\theta_{R,\nu}] \cdot \mathbf{b}^T[k,\theta_{T,\mu}] \cdot g[\kappa-\xi] \quad (5.28)$$

- Correlation matrix

$$\Phi_{\mathcal{HH}} = \mathrm{E}\left\{\mathrm{vec}(\mathbf{H})\mathrm{vec}(\mathbf{H})^H\right\} \quad (5.29)$$

- Construction of correlated channel matrix from matrix $\mathbf{H}_w$ with i.i.d. elements

$$\mathrm{vec}(\mathbf{H}) = \Phi_{\mathcal{HH}}^{1/2} \cdot \mathrm{vec}(\mathbf{H}_w) \quad (5.30)$$

- Simplified model with separated transmitter and receiver correlations

$$\mathbf{H} = \Phi_R^{1/2} \cdot \mathbf{H}_w \cdot \Phi_T^{1/2} \quad (5.31)$$

where Φ_R (Φ_T) are assumed to be the same for all transmit (receive) antennas.

- Relationship between $\Phi_{\mathcal{HH}}$, Φ_T and Φ_R

$$\Phi_{\mathcal{HH}} = \Phi_T^T \otimes \Phi_R \quad (5.32)$$

Figure 5.18: Modelling spatial channels

In practice, only a finite number of propagation paths can be considered. Therefore, the continuously distributed channel impulse response $h(t,\tau,\theta_R)$ will be replaced by a discrete form $h[k,\kappa,\theta_{R,\nu}]$.[2]

A suitable MIMO channel model represented by a set of matrices $\mathbf{H}[k,\kappa]$ can be constructed by using Equation (5.28) in Figure 5.18. The vector $\mathbf{a}[k,\theta_{R,\nu}]$ denotes the steering vector at the receiver which depends on the DoA $\theta_{R,\nu}$ as well as the array geometry. The corresponding counterpart at the transmitter is $\mathbf{b}^T[k,\theta_{T,\mu}]$, where the direction of departure $\theta_{T,\mu}$ is itself a function of $\theta_{R,\nu}$ and the delay κ. Finally, $g[\kappa]$ represents the joint impulse response of transmit and receive filters.

With the assumption that $A(\cdot)$, $\mathbf{a}[\cdot]$ and $\mathbf{b}[\cdot]$ are sufficiently known, an accurate model of the space–time channel can be constructed by using correlation matrices as summarised in

---

[2]The time and delay parameters k and κ are generally aligned to the sampling grid on the entire model. However, we have a finite number of directions of arrival and departure that are not aligned onto a certain grid. In order to indicate their discrete natures, they are indexed by subscripts ν and μ respectively.

Figure 5.18. The exact $N_T N_R \times N_T N_R$ correlation matrix $\Phi_{\mathcal{HH}}$ is given in Equation (5.29), where the operator vec(**A**) stacks all columns of a matrix **A** on top of each other. The channel matrix **H** can be constructed from a matrix $\mathbf{H}_w$ of the same size with i.i.d. elements according to Equation (5.30). To be exact, $\Phi_{\mathcal{HH}}$ should be determined for each delay κ. However, in most cases $\Phi_{\mathcal{HH}}$ is assumed to be identical for all delays.

A frequently used simplified but less general model is obtained if transmitter and receiver correlation are separated. In this case, we have a correlation matrix Φ_T describing the correlations at the transmitter and a matrix Φ_R for the correlations at the receiver. The channel model is now generated by Equation (5.31), as can be verified by[3]

$$E\{\mathbf{HH}^H\} = E\{\Phi_R^{1/2}\mathbf{H}_w\Phi_T^{1/2}\Phi_T^{H/2}\mathbf{H}_w^H\Phi_R^{H/2}\} = \Phi_R$$

and

$$E\{\mathbf{H}^H\mathbf{H}\} = E\{\Phi_T^{H/2}\mathbf{H}_w^H\Phi_R^{H/2}\Phi_R^{1/2}\mathbf{H}_w\Phi_T^{1/2}\} = \Phi_T.$$

A relationship between the separated correlation approach and the optimal one is obtained with the Kronecker product $\otimes$, as shown in Equation (5.32).

This simplification matches reality only if the correlations at the transmit side are identical for all receive antennas, and vice versa. It cannot be applied for pinhole (or keyhole) channels (Gesbert et al., 2003). They describe scenarios where transmitter and receiver may be located in rich scattering environments while the rays between them have to pass a keyhole. Although the spatial fading at transmitter and receiver is mutually independent, we obtain a degenerated channel of unit rank that can be modelled by

$$\mathbf{H} = \mathbf{h}_R \cdot \mathbf{h}_T^H.$$

In order to evaluate the properties of a channel, especially its correlations, the Singular Value Decomposition (SVD) is a suited means. It decomposes **H** into three matrices: a unitary $N_R \times N_R$ matrix **U**, a quasi-diagonal $N_R \times N_T$ matrix Σ and a unitary $N_T \times N_T$ matrix **V**. While **U** and **V** contain the eigenvectors of $\mathbf{HH}^H$ and $\mathbf{H}^H\mathbf{H}$ respectively,

$$\Sigma = \begin{pmatrix} \sigma_1 & & & \\ & \ddots & & \mathbf{0}_{r \times N_T - r} \\ & & \sigma_r & \\ & \mathbf{0}_{N_R - r \times N_T - r} & \end{pmatrix}$$

contains on its diagonal the singular values σ_i of **H**. The number of non-zero singular values is called the rank of a matrix. For i.i.d. elements in **H**, all singular values are identical in the average, and the matrix has full rank as pointed out in Figure 5.19. The higher the correlations, the more energy concentrates on only a few singular values and the rank of the matrix decreases. This rank deficiency can also be expressed by the condition number defined in Equation (5.34) as the ratio of largest to smallest singular value. For unitary (orthogonal) matrices, it amounts to unity and becomes larger for growing correlations. The rank of a matrix will be used in subsequent sections to quantify the diversity degree and the spatial degrees of freedom. The condition number will be exploited in the context of lattice reduced signal detection techniques.

---

[3] Since $\mathbf{H}_w$ consists of i.i.d. elements, $E\{\mathbf{H}_w\Phi\mathbf{H}_w^H\} = \mathbf{I}$ holds.

Analysing MIMO channels

- Singular Value Decomposition (SVD) for flat channel matrix

$$\mathbf{H} = \mathbf{U} \cdot \mathbf{\Sigma} \cdot \mathbf{V}^H \tag{5.33}$$

- Condition number

$$\kappa(\mathbf{H}) = \frac{\sigma_{max}}{\sigma_{min}} = \|\mathbf{H}\|_2 \cdot \|\mathbf{H}^{-1}\|_2 \geq 1 \tag{5.34}$$

  - ℓ_2 norm of a matrix

$$\|\mathbf{H}\|_2 = \sup_{\mathbf{x} \neq 0} \frac{\|\mathbf{A}\mathbf{x}\|}{\|\mathbf{x}\|} \tag{5.35}$$

  - $\|\mathbf{H}\|_2 = \sigma_{max}$
  - $\|\mathbf{H}^{-1}\|_2 = \sigma_{min}^{-1}$

- Rank of a matrix rank($\mathbf{H}$) denotes the number of non-zero singular values.

- Full rank: rank($\mathbf{H}$) = min$\{N_T, N_R\}$.

Figure 5.19: Analysing MIMO channels

5.2.3 Channel Estimation

At the end of this section, some principles of MIMO channel estimation are introduced. Generally, channel knowledge is necessary in order to overcome the disturbing influence of the channel. This holds for many space–time coding and multilayer transmission schemes, especially for those discussed in the next sections. However, there exist some exceptions similar to the concepts of differential and orthogonal modulation schemes for single-input single-output channels which allow an incoherent reception without Channel State Information (CSI). These exceptions are unitary space–time codes (Hochwald and Marzetta, 2000; Hochwald et al., 2000) and differentially encoded space–time modulation (Hochwald and Sweldens, 2000; Hughes, 2000; Schober and Lampe, 2002) which do not require any channel knowledge, either at the transmitter or at the receiver.

While perfect CSI is often assumed for ease of analytical derivations and finding ultimate performance limits, the channel has to be estimated in practice. The required degree of channel knowledge depends on the kind of transmission scheme. The highest spectral efficiency is obtained if both transmitter and receiver have channel knowledge. However, this is the most challenging case. In Time Division Duplex (TDD) systems, reciprocity of the channel is often assumed, so that the transmitter can use its own estimate obtained in the uplink to adjust the transmission parameters for the downlink. By contrast, Frequency Division Duplex (FDD) systems place uplink and downlink in different frequency bands so that reciprocity is not fulfilled. Hence, the transmitter is not able to estimate channel parameters for the downlink transmission directly. Instead, the receiver transmits its estimates over a feedback channel to the transmitter. In many systems, e.g. UMTS (Holma and Toskala, 2004), the data rate of the feedback channel is extremely low and error correction coding is not applied. Hence, the channel state information is roughly quantised, likely to be corrupted by transmission errors and often outdated in fast-changing environments.

Many schemes do not require channel state information at the transmitter but only at the receiver. The loss compared with the perfect case is small for medium and high signal-to-noise ratios and becomes visible only at very low SNRs. Next, we describe the MIMO channel estimation at the receiver.

Principally, reference-based and blind techniques have to be distinguished. The former techniques use a sequence of pilot symbols known to the receiver to estimate the channel. They are inserted into the data stream either as one block at a predefined position in a frame, e.g. as preamble or mid-amble, or they are distributed at several distinct positions in the frame. In order to be able to track channel variations, the sampling theorem of Shannon has to be fulfilled so that the time between successive pilot positions is less than the coherence time of the channel. By contrast, blind schemes do not need a pilot overhead. However, they generally have a much higher computational complexity, require rather large block lengths to converge and need an additional piece of information to overcome the problem of phase ambiguities.

Figure 5.20 gives a brief overview starting with the pilot-assisted channel estimation. The transmitter sends a sequence of N_P pilot symbols over each transmit antenna represented by the $(N_T \times N_P)$ matrix $\mathbf{X}_{\text{pilot}}$. Each row of $\mathbf{X}_{\text{pilot}}$ contains one sequence and each column represents a certain time instant. The received $(N_R \times N_P)$ pilot matrix is denoted by $\mathbf{R}_{\text{pilot}}$ and contains in each row a sequence of one receive antenna. Solving the optimisation problem

$$\hat{\mathbf{H}}_{\text{ML}} = \underset{\tilde{\mathbf{H}}}{\text{argmin}} \left\| \mathbf{R}_{\text{pilot}} - \tilde{\mathbf{H}} \cdot \mathbf{X}_{\text{pilot}} \right\|^2$$

leads to the maximum likelihood estimate $\hat{\mathbf{H}}_{\text{ML}}$ as depicted in Equation (5.37). The Moore–Penrose inverse $\mathbf{X}_{\text{pilot}}^{\dagger}$ is defined (Golub and van Loan, 1996) by

$$\mathbf{X}_{\text{pilot}}^{\dagger} = \mathbf{X}_{\text{pilot}}^{H} \cdot \left(\mathbf{X}_{\text{pilot}} \mathbf{X}_{\text{pilot}}^{H} \right)^{-1}.$$

From the last equation we can conclude that the matrix $\mathbf{X}_{\text{pilot}} \mathbf{X}_{\text{pilot}}^{H}$ has to be invertible, which is fulfilled if $\text{rank}(\mathbf{X}_{\text{pilot}}) = N_P$ holds, i.e. $\mathbf{X}_{\text{pilot}}$ has to be of full rank and the number of pilot symbols in $\mathbf{X}_{\text{pilot}}$ has to be at least as large as the number of transmit antennas N_T.

Depending on the condition number of $\mathbf{X}_{\text{pilot}}$, the inversion of $\mathbf{X}_{\text{pilot}} \mathbf{X}_{\text{pilot}}^{H}$ may lead to large values in $\mathbf{X}_{\text{pilot}}^{\dagger}$. Since the noise $\mathbf{N}$ is also multiplied with $\mathbf{X}_{\text{pilot}}^{\dagger}$, significant noise amplifications can occur. This drawback can be circumvented by choosing long training sequences with $N_P \gg N_T$, which leads to a large pilot overhead and, consequently, to a low overall spectral efficiency. Another possibility is to design appropriate training sequences. If $\mathbf{X}_{\text{pilot}}$ is unitary,

$$\mathbf{X}_{\text{pilot}} \cdot \mathbf{X}_{\text{pilot}}^{H} = \mathbf{I}_{N_T} \quad \Rightarrow \quad \mathbf{X}_{\text{pilot}}^{\dagger} = \mathbf{X}_{\text{pilot}}^{H}$$

holds and no noise amplification disturbs the transmission.

MIMO channel estimation

- Pilot-assisted channel estimation:
  - Received pilot signal

  $$\mathbf{R}_{\text{pilot}} = \mathbf{H} \cdot \mathbf{X}_{\text{pilot}} + \mathbf{N}_{\text{pilot}} \qquad (5.36)$$

  - Pilot-assisted channel estimation

  $$\hat{\mathbf{H}} = \mathbf{R}_{\text{pilot}} \cdot \mathbf{X}_{\text{pilot}}^{\dagger} = \mathbf{H} + \mathbf{N}_{\text{pilot}} \cdot \mathbf{X}_{\text{pilot}}^{\dagger} \qquad (5.37)$$

  - Pilot matrix with unitary $\mathbf{X}_{\text{pilot}}$

  $$\hat{\mathbf{H}} = \mathbf{R}_{\text{pilot}} \cdot \mathbf{X}_{\text{pilot}}^{\text{H}} = \mathbf{H} + \mathbf{N} \cdot \mathbf{X}_{\text{pilot}}^{\text{H}} \qquad (5.38)$$

- Blind channel estimation based on second-order statistics

  $$\hat{\mathbf{H}} = \text{E}\{\mathbf{r}\mathbf{r}^{\text{H}}\} = \text{E}\{\mathbf{H}\mathbf{x}\mathbf{x}^{\text{H}}\mathbf{H}^{\text{H}} + \mathbf{n}\mathbf{n}^{\text{H}}\} = \sigma_{\mathcal{X}}^{2} \cdot \Phi_{\text{R}} + \sigma_{\mathcal{N}}^{2} \cdot \mathbf{I}_{N_{R}} \qquad (5.39)$$

Figure 5.20: MIMO channel estimation

A blind channel estimation approach based on second-order statistics is shown in Equation (5.39). The right-hand side of this equation holds under the assumption of statistically independent transmit signals $\text{E}\{\mathbf{x}\mathbf{x}^{\text{H}}\} = \sigma_{\mathcal{X}}^{2}\mathbf{I}_{N_{T}}$, white noise $\text{E}\{\mathbf{n}\mathbf{n}^{\text{H}}\} = \sigma_{\mathcal{N}}^{2}\mathbf{I}_{N_{R}}$ and the validity of the channel model $\mathbf{H} = \Phi_{\text{R}}^{1/2} \cdot \mathbf{H}_{\text{w}} \cdot \Phi_{\text{T}}^{1/2}$ (cf. Equation (5.31)). It can be observed that this approach does not deliver phase information. Moreover, the estimate only depends on the covariance matrix at the receiver, i.e. the receiver cannot estimate correlations at the transmit antenna array with this method. The same holds for the transmitter in the opposite direction. However, long-term channel characteristics such as directions of arrival that are incorporated in Φ_{R} can be determined by this approach.

5.3 Performance Measures

5.3.1 Channel Capacity

In order to evaluate the quality of a MIMO channel, different performance measures can be used. The ultimate performance limit is represented by the channel capacity indicating the maximum data rate that can be transmitted error free. Fixing the desirable data rate by choosing a specific modulation scheme, the resulting average error probability determines how reliable the received values are. These quantities have been partly introduced for the scalar case and will now be extended for multiple transmit and receive antennas. We start our analysis with a short survey for the scalar case.

Channel Capacity of Scalar Channels

Figure 5.21 starts with the model of the scalar AWGN channel in Equation (5.40) and recalls the well-known results for this simple channel. According to Equation (5.41), the mutual information between the input $x[k]$ and the output $r[k]$ is obtained from the difference in the output and noise entropies. For real continuously Gaussian distributed signals, the differential entropy has the form

$$I_{\text{diff}}(\mathcal{X}) = \text{E}\{-\log_2[p_\mathcal{X}(x)]\} = 0.5 \cdot \log_2(2\pi e \sigma_\mathcal{X}^2),$$

while it amounts to

$$I_{\text{diff}}(\mathcal{X}) = \log_2(\pi e \sigma_\mathcal{X}^2)$$

for circular symmetric complex signals where real and imaginary parts are statistically independent and identically distributed. Inserting these results into Equation (5.41) leads to the famous formulae (5.42b) and (5.42a). It is important to note that the difference between mutual information and capacity is the maximisation of $I(\mathcal{X}; \mathcal{R})$ with respect to the input statistics $p_\mathcal{X}(\mathbf{x})$. For the considered AWGN channel, the optimal continuous distribution of the transmit signal is Gaussian. The differences between real and complex cases can be

---

AWGN channel capacity

- Channel output

$$r[k] = x[k] + n[k] \qquad (5.40)$$

- Mutual information

$$I(\mathcal{X}; \mathcal{R}) = I_{\text{diff}}(\mathcal{R}) - I_{\text{diff}}(\mathcal{R} \mid \mathcal{X}) = I_{\text{diff}}(\mathcal{R}) - I_{\text{diff}}(\mathcal{N}) \qquad (5.41)$$

- Capacity for complex Gaussian input and noise (equivalent baseband)

$$C = \sup_{p_\mathcal{X}(\mathbf{x})} [I(\mathcal{X}; \mathcal{R})] = \log_2\left(1 + \frac{E_s}{N_0}\right) \qquad (5.42a)$$

- Capacity for real Gaussian input (imaginary part not used)

$$C = \frac{1}{2} \cdot \log_2\left(1 + 2\frac{E_s}{N_0}\right) \qquad (5.42b)$$

Figure 5.21: AWGN channel capacity

SPACE–TIME CODES

explained as follows. On the one hand, we do not use the imaginary part and waste half of the available dimensions (factor 1/2 in front of the logarithm). On the other hand, the imaginary part of the noise does not disturb the real $x[k]$, so that the effective SNR is doubled (factor 2 in front of E_s/N_0).

The basic difference between the AWGN channel and a frequency-non-selective fading channel is its time-varying signal-to-noise ratio $\gamma[k] = |h[k]|^2 E_s/N_0$ which depends on the instantaneous fading coefficient $h[k]$. Hence, the instantaneous channel capacity $C[k]$ in Equation (5.43) is a random variable itself and can be described by its statistical properties. For fast-fading channels, a coded frame generally spans over many different fading states so that the decoder exploits diversity by performing a kind of averaging. Therefore, the average capacity $\bar{C}$ among all channel states, termed ergodic capacity and defined in Figure 5.22, is an appropriate means. In Equation (5.44), the expectation is defined as

$$E\{f(\mathcal{X})\} = \int_{-\infty}^{\infty} f(x) \cdot p_{\mathcal{X}}(x) dx .$$

Channel Capacity of Multiple-Input and Multiple-Output Channels

The results in Figure 5.22 can be easily generalised to MIMO channels. The only difference is the handling of vectors and matrices instead of scalar variables, resulting in multivariate distributions of random processes. From the known system description

$$\mathbf{r} = \mathbf{H} \cdot \mathbf{x} + \mathbf{n}$$

of Subsection 5.2.1 we know that $\mathbf{r}$ and $\mathbf{n}$ are N_R dimensional vectors, $\mathbf{x}$ is N_T dimensional and $\mathbf{H}$ is an $N_R \times N_T$ matrix. As for the AWGN channel, Gaussian distributed

---

Scalar fading channel capacities

- Single-Input Single-Output (SISO) fading channel

$$r[k] = h[k] \cdot x[k] + n[k]$$

- Instantaneous channel capacity

$$C[k] = \log_2\left(1 + |h[k]|^2 \cdot \frac{E_s}{N_0}\right) \tag{5.43}$$

- Ergodic channel capacity

$$\bar{C} = E\{C[k]\} = E\left\{\log_2\left(1 + |h[k]|^2 \cdot \frac{E_s}{N_0}\right)\right\} \tag{5.44}$$

Figure 5.22: Scalar fading channel capacities

input alphabets achieve capacity and are considered below. The corresponding multivariate distributions for real and complex random processes with n dimensions are shown in Figure 5.23. The $n \times n$ matrix $\Phi_{\mathcal{AA}}$ denotes the covariance matrix of the n-dimensional process $\mathcal{A}$ and is defined as

$$\Phi_{\mathcal{AA}} = \mathrm{E}\{\mathbf{aa}^{\mathrm{H}}\} = \mathbf{U}_{\mathcal{A}} \cdot \Lambda_{\mathcal{A}} \cdot \mathbf{U}_{\mathcal{A}}^{\mathrm{H}}.$$

The right-handside of the last equation shows the eigenvalue decomposition (see definition (B.0.7) in Appendix B) which decomposes the Hermitian matrix $\Phi_{\mathcal{AA}}$ into the diagonal matrix $\Lambda_{\mathcal{A}}$ with the corresponding eigenvalues $\lambda_{\mathcal{A},i}$, $1 \leq i \leq n$, and the square unitary matrix $\mathbf{U}_{\mathcal{A}}$. The latter contains the eigenvectors of $\mathbf{A}$.

Multivariate distributions and related entropies

- Joint probability density for a real multivariate Gaussian process

$$p_{\mathcal{A}}(\mathbf{a}) = \frac{1}{\sqrt{\det(2\pi \Phi_{\mathcal{AA}})}} \cdot \exp\left[-\mathbf{a}^{\mathrm{T}} \Phi_{\mathcal{AA}}^{-1} \mathbf{a}/2\right]$$

- Joint probability density for complex multivariate Gaussian process

$$p_{\mathcal{A}}(\mathbf{a}) = \frac{1}{\sqrt{\det(\pi \Phi_{\mathcal{AA}})}} \cdot \exp\left[-\mathbf{a}^{\mathrm{H}} \Phi_{\mathcal{AA}}^{-1} \mathbf{a}\right]$$

- Joint entropy of a multivariate process with n dimensions

$$I_{\mathrm{diff}}(\mathcal{A}) = -\mathrm{E}\{\log_2[p_{\mathcal{A}}(\mathbf{a})]\} = -\int_{\mathbb{A}^n} p_{\mathcal{A}}(\mathbf{a}) \cdot \log_2[p_{\mathcal{A}}(\mathbf{a})]\, d\mathbf{a} \qquad (5.45)$$

- Joint entropy of a multivariate real Gaussian process

$$I_{\mathrm{diff}}(\mathcal{A}) = \frac{1}{2} \cdot \log_2[\det(2\pi e \Phi_{\mathcal{AA}})] = \frac{1}{2} \cdot \sum_{i=1}^{n} \log_2(2\pi e \lambda_{\mathcal{A},i}) \qquad (5.46a)$$

- Joint entropy of a multivariate complex Gaussian process

$$I_{\mathrm{diff}}(\mathcal{A}) = \log_2[\det(\pi e \Phi_{\mathcal{AA}})] = \sum_{i=1}^{n} \log_2(\pi e \lambda_{\mathcal{A},i}) \qquad (5.46b)$$

Figure 5.23: Multivariate distributions and related entropies

SPACE-TIME CODES

Following Figure 5.23, we recognize from Equation (5.45) that the differential entropy $I_{\text{diff}}(\mathcal{X})$ is defined in exactly the same way as for scalar processes except for the expectation (integration) over an N_T-dimensional space. Solving the integrals by inserting the multivariate distributions for real and complex Gaussian processes, we obtain the differential entropies in Equations (5.46a) and (5.46b). They both depend on the covariance matrix $\Phi_{\mathcal{AA}}$. Using its eigenvalue decomposition and the fact that the determinant of a matrix equals the product of its eigenvalues, we obtain the right-hand-side expressions of Equations (5.46a) and (5.46b). If the elements of $\mathbf{A}$ are statistically independent and identically distributed (i.i.d.), all with variance $\lambda_{\mathcal{A},i} = \sigma_{\mathcal{A}}^2$, the differential entropy becomes

$$I_{\text{diff}}(\mathcal{A}) = \log_2\left[\prod_{i=1}^n (\pi e \lambda_{\mathcal{A},i})\right] = \sum_{i=1}^n \log_2(\pi e \lambda_{\mathcal{A},i}) \underset{\text{i.i.d.}}{=} n \cdot \log_2(\pi e \sigma_{\mathcal{A}}^2)$$

for the complex case. An equivalent expression is obtained for real processes.

Using the results of Figure 5.23, we can now derive the capacity of a MIMO channel. A look at Equation (5.47) in Figure 5.24 illustrates that the instantaneous mutual information

---

Mutual Information of MIMO systems

- Mutual information of a MIMO channel

$$I(\mathcal{X}; \mathcal{R} \mid \mathbf{H}) = I_{\text{diff}}(\mathcal{R} \mid \mathbf{H}) - I_{\text{diff}}(\mathcal{N}) = \log_2 \frac{\det(\Phi_{\mathcal{RR}})}{\det(\Phi_{\mathcal{NN}})} \quad (5.47)$$

- Inserting SVD $\mathbf{H} = \mathbf{U}_\mathcal{H} \mathbf{\Sigma}_\mathcal{H} \mathbf{V}_\mathcal{H}^H$ and $\Phi_{\mathcal{NN}} = \sigma_\mathcal{N}^2 \mathbf{I}_{N_R}$

$$I(\mathcal{X}; \mathcal{R} \mid \mathbf{H}) = \log_2 \det\left(\mathbf{I}_{N_R} + \frac{1}{\sigma_\mathcal{N}^2} \mathbf{\Sigma}_\mathcal{H} \mathbf{V}_\mathcal{H}^H \Phi_{\mathcal{XX}} \mathbf{V}_\mathcal{H} \mathbf{\Sigma}_\mathcal{H}^H\right) \quad (5.48)$$

- Perfect channel knowledge only at receiver

$$I(\mathcal{X}; \mathcal{R} \mid \mathbf{H}) = \log_2 \det\left(\mathbf{I}_{N_R} + \frac{\sigma_\mathcal{X}^2}{\sigma_\mathcal{N}^2} \mathbf{\Sigma}_\mathcal{H} \mathbf{\Sigma}_\mathcal{H}^H\right) = \sum_{i=1}^{N_T} \log_2\left(1 + \sigma_{\mathcal{H},i}^2 \cdot \frac{\sigma_\mathcal{X}^2}{\sigma_\mathcal{N}^2}\right) \quad (5.49)$$

- Perfect channel knowledge at transmitter and receiver

$$I(\mathcal{X}; \mathcal{R} \mid \mathbf{H}) = \log_2 \det\left(\mathbf{I}_{N_R} + \frac{1}{\sigma_\mathcal{N}^2} \mathbf{\Sigma}_\mathcal{H} \Lambda_\mathcal{X} \mathbf{\Sigma}_\mathcal{H}^H\right) = \sum_{i=1}^{N_T} \log_2\left(1 + \sigma_{\mathcal{H},i}^2 \cdot \frac{\sigma_{\mathcal{X},i}^2}{\sigma_\mathcal{N}^2}\right) \quad (5.50)$$

Figure 5.24: Mutual Information of MIMO systems

for a specific channel $\mathbf{H}$ is similarly defined as the mutual information of the scalar case (Figure 5.21). With the differential entropies

$$I_{\text{diff}}(\mathcal{R} \mid \mathbf{H}) = \log_2\left(\det(\pi e \Phi_{\mathcal{RR}})\right)$$

and

$$I_{\text{diff}}(\mathcal{N}) = \log_2\left(\det(\pi e \Phi_{\mathcal{NN}})\right),$$

we obtain the right-handside of Equation (5.47). With the relation $\mathbf{r} = \mathbf{H}\mathbf{x} + \mathbf{n}$, the covariance matrix $\Phi_{\mathcal{RR}}$ of the channel output $\mathbf{r}$ becomes

$$\Phi_{\mathcal{RR}} = E\{\mathbf{r}\mathbf{r}^H\} = \mathbf{H}\Phi_{\mathcal{XX}}\mathbf{H}^H + \Phi_{\mathcal{NN}}.$$

Moreover, mutually independent noise contributions at the N_R receive antennas are often assumed, resulting in the noise covariance matrix

$$\Phi_{\mathcal{NN}} = \sigma_\mathcal{N}^2 \cdot \mathbf{I}_{N_R}.$$

Inserting these covariance matrices into Equation (5.47) and exploiting the singular value decomposition of the channel matrix $\mathbf{H} = \mathbf{U}_\mathcal{H} \Sigma_\mathcal{H} \mathbf{V}_\mathcal{H}^H$ delivers the result in Equation (5.48). It has to be mentioned that the singular values $\sigma_{\mathcal{H},i}$ of $\mathbf{H}$ are related to the eigenvalues $\lambda_{\mathcal{H},i}$ of $\mathbf{H}\mathbf{H}^H$ by $\lambda_{\mathcal{H},i} = \sigma_{\mathcal{H},i}^2$.

Next, we distinguish two cases with respect to the available channel knowledge. If only the receiver has perfect channel knowledge, the best strategy is to transmit independent data streams over the antenna elements, all with average power $\sigma_\mathcal{X}^2$. This corresponds to the transmit covariance matrix $\Phi_{\mathcal{XX}} = \sigma_\mathcal{X}^2 \cdot \mathbf{I}_{N_T}$ and leads to the result in Equation (5.49). Since the matrix Σ in Equation (5.49) is diagonal, the whole argument of the determinant is a diagonal matrix. Hence, the determinant equals the product of all diagonal elements which is transformed by the logarithm into the sum of the individual logarithms. We recognize from the right-handside of Equation(5.49) that the MIMO channel has been decomposed into a set of parallel (independent) scalar channels with individual signal-to-noise ratios $\sigma_{\mathcal{H},i}^2 \sigma_\mathcal{X}^2 / \sigma_\mathcal{N}^2$. Therefore, the total capacity is simply the sum of the individual capacities of the contributing parallel scalar channels.

If the transmitter knows the channel matrix perfectly, it can exploit the eigenmodes of the channel and, therefore, achieve a higher throughput. In order to accomplish this advantage, the transmitter covariance matrix has to be chosen as

$$\Phi_{\mathcal{XX}} = \mathbf{V}_\mathcal{H} \cdot \Lambda_\mathcal{X} \cdot \mathbf{V}_\mathcal{H}^H,$$

i.e. the eigenvectors have to equal those of $\mathbf{H}$. Inserting the last equation into Equation (5.48), we see that the eigenvector matrices $\mathbf{V}_\mathcal{H}$ eliminate themselves, leading to Equation (5.50). Again, all matrices are diagonal matrices, and we obtain the right-handside of Equation (5.50). The question that still has to be answered is how to choose the eigenvalues $\lambda_{\mathcal{H},i} = \sigma_{\mathcal{H},i}^2$ of $\Phi_{\mathcal{XX}}$, i.e. how to distribute the transmit power over the parallel data streams.

Following the procedure described elsewhere (Cover and Thomas, 1991; Kühn, 2006) using Lagrangian multipliers, the famous waterfilling solution is obtained. It is illustrated in Figure 5.25, where each bin represents one of the scalar channels. We have to imagine the

Waterfilling solution

- Waterfilling solution

$$\sigma_{\mathcal{X},i}^2 = \begin{cases} \theta - \dfrac{\sigma_{\mathcal{N}}^2}{\sigma_{\mathcal{H},i}^2} & \text{for } \theta > \dfrac{\sigma_{\mathcal{N}}^2}{\sigma_{\mathcal{H},i}^2} \\ 0 & \text{else} \end{cases} \quad (5.51)$$

- Total transmit power constraint

$$\sum_{i=1}^{N_T} \sigma_{\mathcal{X},i}^2 \stackrel{!}{=} N_T \cdot \frac{E_s}{N_0} \quad (5.52)$$

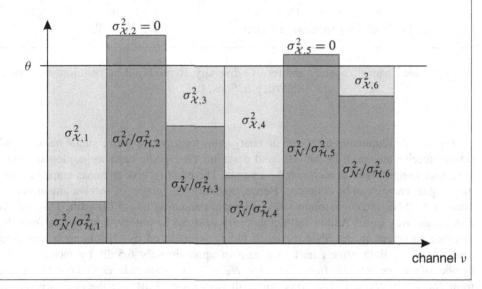

Figure 5.25: Waterfilling solution. Reproduced by permission of John Wiley & Sons, Ltd

diagram as a vessel with a bumpy ground where the height of the ground is proportional to the ratio $\sigma_{\mathcal{N}}^2/\sigma_{\mathcal{H},i}^2$. Pouring water into the vessel is equivalent to distributing transmit power onto the parallel scalar channels. The process is stopped when the totally available transmit power is consumed. Obviously, good channels with a low $\sigma_{\mathcal{N}}^2/\sigma_{\mathcal{H},i}^2$ obtain more transmit power than weak channels. The worst channels whose bins are not covered by the water level θ do not obtain any power, which can also be seen from Equation (5.51). Therefore, we can conclude that much power is spent on good channels transmitting high data rates, while little power is given to bad channels transmitting only very low data rates. This strategy leads to the highest possible data rate.

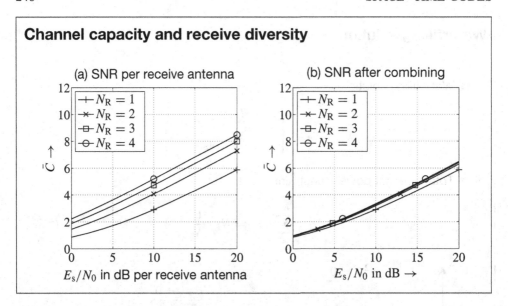

Figure 5.26: Channel capacity and receive diversity. Reproduced by permission of John Wiley & Sons, Ltd

Figure 5.26 illuminates array and diversity gains for a system with a single transmit and several receive antennas. In the left-hand diagram, the ergodic capacity is plotted versus the signal-to-noise ratio at each receive antenna. The more receive antennas employed, the more signal energy can be collected. Hence, doubling the number of receive antennas also doubles the SNR after maximum ratio combining, resulting in a 3 dB gain. This gain is denoted as array gain.[4] Additionally, a diversity gain can be observed, stemming from the fact that variations in the SNR owing to fading are reduced by combining independent diversity paths. Both effects lead to a gain of approximately 6.5 dB by increasing the number of receive antennas from $N_R = 1$ to $N_R = 2$. This gain reduces to 3.6 dB by going from $N_R = 2$ to $N_R = 4$. The array gain still amounts to 3 dB, but the diversity gain is getting smaller if the number of diversity paths is already high.

The pure diversity gains become visible in the right-hand diagram plotting the ergodic capacities versus the SNR after maximum ratio combining. This normalisation removes the array gain, and only the diversity gain remains. We observe that the ergodic capacity increases only marginally owing to a higher diversity degree.

Figure 5.27 shows the ergodic capacities for a system with $N_T = 4$ transmit antennas versus the signal-to-noise ratio E_s/N_0. The MIMO channel matrix consists of i.i.d. complex circular Gaussian distributed coefficients. Solid lines represent the results with perfect channel knowledge only at the receiver, while dashed lines correspond to the waterfilling solution with ideal CSI at transmitter and receiver. Asymptotically for large signal-to-noise ratios, we observe that the capacities increase linearly with the SNR. The slope amounts

---

[4] Certainly, the receiver cannot collect a higher signal power than has been transmitted. However, the channels' path loss has been omitted here so that the average channel gains are normalised to unity.

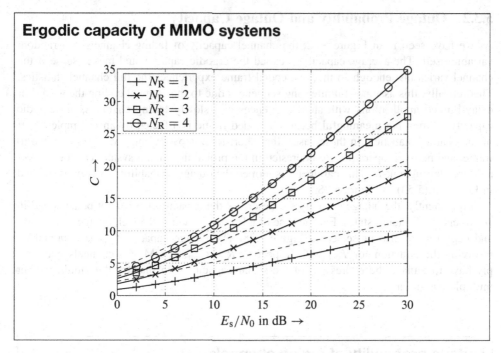

Figure 5.27: Ergodic capacity of MIMO systems with $N_T = 4$ transmit antennas and varying N_R (solid lines: CSI only at receiver; dashed lines: waterfilling solution with perfect CSI at transmitter and receiver)

to 1 bit/s/Hz for $N_R = 1$, 2 bit/s/Hz for $N_R = 2$, 3 bit/s/Hz for $N_R = 3$ and 4 bit/s/Hz for $N_R = 4$, and therefore depends on the rank r of $\mathbf{H}$. For fixed $N_T = 4$, the number of non-zero eigenmodes r grows with the number of receive antennas up to $r_{\max} = \text{rank}(\mathbf{H}) = 4$. These data rate enhancements are called multiplexing gains because we can transmit up to $r_{\max}$ parallel data streams over the MIMO channel. This confirms our theoretical results that the capacity increases linearly with the rank of $\mathbf{H}$ while it grows only logarithmically with the SNR.

Moreover, we observe that perfect CSI at the transmitter leads to remarkable improvements for $N_R < N_T$. For these configurations, the rank of our system is limited by the number of receive antennas. Exploiting the non-zero eigenmodes requires some kind of beamforming which is only possible with appropriate channel knowledge at the transmitter. For $N_R = 1$, only one non-zero eigenmode exists. With transmitter CSI, we obtain an array gain of $10 \log_{10}(4) \approx 6$ dB which is not achievable without channel knowledge. For $N_R = N_T = 4$, the additional gain due to channel knowledge at the transmitter is visible only at low SNR where the waterfilling solution drops the weakest eigenmodes and concentrates the transmit power only on the strongest modes, whereas this is impossible without transmitter CSI. At high SNR, the water level in Figure 5.25 is so high that all eigenmodes are active and a slightly different distribution of the transmit power has only a minor impact on the ergodic capacity.

5.3.2 Outage Probability and Outage Capacity

As we have seen from Figure 5.22, the channel capacity of fading channels is a random variable itself. The average capacity is called the ergodic capacity and makes sense if the channel varies fast enough so that one coded frame experiences the full channel statistics. Theoretically, this assumes infinite long sequences due to the channel coding theorem. For delay-limited applications with short sequences and slowly fading channels, the ergodic capacity is often not meaningful because a coded frame is affected by an incomplete part of the channel statistics. In these cases, the 'short-term capacity' may vary from frame to frame, and network operators are interested in the probability that a system cannot support a desired throughput R. This parameter is termed the outage probability P_out and is defined in Equation (5.53) in Figure 5.28.

Equivalently, the outage capacity C_p describes the capacity that cannot be achieved in p percent of all fading states. For the case of a Rayleigh fading channel, outage probability and capacity are also presented in Figure 5.28. The outage capacity C_out is obtained by resolving the equation for P_out with respect to $R = C_\text{out}$. For MIMO channels, we simply have to replace the expression of the scalar capacity with that for the multiple-input multiple-output case.

Outage probability of fading channels

- Outage probability of a scalar channel

$$P_\text{out} = \Pr\{C[k] < R\} = \Pr\{|h[k]|^2 < \frac{2^R - 1}{E_s/N_0}\} \qquad (5.53)$$

  – Outage probability for scalar Rayleigh fading channels (Kühn, 2006)

$$P_\text{out} = 1 - \exp\left(\frac{1 - 2^R}{E_s/N_0}\right)$$

  – Outage capacity for scalar Rayleigh fading channel

$$C_\text{out} = \log_2\left(1 - E_s/N_0 \cdot \log(1 - P_\text{out})\right).$$

- Outage probability for MIMO channel with singular values $\sigma_{\mathcal{H},i}$

$$P_\text{out} = \Pr\{C[k] < R\} = \Pr\{\sum_{\mu=1}^{r} \log_2\left[1 + \sigma_{\mathcal{H},i}^2 \cdot \frac{\sigma_{\mathcal{X},i}^2}{\sigma_\mathcal{N}^2}\right] < R\} \qquad (5.54)$$

Figure 5.28: Outage probability of fading channels

SPACE–TIME CODES

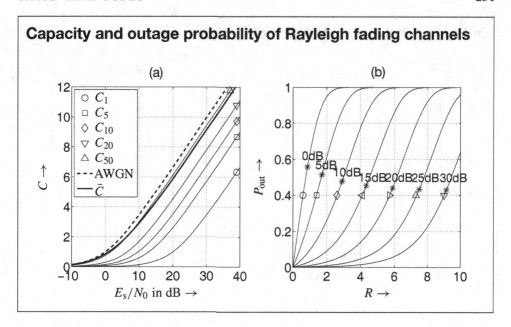

Figure 5.29: Capacity and outage probability of Rayleigh fading channels

The left-hand diagram in Figure 5.29 shows a comparison between the ergodic capacities of AWGN and flat Rayleigh fading channels (bold lines). For sufficiently large SNR, the curves are parallel and we can observe a loss due to fading of roughly 2.5 dB. Compared with the loss of approximately 17 dB at a bit error rate (BER) of $P_b = 10^{-3}$ in the uncoded case, the observed difference is rather small. This discrepancy can be explained by the fact that the channel coding theorem presupposes infinite long code words allowing the decoder to exploit a high diversity gain. Therefore, the loss in capacity compared with the AWGN channel is relatively small. Astonishingly, the ultimate limit of $10 \log_{10}(E_b/N_0) = -1.59$ dB is the same for AWGN and Rayleigh fading channels.

Additionally, the left-hand diagram shows the outage capacities for different values of P_{out}. For example, the capacity C_{50} can be ensured with a probability of 50% and is close to the ergodic capacity $\bar{C}$. The outage capacities C_p decrease dramatically for smaller P_{out}, i.e. the higher the requirements, the higher is the risk of an outage event. At a spectral efficiency of 6 bit/s/Hz, the loss compared with the AWGN channel in terms of E_b/N_0 amounts to nearly 8 dB for $P_{\text{out}} = 0.1$ and roughly 18 dB for $P_{\text{out}} = 0.01$.

The right-hand diagram depicts the outage probability versus the target throughput R for different values of E_s/N_0. As expected for large signal-to-noise ratios, high data rates can be guaranteed with very low outage probabilities. However, P_{out} grows rapidly with decreasing E_s/N_0. The asterisks denote the outage probability of the ergodic capacity $R = \bar{C}$. As could already be observed in the left-hand diagram, it is close to a probability of 0.5.

Finally, Figure 5.30 shows the outage probabilities for $1 \times N_R$ and $4 \times N_R$ MIMO systems at an average signal-to-noise ratio of 10 dB. From figure (a) it becomes obvious

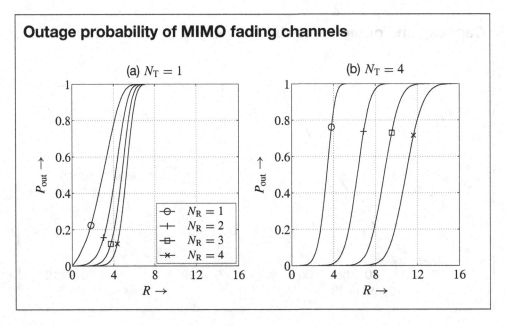

Figure 5.30: Outage probability of MIMO fading channels

that an increasing number of receive antennas enlarges the diversity degree and, hence, minimises the risk of an outage event. However, there is no multiplexing gain with only a single transmit antenna, and the gains for additional receive antennas become smaller if the number of receiving elements is already large. This is a well-known effect from diversity, assuming an appropriate scaling of the signal-to-noise ratio. In figure (b), the system with $N_T = 4$ transmit antennas is considered. Here, we observe larger gains with each additional receive antenna, since the number of eigenmodes increases so that we obtain a multiplexing gain besides diversity enhancements.

5.3.3 Ergodic Error Probability

Having analysed MIMO systems on an information theory basis, we will now have a look at the error probabilities. The following derivation was first introduced for space–time codes (Tarokh *et al.*, 1998). However, it can be applied to general MIMO systems. It assumes an optimal maximum likelihood detection and perfect channel knowledge at the receiver and a block-wise transmission, i.e. L consecutive vectors $\mathbf{x}[k]$ are written into the $N_T \times L$ transmit matrix

$$\mathbf{X} = \begin{bmatrix} \mathbf{x}[0] \, \mathbf{x}[1] \, \cdots \, \mathbf{x}[L-1] \end{bmatrix} = \begin{pmatrix} x_1[0] & x_1[1] & \cdots & x_1[L-1] \\ x_2[0] & x_2[1] & \cdots & x_2[L-1] \\ \vdots & & \ddots & \vdots \\ x_{N_T}[0] & x_{N_T}[1] & \cdots & x_{N_T}[L-1] \end{pmatrix}.$$

SPACE–TIME CODES

The MIMO channel is assumed to be constant during L time instants so that we receive an $N_R \times L$ matrix $\mathbf{R}$

$$\mathbf{R} = [\mathbf{r}[0]\,\mathbf{r}[1]\,\cdots\,\mathbf{r}[L-1]] = \begin{pmatrix} r_1[0] & r_1[1] & \cdots & r_1[L-1] \\ r_2[0] & r_2[1] & \cdots & r_2[L-1] \\ \vdots & & \ddots & \vdots \\ r_{N_R}[0] & r_{N_R}[1] & \cdots & r_{N_R}[L-1] \end{pmatrix} = \mathbf{H} \cdot \mathbf{X} + \mathbf{N},$$

where $\mathbf{N}$ contains the noise vectors $\mathbf{n}[k]$ within the considered block. The set of all possible matrices $\mathbf{X}$ is termed $\mathbb{X}$. In the case of a space–time block code, certain constraints apply to $\mathbf{X}$, limiting the size of $\mathbb{X}$ (see Section 5.4). By contrast, for the well-known Bell Labs Layered Space–Time (BLAST) transmission, there are no constraints on $\mathbf{X}$, leading to a size $|\mathbb{X}| = M^{N_T L}$, where M is the size of the modulation alphabet.

Calculating the average error probability generally starts with the determination of the pairwise error probability $\Pr\{\mathbf{B} \to \tilde{\mathbf{B}} \mid \mathbf{H}\}$. Equivalently to the explanation in Chapter 3, it denotes the probability that the detector decides in favour of a code matrix $\tilde{\mathbf{X}}$ although $\mathbf{X}$ was transmitted. Assuming uncorrelated noise contributions at each receive antenna, the optimum detector performs a maximum likelihood estimation, i.e. it determines that code matrix $\tilde{\mathbf{X}}$ which minimises the squared Frobenius distance $\|\mathbf{R} - \mathbf{H}\tilde{\mathbf{X}}\|_F^2$ (see Appendix B on page 312). Hence, we have to consider not the difference $\|\mathbf{X} - \tilde{\mathbf{X}}\|_F^2$, but the difference in the noiseless received signals $\|\mathbf{H}\mathbf{X} - \mathbf{H}\tilde{\mathbf{X}}\|_F^2$, as done in Equation (5.55) in Figure 5.31.

In order to make the squared Frobenius norm independent of the average power E_s of a symbol, we normalise the space–time code words by the average power per symbol to

$$\mathbf{B} = \frac{\mathbf{X}}{\sqrt{E_s/T_s}} \quad \text{and} \quad \tilde{\mathbf{B}} = \frac{\tilde{\mathbf{X}}}{\sqrt{E_s/T_s}}.$$

If the μth row of $\mathbf{H}$ is denoted by $\underline{\mathbf{h}}_\mu$, the squared Frobenius norm can be written as

$$\|\mathbf{H}(\mathbf{B} - \tilde{\mathbf{B}})\|_F^2 = \|\underline{\mathbf{h}}_\mu(\mathbf{B} - \tilde{\mathbf{B}})\|^2 = \sum_{\mu=1}^{N_R} \underline{\mathbf{h}}_\mu (\mathbf{B} - \tilde{\mathbf{B}})(\mathbf{B} - \tilde{\mathbf{B}})^H \underline{\mathbf{h}}_\mu^H.$$

This rewriting and the normalisation lead to the form given in Equation (5.56). We now apply the eigenvalue decomposition on the Hermitian matrix

$$(\mathbf{B} - \tilde{\mathbf{B}})(\mathbf{B} - \tilde{\mathbf{B}})^H = \mathbf{U}\Lambda\mathbf{U}^H.$$

The matrix Λ is diagonal and contains the eigenvalues λ_ν of $(\mathbf{B} - \tilde{\mathbf{B}})(\mathbf{B} - \tilde{\mathbf{B}})^H$, while $\mathbf{U}$ is unitary and consists of the corresponding eigenvectors. Inserting the eigenvalue decomposition into Equation (5.56) yields

$$\|\mathbf{H}(\mathbf{B} - \tilde{\mathbf{B}})\|_F^2 = \sum_{\mu=1}^{N_R} \underline{\mathbf{h}}_\mu \mathbf{U} \cdot \Lambda \cdot \mathbf{U}^H \underline{\mathbf{h}}_\mu^H = \sum_{\mu=1}^{N_R} \underline{\boldsymbol{\beta}}_\mu \Lambda \underline{\boldsymbol{\beta}}_\mu^H$$

with $\underline{\boldsymbol{\beta}}_\mu = [\beta_{\mu,1}, \ldots, \beta_{\mu,L}]$. Since Λ is diagonal, its multiplication with $\underline{\boldsymbol{\beta}}_\mu$ and $\underline{\boldsymbol{\beta}}_\mu^H$ from the left- and the right-handside respectively reduces to

$$\|\mathbf{H}(\mathbf{B} - \tilde{\mathbf{B}})\|_F^2 = \sum_{\mu=1}^{N_R} \sum_{\nu=1}^{L} |\beta_{\mu,\nu}|^2 \cdot \lambda_\nu = \sum_{\mu=1}^{N_R} \sum_{\nu=1}^{r} |\beta_{\mu,\nu}|^2 \cdot \lambda_\nu.$$

Pairwise error probability

- Pairwise error probability for code matrices $\mathbf{X}$ and $\tilde{\mathbf{X}}$

$$\Pr\{\mathbf{X} \to \tilde{\mathbf{X}} \mid \mathbf{H}\} = \frac{1}{2} \cdot \text{erfc}\left(\sqrt{\frac{\|\mathbf{HX} - \mathbf{H\tilde{X}}\|_F^2}{4\sigma_\mathcal{N}^2}}\right) \qquad (5.55)$$

- Normalisation to unit average power per symbol

$$\|\mathbf{H} \cdot (\mathbf{X} - \tilde{\mathbf{X}})\|_F^2 = \frac{E_s}{T_s} \cdot \sum_{\mu=1}^{N_R} \mathbf{h}_\mu \cdot (\mathbf{B} - \tilde{\mathbf{B}})(\mathbf{B} - \tilde{\mathbf{B}})^H \cdot \mathbf{h}_\mu^H \qquad (5.56)$$

- Substitution of $\underline{\boldsymbol{\beta}}_\mu = \mathbf{h}_\mu \mathbf{U}$

$$\|\mathbf{H} \cdot (\mathbf{X} - \tilde{\mathbf{X}})\|_F^2 = \frac{E_s}{T_s} \cdot \sum_{\mu=1}^{N_R} \underline{\boldsymbol{\beta}}_\mu \cdot \Lambda \cdot \underline{\boldsymbol{\beta}}_\mu^H = \frac{E_s}{T_s} \cdot \sum_{\mu=1}^{N_R} \sum_{\nu=1}^{r} |\beta_\mu|^2 \cdot \lambda_\mu \qquad (5.57)$$

- With $\sigma_\mathcal{N}^2 = N_0/T_s$, pairwise error probability becomes

$$\Pr\{\mathbf{X} \to \tilde{\mathbf{X}} \mid \mathbf{H}\} = \frac{1}{2} \cdot \text{erfc}\left(\sqrt{\frac{E_s}{4N_0} \cdot \sum_{\mu=1}^{N_R} \sum_{\nu=1}^{r} |\beta_{\mu,\nu}|^2 \cdot \lambda_\nu}\right) \qquad (5.58)$$

Figure 5.31: Pairwise error probability for space–time code words

Assuming that the rank of Λ equals rank$\{\Lambda\} = r \leq L$, i.e. r eigenvalues λ_μ are non-zero, the inner sum can be restricted to run from $\nu = 1$ only to $\nu = r$ because $\lambda_{\nu>r} = 0$ holds. The last equality is obtained because Λ is diagonal. Inserting the new expression for the squared Frobenius norm into the pairwise error probability of Equation (5.55) delivers the result in Equation (5.58).

The last step in our derivation starts with the application of the upper bound $\text{erfc}(\sqrt{x}) < e^{-x}$ on the complementary error function. Rewriting the double sum in the exponent into the product of exponential functions leads to the result in inequality (5.59) in Figure 5.32. In order to obtain a pairwise error probability averaged over all possible channel observations, the expectation of Equation (5.58) with respect to $\mathbf{H}$ has to be determined. This expectation is calculated over all channel coefficients $h_{\mu,\nu}$ of $\mathbf{H}$. At this point it has to be mentioned that the multiplication of a vector $\mathbf{h}_\mu$ with $\mathbf{U}$ performs just a rotation in the N_T-dimensional

SPACE-TIME CODES

Determinant and rank criteria

- Upper bound on pairwise error probability by erfc($\sqrt{x}$) < e^{-x}

$$\Pr\{\mathbf{B} \to \tilde{\mathbf{B}} \mid \mathbf{H}\} \leq \frac{1}{2} \cdot \prod_{\mu=1}^{N_R} \prod_{\nu=1}^{r} \exp\left[-|\beta_\mu|^2 \lambda_\mu \frac{E_s}{4N_0}\right] \quad (5.59)$$

- Expectation with respect to **H** yields

$$\Pr\{\mathbf{B} \to \tilde{\mathbf{B}}\} \leq \frac{1}{2} \cdot \left[\frac{E_s}{4N_0} \cdot \left(\prod_{\nu=1}^{r} \lambda_\nu\right)^{1/r}\right]^{-rN_R} \quad (5.60)$$

- Rank criterion: maximise the minimum rank of $(\mathbf{B} - \tilde{\mathbf{B}})$

$$g_d = N_R \cdot \min_{(\mathbf{B},\tilde{\mathbf{B}})} \mathrm{rank}\,(\mathbf{B} - \tilde{\mathbf{B}}) \quad (5.61)$$

- Determinant criterion: maximise the minimum of $(\prod_{\nu=1}^{r} \lambda_\nu)^{1/r}$

$$g_c = \min_{(\mathbf{B},\tilde{\mathbf{B}})} \left(\prod_{\nu=1}^{r} \lambda_\nu\right)^{1/r} \quad (5.62)$$

Figure 5.32: Determinant and rank criteria

---

space. Hence the coefficients $\beta_{\mu,\nu}$ of the vector $\underline{\beta}_\mu$ have the same statistics as the channel coefficients $h_{\mu,\nu}$ (Naguib et al., 1997).

Assuming the frequently used case of independent Rayleigh fading, the coefficients $h_{\mu,\nu}$, and, consequently also $\beta_{\mu,\nu}$, are complex rotationally invariant Gaussian distributed random variables with unit power $\sigma_\mathcal{H}^2 = 1$. Hence, their squared magnitudes are chi-squared distributed with two degrees of freedom, i.e.

$$p_{\beta_{\mu,\nu}}(\xi) = e^{-\xi}$$

holds. The expectation of inequality (5.59) now results in

$$\Pr\{\mathbf{B} \to \tilde{\mathbf{B}}\} = E_\mathcal{H}\left\{\Pr\{\mathbf{B} \to \tilde{\mathbf{B}} \mid \mathbf{H}\}\right\}$$

$$\leq \frac{1}{2} \cdot \prod_{\mu=1}^{N_R} \prod_{\nu=1}^{r} E_\beta\left\{\exp\left[-\lambda_\nu \cdot |\beta_{\mu,\nu}|^2 \cdot \frac{E_s}{4N_0}\right]\right\}$$

$$\leq \frac{1}{2} \cdot \prod_{\mu=1}^{N_R} \prod_{\nu=1}^{r} \int_0^\infty e^{-\xi} \cdot \exp\left[-\xi \cdot \lambda_\nu \frac{E_s}{4N_0}\right] d\xi$$

$$\leq \frac{1}{2} \cdot \left[\prod_{\nu=1}^{r} \frac{1}{1+\lambda_\nu \cdot \frac{E_s}{4N_0}}\right]^{N_R}.$$

The upper bound can be relaxed a little by dropping the $+1$ in the denominator. It is still tight for large SNR. Slight rewriting of the last inequality leads to inequality (5.60).

We can draw some important conclusions from the result in inequality (5.60). It resembles the error probability of a transmission with D-fold diversity which is proportional to

$$P_s \propto \left(\frac{4E_s}{N_0}\right)^{-D}.$$

A comparison shows that the exponent $r N_R$ in inequality (5.60) is equivalent to the diversity degree D. Hence, in order to achieve the maximum possible diversity degree, the minimum rank r among all pairwise differences $\mathbf{B} - \tilde{\mathbf{B}}$ should be maximised, leading to the rank criterion presented in Equation (5.61). The maximum diversity degree equals $N_T N_R$. Since the error probabilities are generally scaled logarithmically when depicted versus the SNR, the diversity degree in the exponent determines the slope of the error probability curves due to

$$\log\left[\left(\frac{4E_s}{N_0}\right)^{-D}\right] = D \cdot \log\left(\left[\frac{4E_s}{N_0}\right]^{-1}\right)$$

By contrast, the product $\left(\prod_{\nu=1}^{r} \lambda_\nu\right)^{1/r}$ multiplies the signal-to-noise ratio E_s/N_0 in inequality (5.60). The logarithm turns this multiplication into an addition and, therefore, into a horizontal shift of the corresponding error probability curve. By analogy with coding theory, this shift is called the coding gain. The largest coding gain is obtained if the minimum of all possible products is maximised as stated in Equation (5.62). If the design of $\mathbf{B}$ ensures a full rank $r = \text{rank}\{\mathbf{B} - \tilde{\mathbf{B}}\} = N_T$, the product of the eigenvalues equals the determinant $\det(\mathbf{B} - \tilde{\mathbf{B}})$

$$g_c = \min_{(\mathbf{B},\tilde{\mathbf{B}})} \left(\prod_{\nu=1}^{N_T} \lambda_\nu\right)^{1/N_T} = \min_{(\mathbf{B},\tilde{\mathbf{B}})} \left(\det(\mathbf{B} - \tilde{\mathbf{B}})\right)^{1/N_T}.$$

Therefore, the criterion is termed the determinant criterion, as presented on Figure 5.32.

Some comments have to be made on these criteria. If no constraints are imposed on the matrix $\mathbf{B}$, i.e. it contains arbitrary symbols as in the case of Bell Labs Layered Space–Time (BLAST)-like layered space–time transmissions, the maximum minimal rank equals N_R and only receive diversity is gained. Consequently, no coding gain can be expected. However, N_T data streams can be transmitted in parallel (one over each antenna) boosting the achievable data rate. If appropriate space–time encoding is applied at the transmitter, the code design may ensure a higher minimum rank, at most $r = N_T N_R$, as well as a coding gain. However, the data rate is lower, as in the first case. Obviously, a trade-off between reliability and data rate can be achieved (Zheng and Tse, 2003). In the next section, a general description of space–time coding for the cases just discussed will be introduced.

5.4 Orthogonal Space–Time Block Codes

In this chapter we pursue the goal of obtaining spatial diversity by deploying several antennas at the transmitter but only a single antenna at the receiver. However, a generalisation to multiple receive antennas is straightforward (Kühn, 2006). Furthermore, it is assumed that the average energy E_s per transmitted symbol is constant and, in particular, independent of N_T and the length of a space–time block. Since aiming for diversity is mostly beneficial if the channel between one transmit and one receive antenna provides no diversity, we consider frequency-non-selective channels. Moreover, the channel is assumed to be constant during one space–time code word.

We saw from Section 5.1 that spatial receive diversity is simply achieved by maximum ratio combining the received samples, resulting in a coherent (constructive) superposition, i.e. the squared magnitudes have been summed. Unfortunately, transmitting the same symbol $s[\ell]$ from all N_T transmit antennas generally leads to an incoherent superposition

$$r[\ell] = \sum_{\nu=1}^{N_T} h_\nu[\ell] \cdot \frac{s[\ell]}{\sqrt{N_T}} + n[\ell] = s[\ell] \cdot \frac{1}{\sqrt{N_T}} \cdot \sum_{\nu=1}^{N_T} h_\nu[\ell] + n[\ell] = s[\ell] \cdot \tilde{h}[\ell] + n[\ell]$$

at the receive antenna. The factor $1/\sqrt{N_T}$ ensures that the average transmit power per symbol is independent of N_T. In the case of i.i.d. Rayleigh fading coefficients, the new channel coefficient $\tilde{h}[\ell]$ has the same distribution as each single coefficient $h_\nu[\ell]$ and nothing has been won. If the transmitter knows the channel coefficients, it can predistort the symbols so that the superposition at the receiver becomes coherent. This strategy is known as beamforming and is not considered in this book. Therefore, more sophisticated signalling schemes are required in order to achieve a diversity gain.

Although orthogonal space–time block codes do not provide a coding gain, they have the great advantage that decoding simply requires some linear combinations of the received symbols. Moreover, they provide the full diversity degree achievable with a certain number of transmit and receive antennas. In order to have an additional coding gain, they can be easily combined with conventional channel coding concepts, as discussed in the previous chapters.

5.4.1 Alamouti's Scheme

Before we give a general description of space–time block codes, a famous but simple example should illustrate the basic principle. We consider the approach introduced by Alamouti (Alamouti, 1998) using two transmit antennas and a single receive antenna. The original structure is depicted in Figure 5.33. As we will see, each symbol $s[\ell]$ is transmitted twice. In order to keep the transmit power per symbol constant, each instance of the symbol is normalised by the factor $1/\sqrt{2}$.

Two consecutive symbols $s[\ell - 1]$ and $s[\ell]$ are collected from the input sequence. They are denoted as $s_1 = s[\ell - 1]$ and $s_2 = s[\ell]$ respectively and are mapped onto the $N_T = 2$ transmit antennas as follows. At the first time instant, $x_1[\ell] = s_1/\sqrt{2}$ is sent over the first antenna and $x_2[\ell] = s_2/\sqrt{2}$ over the second one. At the receiver, we obtain the superposition

$$r[\ell] = h_1 x_1[\ell] + h_2 x_2[\ell] + n[\ell] = \frac{1}{\sqrt{2}} \cdot (h_1 s_1 + h_2 s_2) + n[\ell] \,.$$

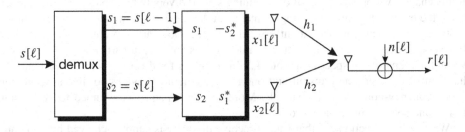

Figure 5.33: Orthogonal space–time block code by Alamouti with $N_T = 2$ transmit antennas

Next, both symbols are exchanged and the first antenna transmits $x_1[\ell+1] = -s_2^*/\sqrt{2}$ while the second antenna emits $x_2[\ell+1] = s_1^*/\sqrt{2}$. This leads to

$$r[\ell+1] = h_1 x_2[\ell+1] + h_2 x_2[\ell+1] + n[\ell+1] = \frac{1}{\sqrt{2}} \cdot \left(h_1 - s_2^* + h_2 s_1^* \right) + n[\ell+1].$$

Using vector notations, we can write the two received symbols and the two noise samples into vectors $\mathbf{r} = [r[\ell]\ r[\ell+1]]^T$ and $\mathbf{n} = [n[\ell]\ n[\ell+1]]^T$ respectively. This yields the compact description

$$\mathbf{r} = \frac{1}{\sqrt{2}} \cdot \begin{pmatrix} s_1 & s_2 \\ -s_2^* & s_1^* \end{pmatrix} \cdot \begin{pmatrix} h_1 \\ h_2 \end{pmatrix} + \mathbf{n} = \mathbf{X}_2 \cdot \mathbf{h} + \mathbf{n} \tag{5.65}$$

The columns of the space–time code word

$$\mathbf{X}_2 = \begin{pmatrix} \mathbf{x}[\ell] & \mathbf{x}[\ell+1] \end{pmatrix} = \frac{1}{\sqrt{2}} \cdot \begin{pmatrix} s_1 & -s_2^* \\ s_2 & s_1^* \end{pmatrix} \tag{5.66}$$

represent the symbols transmitted at a certain time instant, while the rows represent the symbols transmitted over a certain antenna. The entire set of all code words is denoted by $\mathbb{X}_2$. Since $K = 2$ symbols s_1 and s_2 are transmitted during $L = 2$ time slots, the rate of this

SPACE–TIME CODES

code is $R = K/L = 1$. It is important to mention that the columns in $\mathbf{X}_2$ are orthogonal, so that Alamouti's scheme does not provide a coding gain.

Taking the conjugate complex of the second line in (5.65), we can rewrite this equation and obtain Equation (5.63) in Figure 5.33. Obviously, this slight modification has transformed the Multiple-Input Single-Output (MISO) channel $\mathbf{h}$ into an equivalent MIMO channel

$$\tilde{\mathbf{H}} = \mathbf{H}[\mathbf{X}_2] = \frac{1}{\sqrt{2}} \cdot \begin{pmatrix} h_1 & h_2 \\ -h_2^* & h_1^* \end{pmatrix}.$$

This matrix has orthogonal columns, so that the matched filter $\tilde{\mathbf{H}}$ represents the optimum receive filter according to Section 5.1. The matched filter output is given in Equation (5.64).

A comparison of the diagonal elements with the results of the receive diversity concept on page 224 illustrates the equivalence of both concepts. The multiple antenna side has simply been moved from the receiver to the transmitter, leading to similar results. In both cases, the squared magnitudes of the contributing channel coefficients h_ν are summed. Hence, the full diversity degree of $D = 2$ is obtained which is the largest possible degree for $N_T = 2$.

However, there exists a major difference between transmit and receive diversity which can be illuminated by deriving the signal-to-noise ratio. For Alamouti's scheme, the signal power after matched filtering becomes

$$S = \frac{1}{4} \cdot \left(|h_1|^2 + |h_2|^2\right)^2 \cdot \sigma_S^2$$

while

$$N = \frac{1}{2} \cdot \left(|h_1|^2 + |h_2|^2\right) \cdot \sigma_\mathcal{N}^2$$

holds for the received noise power. Consequently, we obtain

$$\gamma = \frac{S}{N} = \frac{1}{2} \cdot \left(|h_1|^2 + |h_2|^2\right) \cdot \frac{\sigma_S^2}{\sigma_\mathcal{N}^2} = \frac{1}{2} \cdot \left(|h_1|^2 + |h_2|^2\right) \cdot \frac{E_s}{N_0} \quad (5.67)$$

instead of $(|h_1|^2 + |h_2|^2)E_s/N_0$ for the receive diversity concept. Comparing both SNRs, we observe that they differ by the factor $1/\sqrt{2}$. This corresponds to an SNR loss of 3 dB because no array gain is possible without channel knowledge at the transmitter. The reason for this difference is that we assumed perfect channel knowledge at the receiver, so that the matched filter delivered an array gain of $10\log_{10}(N_R) \approx 3$ dB. However, we have no channel knowledge at the transmitter, and hence no array gain. With perfect channel knowledge at the transmitter, both results would have been identical.

Moreover, the off-diagonal elements are zero, so that no interference between s_1 and s_2 disturbs the decision. Additionally, the noise remains white when multiplied with a matrix consisting of orthogonal columns. Hence, the symbol-by symbol detection

$$\hat{s}_\mu = \underset{a \in \mathbb{S}}{\operatorname{argmin}} \left| y_\mu - (|h_1|^2 + |h_2|^2)a \right|^2. \quad (5.68)$$

is optimum.

Application to UMTS

In the UMTS standard (release 99) (3GPP, 1999), a slightly different implementation was chosen because the compatibility with one-antenna devices should be preserved. This modification does not change the achievable diversity gain. Instead of setting up the space–time code word according to Equation (5.66), the code matrix has the form

$$\mathbf{X}_2 = \begin{pmatrix} \mathbf{x}[\ell-1] & \mathbf{x}[\ell] \end{pmatrix} = \frac{1}{\sqrt{2}} \cdot \begin{pmatrix} s_1 & s_2 \\ -s_2^* & s_1^* \end{pmatrix} \qquad (5.69)$$

This implementation transmits both original symbols s_1 and s_2 over the first antenna, whereas the complex conjugate symbols s_1^* and $-s_2^*$ are emitted over the second antenna. If a transmitter has only a single antenna, we simply have to remove the second row of $\mathbf{X}_2$; the signalling in the first row is not affected at all. On the other hand, switching from $N_T = 1$ to $N_T = 2$ just requires the activation of the second antenna without influencing the data stream $x_1[\ell]$. The original approach of Alamouti would require a complete change of $\mathbf{X}_2$.

5.4.2 Extension to More than Two Transmit Antennas

Figure 5.34 shows the basic structure of a space–time block coding system for $N_R = 1$ receive antenna. The space–time encoder collects a block of K successive symbols s_μ and assigns them onto a sequence of L consecutive vectors

$$\mathbf{x}[k] = \begin{pmatrix} x_1[k] & \cdots & x_{N_T}[k] \end{pmatrix}^T$$

with $0 \leq k < L$. Therefore, the code matrix $\mathbf{X}_{N_T}$ consists of K symbols $s_1, \ldots, s_K$ as well as their conjugate complex counterparts $s_1^*, \ldots, s_K^*$ that are arranged in N_T rows and L columns. Since the matrix occupies L time instants, the code rate is given in Equation (5.70). When comparing different space–time coding schemes, one important parameter is the spectral efficiency η in Equation (5.71). It equals the product of R and the number m of bits per modulation symbol $s[\ell]$. Therefore, it determines the average number of bits that are transmitted per channel use. In order to obtain orthogonal space–time block codes, the rows in $\mathbf{X}_{N_T}$ have to be orthogonal, resulting in the orthogonality constraint given in Equation(5.72).

The factor in front of the identity matrix ensures that the average transmit power per symbol equals E_s/T_s and is independent of N_T and L. Since $\mathbf{X}_{N_T}^H \mathbf{X}_{N_T}$ is a square $N_T \times N_T$ matrix, the equality

$$\text{tr}\left\{ \mathbf{X}_{N_T} \mathbf{X}_{N_T}^H \right\} = K \cdot \frac{E_s}{T_s} \qquad (5.73)$$

holds. In order to illustrate the condition in Equation (5.73), we will first look at Alamouti's scheme. Each of the two symbols is transmitted twice during one block, once as the original version and a second time as the complex conjugate version. As the total symbol power should be fixed to E_s/T_s, a scaling factor of $1/\sqrt{2}$ in front of $\mathbf{X}_2$ is required, as already used on page 258. For general codes where a symbol is transmitted N times (either the original symbol or its complex conjugate) with full power, the scaling factor $1/\sqrt{N}$ is obtained. As we will see, some schemes attenuate symbols in $\mathbf{X}_{N_T}$ differently, i.e. they are not transmitted each time with full power. Consequently, this has to be considered when determining an appropriate scaling factor.

SPACE–TIME CODES

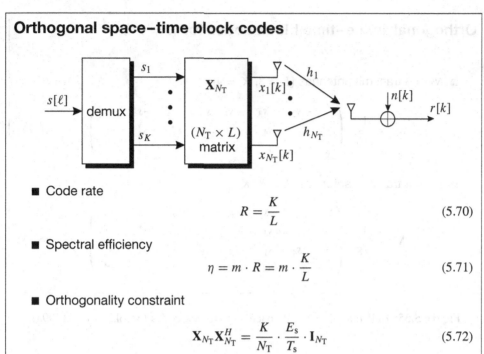

Figure 5.34: General structure of orthogonal space–time block codes with N_T transmit antennas

Unfortunately, Alamouti's scheme is the only orthogonal space–time code with rate $R = 1$. For $N_T > 2$, orthogonal codes exist only for rates $R < 1$. This loss in spectral efficiency can be compensated for by using larger modulation alphabets, e.g. replacing Quaternary Phase Shift Keying (QPSK) with 16-QAM, so that more bits per symbol can be transmitted. However, we know from Subsection 5.1.1 that increasing the modulation size M leads to higher error rates. Hence, we have to answer the question as to whether the achievable diversity gain will be larger than the SNR loss due to a change of the modulation scheme.

Half-rate codes exist for an arbitrary number of transmit antennas. Code matrices for $N_T = 3$ and $N_T = 4$ are presented and summarised in Figure 5.35. Both codes achieve the full diversity degree that equals N_T. For $N_T = 3$, each of the $K = 4$ symbols occurs 6 times in $\mathbf{X}_3$, resulting in a scaling factor of $1/\sqrt{6}$. With $N_T = 4$ transmit antennas, again four symbols are mapped onto $L = 8$ time slots, where each symbol is used 8 times, leading to a factor $1/\sqrt{8}$.

Figure 5.36 shows two codes with $N_T = 3$ and $N_T = 4$ (Tarokh et al., 1999a,b). Both codes again have the full diversity degree and map $K = 3$ symbols onto $L = 4$ time slots, leading to a rate of $R = 3/4$. In order to distinguish them from the codes presented so far, we use the notation $\mathbf{T}_3$ and $\mathbf{T}_4$. For $N_T = 3$, the orthogonal space–time code word is presented in Equation (5.76). Summing the squared magnitudes for each symbol in $\mathbf{T}_3$

Orthogonal space–time block codes for $R = 1/2$

- $N_T = 3$ transmit antennas ($L = 8$, $K = 4$)

$$\mathbf{X}_3 = \frac{1}{\sqrt{6}} \cdot \begin{pmatrix} s_1 & -s_2 & -s_3 & -s_4 & s_1^* & -s_2^* & -s_3^* & -s_4^* \\ s_2 & s_1 & s_4 & -s_3 & s_2^* & s_1^* & s_4^* & -s_3^* \\ s_3 & -s_4 & s_1 & s_2 & s_3^* & -s_4^* & s_1^* & s_2^* \end{pmatrix} \quad (5.74)$$

- $N_T = 4$ transmit antennas ($L = 8$, $K = 4$)

$$\mathbf{X}_4 = \frac{1}{\sqrt{8}} \cdot \begin{pmatrix} s_1 & -s_2 & -s_3 & -s_4 & s_1^* & -s_2^* & -s_3^* & -s_4^* \\ s_2 & s_1 & s_4 & -s_3 & s_2^* & s_1^* & s_4^* & -s_3^* \\ s_3 & -s_4 & s_1 & s_2 & s_3^* & -s_4^* & s_1^* & s_2^* \\ s_4 & s_3 & -s_2 & s_1 & s_4^* & s_3^* & -s_2^* & s_1^* \end{pmatrix} \quad (5.75)$$

Figure 5.35: Half-rate orthogonal space–time block codes (Tarokh et al., 1999a)

Orthogonal space–time block codes for $R = 3/4$

- $N_T = 3$ transmit antennas ($L = 4$, $K = 3$)

$$\mathbf{T}_3 = \frac{1}{\sqrt{12}} \cdot \begin{pmatrix} 2s_1 & -2s_2^* & \sqrt{2}s_3^* & \sqrt{2}s_3^* \\ 2s_2 & 2s_1^* & \sqrt{2}s_3^* & -\sqrt{2}s_3^* \\ \sqrt{2}s_3 & \sqrt{2}s_3 & -s_1 - s_1^* + s_2 - s_2^* & s_1 - s_1^* + s_2 + s_2^* \end{pmatrix} \quad (5.76)$$

- $N_T = 4$ transmit antennas ($L = 4$, $K = 3$)

$$\mathbf{T}_4 = \frac{1}{4} \begin{pmatrix} 2s_1 & -2s_2^* & \sqrt{2}s_3^* & \sqrt{2}s_3^* \\ 2s_2 & 2s_1^* & \sqrt{2}s_3^* & -\sqrt{2}s_3^* \\ \sqrt{2}s_3 & \sqrt{2}s_3 & -s_1 - s_1^* + s_2 - s_2^* & s_1 - s_1^* + s_2 + s_2^* \\ \sqrt{2}s_3 & -\sqrt{2}s_3 & -s_1 - s_1^* - s_2 - s_2^* & -(s_1 + s_1^* + s_2 + s_2^*) \end{pmatrix} \quad (5.77)$$

Figure 5.36: Orthogonal space–time block codes with rate $R = 3/4$ (Tarokh et al., 1999a)

SPACE–TIME CODES

results in a value of 12. Hence, the scaling factor amounts to $1/\sqrt{12}$. For the code $\mathbb{T}_4$, the summation yields a value of 16, and thus a scaling factor of 0.25.

The detection at the receiver works in the same way as for Alamouti's scheme. Kühn derived the equivalent channel matrices $\mathbf{H}[\mathbf{X}_{N_T}]$ for each space–time block code (Kühn, 2006). They make it possible to describe the received vector as the product of $\mathbf{H}[\mathbf{X}_{N_T}]$ and the transmit vector $\mathbf{x}$.[5] Since the columns of these matrices are orthogonal, a simple multiplication with their Hermitian delivers the desired data symbol estimates $\tilde{s}_\mu$.

5.4.3 Simulation Results

After the introduction of several orthogonal space–time block codes, we now analyse their error rate performance. A first comparison regards all discussed schemes with BPSK modulation. Hence, the spectral efficiencies are different, as shown in Figure 5.37. In the left-hand diagram, the error rates are depicted versus E_s/N_0. Obviously, codes with identical diversity degrees such as $\mathbb{X}_3$ and $\mathbb{T}_3$ achieve the same error rates because their differing code rates are not considered. This somehow leads to an unfair comparison. Instead, one

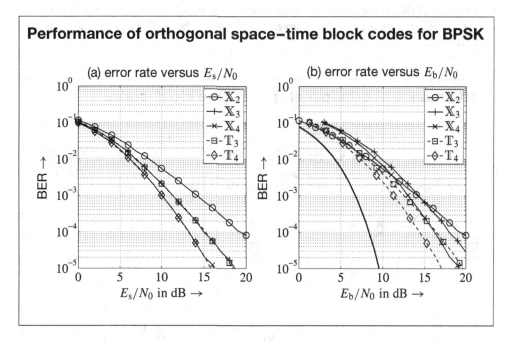

Figure 5.37: Bit error rates for different orthogonal STBCs with BPSK and code rates $R(\mathbb{X}_2) = 1$, $R(\mathbb{X}_3) = R(\mathbb{X}_4) = 0.5$ and $R(\mathbb{T}_3) = R(\mathbb{T}_4) = 0.75$ (AWGN reference: bold line). Reproduced by permission of John Wiley & Sons, Ltd

---

[5]For the code matrices $\mathbf{T}_3$ and $\mathbf{T}_4$, a real-valued description is required, i.e. the linear equation system does not contain complex symbols and their conjugates, but only real and imaginary parts of all components.

should depict the error rates versus E_b/N_0, where $E_s = RE_b$ holds. First, the corresponding results in the right-hand diagram illustrate that the slopes of all curves are still the same because using E_b/N_0 does not change the diversity degree. However, horizontal shifts can be observed, so that $\mathbb{T}_4$ now yields the best results. The higher diversity degree of 4 overcompensates for the lower code rate compared with Alamouti's scheme. On the other hand, the half-rate codes lose most, and Alamouti's code outperforms $\mathbb{X}_3$ over a wide range of E_b/N_0 values. As a reference, the AWGN curve is plotted as a bold line. Certainly, it cannot be reached by any of the codes owing to a maximum diversity degree of only 4.

Next, we would like to compare different space–time coding schemes under the constraint of identical spectral efficiencies. Therefore, the modulation schemes have to be adapted. In order to achieve a spectral efficiency $\eta = 2$ bit/s/Hz, Alamouti's scheme has to employ QPSK, while the codes $\mathbb{X}_3$ and $\mathbb{X}_4$ with $R = 1/2$ have to use 16-QAM or 16-PSK. Owing to the better performance of 16-QAM, we confine ourselves to that modulation scheme. For $\eta = 3$ bit/s/Hz, 8-PSK is chosen for $\mathbb{X}_2$ and 16-QAM for $\mathbb{T}_3$ and $\mathbb{T}_4$.

The results are depicted in Figure 5.38. On account of to the higher robustness of QPSK compared with 16-QAM the code $\mathbb{X}_2$ performs better than $\mathbb{X}_3$ and $\mathbb{X}_4$ for low and medium signal-to-noise ratios (left-hand diagram). Asymptotically, the diversity gain becomes dominating owing to the larger slope of the curves, so that $\mathbb{X}_3$ and $\mathbb{X}_4$ are better for large SNR. The error rates of all space–time codes are significantly better than simple QPSK transmission without diversity, although the AWGN performance is not reached.

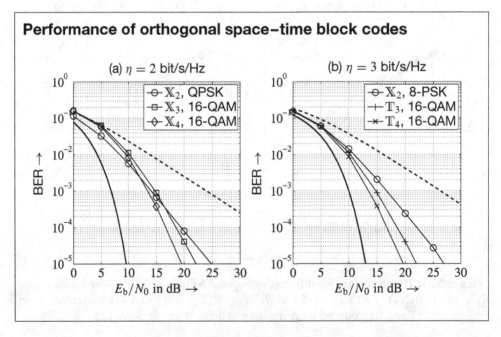

Figure 5.38: Bit error rates for different orthogonal STBCs with $\eta = 2$ bit/s/Hz and $\eta = 3$ bit/s/Hz (AWGN reference: bold solid line; Rayleigh reference: bold dashed line)

SPACE-TIME CODES

Diagram (b) illustrates results for a spectral efficiency $\eta = 3$ bit/s/Hz. Owing to the relative high code rate $R = 0.75$ for $\mathbb{T}_3$ and $\mathbb{T}_4$, only four instead of three bits per symbol have to be transmitted, leading to the modulation schemes 8-PSK and 16-QAM. As we know from Subsection 5.1.1, 16-QAM performs nearly as well as 8-PSK. Hence, the larger diversity gain for $\mathbb{T}_3$ and $\mathbb{T}_4$ is visible over the whole range of SNRs, and the weaker performance of 8-PSK compared with 16-QAM is negligible.

5.5 Spatial Multiplexing

5.5.1 General Concept

In the previous section, coding in space and time has been discussed, mainly for systems with multiple transmit but only a single receive antenna. This approach aims to increase the diversity degree and, hence, the reliability of a communication link. The achievable data rate is not changed or even decreased by code rates $R < 1$. By Contrast, multiple antennas at transmitter and receiver make it possible to increase the data rate by exploiting the resource space without increasing the bandwidth. This multiplexing gain is possible by transmitting independent data streams over spatially separated channels. We will see later that diversity and multiplexing gains can be obtained simultaneously by using a more general description such as linear dispersion codes.

In this section, we assume frequency-non-selective fading channels between each pair of transmit and receive antennas that can be modelled by complex-valued coefficients whose real and imaginary parts are independent and Gaussian distributed. Linear modulation schemes such as M-PSK or M-QAM are used at the transmitter. The improved throughput by increasing M comes along with a larger sensitivity to noise and interference. Moreover, the computational complexity of some detection schemes grows exponentially with the number of the possible transmit vectors and, thus, with M. As stated in Section 5.2, the general system model can be described by

$$\mathbf{r} = \mathbf{H} \cdot \mathbf{x} + \mathbf{n} \tag{5.78}$$

Principally, two cases have to be distinguished.

Exploiting Eigenmodes of Spatial Channel

If channel knowledge is available at transmitter and receiver side, we saw in Subsection 5.3.1 that capacity can be achieved by exploiting the channel's eigenmodes and applying the waterfilling principle. The maximum number of non-zero eigenmodes equals the rank of $\mathbf{H}$ and is upper bounded by $\min\{N_T, N_R\}$. However, it has to be mentioned that the number of parallel data streams that can be supported also depends on the available transmit power $\sigma_\mathcal{X}^2 > 0$ and may be smaller than the rank of $\mathbf{H}$.

The transmission strategy is accomplished by performing a singular value decomposition of the channel matrix $\mathbf{H} = \mathbf{U}_\mathcal{H} \mathbf{\Sigma}_\mathcal{H} \mathbf{V}_\mathcal{H}^H$. If the waterfilling solution distributes the total transmit power onto r data streams with powers λ_i, the transmit vector is constructed by

$$\mathbf{x} = \mathbf{V}_\mathcal{X} \cdot \mathbf{\Lambda}_\mathcal{X}^{1/2} \cdot \mathbf{s},$$

where $\Lambda_\mathcal{X} = \text{diag}[\lambda_1 \cdots \lambda_r]$ is a diagonal matrix containing the different power levels. The transmit filter $\mathbf{V}_\mathcal{X}$ comprises those columns of $\mathbf{V}_\mathcal{H}$ that correspond to the r largest singular values of $\mathbf{H}$. Equivalently, the receive filter $\mathbf{U}_\mathcal{X}$ contains those columns of $\mathbf{U}_\mathcal{H}$ that are associated with the used eigenmodes. Using $\mathbf{V}_\mathcal{X}$ and $\mathbf{U}_\mathcal{X}$ leads to

$$\mathbf{y} = \mathbf{U}_\mathcal{X}^H \cdot \mathbf{r} = \Sigma_\mathcal{H} \cdot \Lambda_\mathcal{X} \cdot \mathbf{s} + \tilde{\mathbf{n}}.$$

Since $\Sigma_\mathcal{H}$ and $\Lambda_\mathcal{X}$ are diagonal matrices, the data streams are perfectly separated into parallel channels whose signal-to-noise ratios amount to

$$\gamma_i = \sigma_{\mathcal{H},i}^2 \cdot \lambda_i \cdot \frac{\sigma_\mathcal{X}^2}{\sigma_\mathcal{N}^2} = \sigma_{\mathcal{H},i}^2 \cdot \lambda_i \cdot \frac{E_s}{N_0}.$$

They depend on the squared singular values $\sigma_{\mathcal{H},i}^2$ and the specific transmit power levels λ_i. For digital communications, the input alphabet is not Gaussian distributed as assumed in Section 5.3 but consists of discrete symbols. Hence, finding the optimal bit and power allocation is a combinatorial problem and cannot be found by gradient methods as discussed on 241. Instead, algorithms presented elsewhere (Fischer and Huber, 1996; Hughes-Hartogs, 1989; Krongold et al., 1999; Mutti and Dahlhaus, 2004) have to be used. Different optimisation strategies are possible. One target may be to maximise the throughput at a given average error rate. Alternatively, we can minimise the error probability at a given total throughput or minimise the transmit power for target error and data rates.

No Channel Knowledge at Transmitter

In Section 5.3 it was shown that the resource space can be used even in the absence of channel knowledge at the transmitter. The loss compared with the optimal waterfilling solution is rather small. However, the price to be paid is the application of advanced signal processing tools at the receiver. A famous example is the Bell Labs Layered Space–Time (BLAST) architecture (Foschini, 1996; Foschini and Gans, 1998; Foschini et al., 1999). Since no channel knowledge is available at the transmitter, the best strategy is to transmit independent equal power data streams called layers. At least two BLAST versions exist. The first, termed diagonal BLAST, distributes the data streams onto the transmit antennas according to a certain permutation pattern. This kind of interleaving ensures that the symbols within each layer experience more different fading coefficients, leading to a higher diversity gain during the decoding process.

In this section, we focus only on the second version, termed vertical BLAST, which is shown in Figure 5.39. Hence, no interleaving between layers is performed and each data stream is solely assigned to a single antenna, leading to $\mathbf{x} = \mathbf{s}$. The name stems from the vertical arrangement of the layers. This means that channel coding is applied per layer. Alternatively, it is also possible to distribute a single coded data stream onto the transmit antennas. There are slight differences between the two approaches, especially concerning the detection at the receiver. As we will soon see, per-layer encoding makes it possible to include the decoder in an iterative turbo detection, while this is not directly possible for a single code stream.

From the mathematical description in Equation (5.78) we see that a superposition

$$r_\mu = \sum_{\nu=1}^{N_T} h_{\mu,\nu} \cdot x_\nu + n_\mu$$

SPACE–TIME CODES

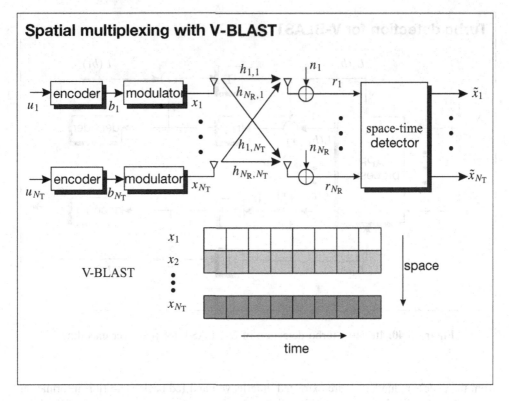

Figure 5.39: Multilayer transmission with vertical BLAST (V-BLAST) and per-layer encoding

of all transmitted layers is obtained at each receive antenna $1 \leq \mu \leq N_R$. The task of the space–time detector is to separate the data streams again. In coded systems, the space–time detector and decoder are generally implemented separately for complexity reasons. However, both components can exchange iteratively information equivalent to the turbo principle for concatenated codes discussed in Chapter 4. We will continue the discussion with the optimum bit-by-bit space–time detector that delivers LLRs. Since the computational complexity becomes demanding for high modulation levels and many layers, alternative detection strategies are discussed subsequently.

5.5.2 Iterative APP Preprocessing and Per-layer Decoding

As mentioned above, we will now consider a receiver structure as depicted in Figure 5.40. A joint preprocessor calculates LLRs $L(\hat{b}_\nu \mid \mathbf{r})$ for each layer ν on the basis of the received samples r_1 up to r_{N_R}. As we assume a memoryless channel $\mathbf{H}$, the preprocessor can work in a symbol-by-symbol manner without performance loss. After de-interleaving, the FEC decoders deliver independently for each layer estimates $\hat{u}_\nu$ of the information bits as well as LLRs $L(\hat{b}_\nu)$ of the coded bits. From the LLRs, the extrinsic parts, i.e. the parts generated

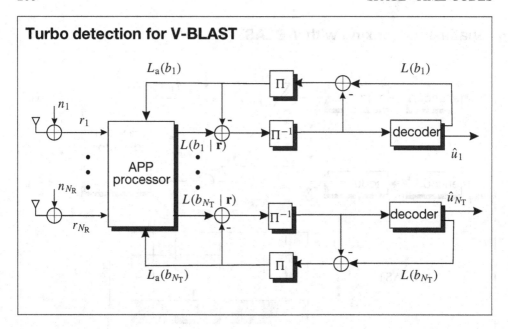

Figure 5.40: Iterative turbo detection of V-BLAST for per-layer encoding

from the code's redundancy, are extracted, interleaved and fed back as a-priori information to the preprocessor. Now, a second iteration can start with an improved output of the joint preprocessor owing to the available side information from the previous iteration. As already described in Chapter 4, an iterative detection process is obtained that aims to approach the maximum likelihood solution (Liew and Hanzo, 2002).

We will now have a more thorough look at the joint preprocessor. The major steps are summarised in Figure 5.41. From Equation (5.79) we see that the LLR depends on the ratio of conditional probabilities $\Pr\{b_\nu = \xi \mid \mathbf{r}\}$ which are defined as

$$\Pr\{b_\nu = \xi \mid \mathbf{r}\} = \frac{p(b_\nu, \mathbf{r})}{p(\mathbf{r})} .$$

Since $p(\mathbf{r})$ is identical in numerator and denominator of the log likelihood ratio, it can be skipped, leading to the right-hand side of Equation (5.79). The joint probability densities have to be further processed to obtain a tractable form. For multilevel modulation schemes such as QAM and PSK, each symbol s is assigned to several bits b_ν with $1 \leq \nu \leq \mathrm{ld}(M)$. The probability that a certain bit b_ν takes a value ξ can be calculated by summing the probabilities $\Pr\{s\}$ over those symbols s whose νth bit equals ξ

$$\Pr\{b_\nu = \xi\} = \sum_{s, b\nu = \xi} \Pr\{s\}.$$

At this point it has to be emphasised that $\mathbf{r}$ comprises corrupted versions of all N_T transmitted symbols. Therefore, it is not sufficient to consider a single symbol s. Instead, the set

SPACE-TIME CODES

APP-preprocessor for V-BLAST

- LLR of APP preprocessor

$$L(b_\nu \mid \mathbf{r}) = \log \frac{\Pr\{b_\nu = 0 \mid \mathbf{r}\}}{\Pr\{b_\nu = 1 \mid \mathbf{r}\}} = \log \frac{p(b_\nu = 0, \mathbf{r})}{p(b_\nu = 1, \mathbf{r})} \qquad (5.79)$$

- Joint probability density

$$p(b_\nu = \xi, \mathbf{r}) = \sum_{\mathbf{s} \in \mathbb{S}_\nu(\xi)} p(\mathbf{s}, \mathbf{r}) = \sum_{\mathbf{s} \in \mathbb{S}_\nu(\xi)} p(\mathbf{r} \mid \mathbf{s}) \cdot \Pr\{\mathbf{s}\} \qquad (5.80)$$

- Conditional probability density function

$$p(\mathbf{r} \mid \mathbf{s}) \propto \exp\left[-\frac{\|\mathbf{r} - \mathbf{H}\mathbf{s}\|^2}{\sigma_\mathcal{N}^2}\right] \qquad (5.81)$$

- A-priori probability

$$\Pr\{\mathbf{s}\} = \prod_{\nu=1}^{N_T \,\mathrm{ld}(M)} \Pr\{b_\nu(\mathbf{s})\} \qquad (5.82)$$

Figure 5.41: APP calculation steps for the joint preprocessor of V-BLAST systems

of N_T symbols and, correspondingly, all bits b_ν with $1 \le \nu \le N_T \,\mathrm{ld}(M)$ have to be taken into account. For this purpose, we define a kind of vector modulation that maps a vector

$$\mathbf{b} = \begin{pmatrix} b_1 & b_2 & \cdots & b_{N_T \,\mathrm{ld}(M)} \end{pmatrix}^\mathrm{T}$$

of $N_T \,\mathrm{ld}(M)$ bits onto a vector

$$\mathbf{s} = \begin{pmatrix} s_1 & s_2 & \cdots & s_{N_T} \end{pmatrix}^\mathrm{T}$$

of N_T symbols. The set of vectors is divided into subsets $\mathbb{S}_\nu(\xi)$ containing those symbol ensembles for which the νth bit of their binary representation $\mathbf{b}$ takes the value $b_\nu = \xi$, leading to Equation (5.80). This transforms the generally unknown joint probability $p(b_\nu, \mathbf{r})$ into an expression depending on the conditional densities $p(\mathbf{r} \mid \mathbf{s})$ and the a-priori probabilities $\Pr\{\mathbf{s}\}$. The former determine the channel statistics and are proportional to the expression in Equation (5.81). Assuming that all bits b_ν are statistically independent, the a-priori probability $\Pr\{\mathbf{s}\}$ can be factorised into the marginal probabilities according to Equation (5.82) where the bits $b_\nu(\mathbf{s})$ are determined by the specific symbol vector $\mathbf{s}$.

From the above equations we can already conclude that the computational complexity grows linearly with the number of possible hypotheses and, hence, exponentially with the number of layers as well as the number of bits per symbol. Thus, this optimal approach is feasible only for small systems and small modulation alphabets.

Turbo Iterations

In the first iteration, no a-priori information from the decoders is available. Hence, the a-priori probabilities are constant, $\Pr\{s\} = 2^{-N_T \operatorname{ld}(M)}$, and can be dropped. The insertion of Equation (5.80) and Equation (5.81) into Equation (5.79) then leads to Equation (5.83) in Figure 5.42. We observe that the numerator and denominator only differ by the subsets $\$_\nu(\xi)$ which distinguish the two hypotheses $b_\nu = 0$ and $b_\nu = 1$. Subsequent per-layer soft-output decoding according to Bahl, Cocke, Jelinek, Raviv (BCJR) or max-log MAP algorithms (rf. to Section 3.4) provides LLRs for each code bit and layer.

APP preprocessor and turbo iterations

■ LLR of APP preprocessor in first iteration

$$L(b_\nu \mid \mathbf{r}) = \log \frac{\sum_{\mathbf{s} \in \$_\nu(0)} \exp\left[-\frac{\|\mathbf{r}-\mathbf{Hs}\|^2}{\sigma_\mathcal{N}^2}\right]}{\sum_{\mathbf{s} \in \$_\nu(1)} \exp\left[-\frac{\|\mathbf{r}-\mathbf{Hs}\|^2}{\sigma_\mathcal{N}^2}\right]} \qquad (5.83)$$

■ Calculating a-priori probabilities from decoder LLRs

$$\Pr\{b_\nu = \xi\} = \frac{\exp[-\xi L_a(b_\nu)]}{1 + \exp[-L_a(b_\nu)]} \qquad (5.84)$$

■ A-priori information per symbol

$$\Pr\{\mathbf{s}\} = \prod_{\nu=1}^{N_T \operatorname{ld}(M)} \frac{\exp[-b_\nu(\mathbf{s}) L_a(b_\nu)]}{1 + \exp[-L_a(b_\nu)]} \to \prod_{\nu=1}^{N_T \operatorname{ld}(M)} \exp[-b_\nu(\mathbf{s}) L_a(b_\nu)] \qquad (5.85)$$

■ LLR of APP preprocessor after first iteration

$$L(b_\nu \mid \mathbf{r}) = \log \frac{\sum_{\mathbf{s} \in \$_\nu(0)} \exp\left[-\frac{\|\mathbf{r}-\mathbf{Hs}\|^2}{\sigma_\mathcal{N}^2} - \sum_{\nu=1}^{N_T \operatorname{ld}(M)} b_\nu(\mathbf{s}) L_a(b_\nu)\right]}{\sum_{\mathbf{s} \in \$_\nu(1)} \exp\left[-\frac{\|\mathbf{r}-\mathbf{Hs}\|^2}{\sigma_\mathcal{N}^2} - \sum_{\nu=1}^{N_T \operatorname{ld}(M)} b_\nu(\mathbf{s}) L_a(b_\nu)\right]} \qquad (5.86)$$

Figure 5.42: APP preprocessor and turbo iterations for the V-BLAST system

SPACE–TIME CODES

They are fed back into the APP processor and have to be converted into a-priori probabilities Pr{s}. Using the results from the introduction of APP decoding given in Section 3.4, we can easily derive Equation (5.83) where $\xi \in \{0, 1\}$ holds. Inserting the expression for the bit-wise a-priori probability in Equation (5.84) into Equation (5.82) directly leads to Equation (5.85). Since the denominator of Equation (5.84) does not depend on the value of b_ν itself but only on the decoder output $L(b_\nu)$, it is independent of the specific symbol vector s represented by the bits b_ν. Hence, it becomes a constant factor regarding s and can be dropped, as done on the right-hand side.

Replacing the a-priori probabilities in Equation (5.79) with the last intermediate results leads to the final expression in Equation (5.86). On account of $b_\nu \in \{0, 1\}$, the a-priori LLRs only contribute to the entire result if a symbol vector s with $b_\nu = 1$ is considered. For these cases, the $L(b_\nu)$ contains the correct information only if it is negative, otherwise its information is wrong. Therefore, true a-priori information increases the exponents in Equation (5.86) which is consistent with the negative squared Euclidean distance which should also be maximised.

Max-Log MAP Solution

As already mentioned above, the complexity of the joint preprocessor still grows exponentially with the number of users and the number of bits per symbol. In Section 3.4 a suboptimum derivation of the BCJR algorithm has been introduced. This max-log MAP approach works in the logarithmic domain and uses the Jacobian logarithm

$$\log(e^{x_1} + e^{x_2}) = \log\left[e^{\max\{x_1, x_2\}}\left(1 + e^{-|x_1 - x_2|}\right)\right]$$
$$= \max\{x_1, x_2\} + \log\left[1 + e^{-|x_1 - x_2|}\right]$$

Obviously, the right-hand side depends on the maximum of x_1 and x_2 as well as on the absolute difference. If the latter is large, the logarithm is close to zero and can be dropped. We obtain the approximation

$$\log(e^{x_1} + e^{x_2}) \approx \max\{x_1, x_2\} \tag{5.87}$$

Applying approximation (5.87) to Equation (5.86) leads to

$$L(b_\nu \mid \mathbf{r}) \approx \min_{\mathbf{s} \in \mathbb{S}_\nu(1)} \left\{ \frac{\|\mathbf{r} - \mathbf{H}\mathbf{s}\|^2}{\sigma_\mathcal{N}^2} + \sum_{\nu=1}^{N_T \operatorname{ld}(M)} b_\nu(\mathbf{s}) L_a(b_\nu) \right\}$$
$$- \min_{\mathbf{s} \in \mathbb{S}_\nu(0)} \left\{ \frac{\|\mathbf{r} - \mathbf{H}\mathbf{s}\|^2}{\sigma_\mathcal{N}^2} + \sum_{\nu=1}^{N_T \operatorname{ld}(M)} b_\nu(\mathbf{s}) L_a(b_\nu) \right\} \tag{5.88}$$

We observe that the ratio has become a difference and the sums in numerator and denominator have been exchanged by the minima searches. The latter can be performed pairwise between the old minimum and the new hypothesis. It has to be mentioned that the first minimisation runs over all $\mathbf{s} \in \mathbb{S}_\nu(0)$, while the second uses $\mathbf{s} \in \mathbb{S}_\nu(1)$.

Using Expected Symbols as A-Priori Information

Although the above approximation slightly reduces the computational costs, often they may be still too high. The prohibitive complexity stems from the sum over all possible hypotheses, i.e. symbol vectors **s**. A further reduction can be obtained by replacing the explicit consideration of each hypothesis with an average symbol vector $\bar{\mathbf{s}}$. To be more precise, the sum in Equation (5.86) is restricted to the M hypotheses of a single symbol s_μ in **s** containing the processed bit b_ν. This reduces the number of hypotheses markealy from M^{N_T} to M. For the remaining $N_T - 1$ symbols in **s**, their expected values

$$\bar{s}_\mu = \sum_{\xi \in \mathbb{S}} \xi \cdot \Pr\{\xi\} \propto \sum_{\xi \in \mathbb{S}} \xi \cdot \prod_{\nu=1}^{\mathrm{ld}(M)} e^{-b_\nu(\xi) L_a(b_\nu)} \qquad (5.89)$$

are used. They are obtained from the a-priori information of the FEC decoders. As already mentioned before, the bits b_ν are always associated with the current symbol ξ of the sum. In the first iteration, no a-priori information from the decoders is available, so that no expectation can be determined. Assuming that all symbols are equally likely would result in $\bar{s}_\mu \equiv 0$ for all μ. Hence, the influence of interfering symbols is not considered at all and the tentative result after the first iteration can be considered to be very bad. In many cases, convergence of the entire iterative process cannot be achieved. Instead, either the full-complexity algorithm or alternatives that will be introduced in the next subsections have to be used in this first iteration.

5.5.3 Linear Multilayer Detection

A reduction in the computational complexity can be achieved by separating the layers with a linear filter. These techniques are already well known from multiuser detection strategies in CDMA systems (Honig and Tsatsanis, 2000; Moshavi, 1996). In contrast to optimum maximum likelihood detectors, linear approaches do not search for a solution in the finite signal alphabet but assume continuously distributed signals. This simplifies the combinatorial problem to an optimisation task that can be solved by gradient methods. This leads to a polynomial complexity with respect to the number of layers instead of an exponential dependency. Since channel coding and linear detectors can be treated separately, this section considers an uncoded system. The MIMO channel output can be expressed by the linear equation system

$$\mathbf{r} = \mathbf{Hs} + \mathbf{n}$$

which has to be solved with respect to **s**.

Zero-Forcing Solution

The zero-forcing filter totally suppresses the interfering signals in each layer. It delivers the vector $\mathbf{s}_{\mathrm{ZF}} \in \mathbb{C}^{N_T}$ which minimises the squared Euclidean distance to the received vector **r**

$$\mathbf{s}_{\mathrm{ZF}} = \underset{\tilde{\mathbf{s}} \in \mathbb{C}^{N_T}}{\operatorname{argmin}} \|\mathbf{r} - \mathbf{H}\tilde{\mathbf{s}}\|^2 \qquad (5.90)$$

Although Equation (5.90) resembles the maximum likelihood approach, it significantly differs from it by the unconstraint search space $\mathbb{C}^{N_T}$ instead of $\mathbb{S}^{N_T}$. Hence, the result $\mathbf{s}_{\mathrm{ZF}}$ is

SPACE-TIME CODES

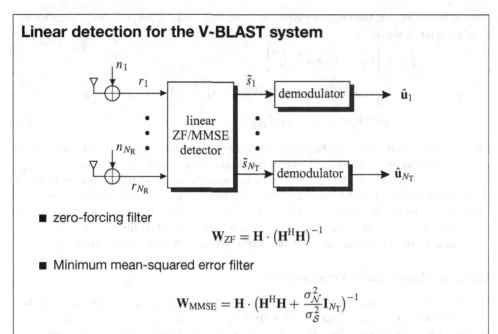

Figure 5.43: Linear detection for the V-BLAST system

generally not an element of $\mathbb{S}^{N_T}$. However, this generalisation transforms the combinatorial problem into one that can be solved by gradient methods. Hence, the squared Euclidean distance in Equation (5.90) is partially differentiated with respect to $\tilde{\mathbf{s}}^H$. Setting this derivative to zero

$$\frac{\partial}{\partial \tilde{\mathbf{s}}^H} \|\mathbf{r} - \mathbf{H}\tilde{\mathbf{s}}\|^2 = \frac{\partial}{\partial \tilde{\mathbf{s}}^H} (\mathbf{r} - \mathbf{H}\tilde{\mathbf{s}})^H \cdot (\mathbf{r} - \mathbf{H}\tilde{\mathbf{s}}) = -\mathbf{H}^H \mathbf{r} + \mathbf{H}^H \mathbf{H}\tilde{\mathbf{s}} \stackrel{!}{=} 0 \quad (5.91)$$

yields the solution

$$\mathbf{s}_{ZF} = \mathbf{W}_{ZF}^H \cdot \mathbf{r} = \mathbf{H}^\dagger \cdot \mathbf{r} = (\mathbf{H}^H \mathbf{H})^{-1} \cdot \mathbf{H}^H \cdot \mathbf{r} \quad (5.92)$$

The filter matrix $\mathbf{W}_{ZF} = \mathbf{H}^\dagger = \mathbf{H}(\mathbf{H}^H\mathbf{H})^{-1}$ is called the Moore–Penrose, or pseudo, inverse and can be expressed by the right-hand side of Equation (5.92) if $\mathbf{H}$ has full rank. In this case, the inverse of $\mathbf{H}^H\mathbf{H}$ exists and the filter output becomes

$$\mathbf{s}_{ZF} = (\mathbf{H}^H\mathbf{H})^{-1}\mathbf{H}^H \cdot (\mathbf{H}\mathbf{s} + \mathbf{n}) = \mathbf{s} + \mathbf{W}_{ZF}^H \cdot \mathbf{n} \quad (5.93)$$

Since the desired data vector $\mathbf{s}$ is only disturbed by noise, the final detection is obtained by a scalar demodulator $\hat{\mathbf{s}}_{ZF} = \mathcal{Q}(\mathbf{s}_{ZF})$ of the filter outputs $\mathbf{s}_{ZF}$. The non-linear function $\mathcal{Q}(\cdot)$ represents the hard-decision demodulation.

Although the filter output does not suffer from interference, the resulting signal-to-noise ratios per layer may vary significantly. This effect can be explained by the fact that the total suppression of interfering signals is achieved by projecting the received vector into the null space of all interferers. Since the desired signal may have only a small component lying in this subspace, the resulting signal-to-noise ratio is low. It can be expressed by the error covariance matrix

$$\Phi_{ZF} = E\left\{(s_{ZF} - s)(s_{ZF} - s)^H\right\}$$
$$= E\left\{(s + W_{ZF}^H n - s)(s + W_{ZF}^H n - s)^H\right\} = W_{ZF}^H E\{nn^H\} W_{ZF}$$
$$= \sigma_\mathcal{N}^2 \cdot W_{ZF}^H W_{ZF} = \sigma_\mathcal{N}^2 \cdot (H^H H)^{-1} \quad (5.94)$$

which contains on its diagonal the mean-squared error for each layer. The last row holds if the covariance matrix of the noise equals a diagonal matrix containing the noise power $\sigma_\mathcal{N}^2$.

The main drawback of the zero-forcing solution is the amplification of the background noise. If the matrix $H^H H$ has very small eigenvalues, its inverse may contain very large values that enhance the noise samples. At low signal-to-noise ratios, the performance of the Zero-Forcing filter may be even worse than a simple matched filter. A better solution is obtained by the Minimum Mean-Square Error (MMSE) filter described next.

Minimum Mean-Squared Error Solution

Looking back to Equation (5.90), we observe that the zero-forcing solution s_{ZF} does not consider that the received vector r is disturbed by noise. By contrast, the MMSE detector W_{MMSE} does not minimise the squared Euclidean distance between the estimate and the r, but between the estimate

$$s_{MMSE} = W_{MMSE}^H \cdot r \quad \text{with} \quad W_{MMSE} = \underset{W \in \mathbb{C}^{N_T \times N_R}}{\operatorname{argmin}} E\left\{\|W^H r - s\|^2\right\} \quad (5.95)$$

and the true data vector s. Similarly to the ZF solution, the partial derivative of the squared Euclidean distance with respect to W is determined and set to zero. With the relation $\partial W^H / \partial W = 0$ (Fischer, 2002), the approach

$$\frac{\partial}{\partial W} E\left\{\operatorname{tr}\left[(W^H r - x)(W^H r - x)^H\right]\right\} = W^H \Phi_{\mathcal{RR}} - \Phi_{\mathcal{SR}} \stackrel{!}{=} 0 \quad (5.96)$$

leads to the well-known Wiener solution

$$W_{MMSE}^H = \Phi_{\mathcal{SR}} \cdot \Phi_{\mathcal{RR}}^{-1} \quad (5.97)$$

The covariance matrix of the received samples has the form

$$\Phi_{\mathcal{RR}} = E\{rr^H\} = H \Phi_{\mathcal{XX}} H^H + \Phi_{\mathcal{NN}} \quad (5.98)$$

while the cross-covariance matrix becomes

$$\Phi_{\mathcal{SR}} = E\{sr^H\} = \Phi_{\mathcal{SS}} H^H + \Phi_{\mathcal{SN}} \quad (5.99)$$

Assuming that noise samples and data symbols are independent of each other and identically distributed, we obtain the basic covariance matrices

$$\Phi_{\mathcal{NN}} = E\{nn^H\} = \sigma_\mathcal{N}^2 \cdot I_{N_R} \quad (5.100a)$$
$$\Phi_{\mathcal{SS}} = E\{ss^H\} = \sigma_\mathcal{S}^2 \cdot I_{N_T} \quad (5.100b)$$
$$\Phi_{\mathcal{SN}} = E\{sn^H\} = 0 \quad (5.100c)$$

Inserting them into Equations (5.98) and (5.99) yields the final MMSE filter matrix

$$\mathbf{W}_{\text{MMSE}} = \mathbf{H}\left(\mathbf{H}^H\mathbf{H} + \frac{\sigma_N^2}{\sigma_S^2}\mathbf{I}_{N_R}\right)^{-1} \quad (5.101)$$

The MMSE detector does not suppress the multiuser interference perfectly, and some residual interference still disturbs the transmission. Moreover, the estimate is biased. The error covariance matrix, with Equation (5.101), now becomes

$$\Phi_{\text{MMSE}} = \mathrm{E}\left\{(\mathbf{s}_{\text{MMSE}} - \mathbf{s})(\mathbf{s}_{\text{MMSE}} - \mathbf{s})^H\right\} = \sigma_N^2\left(\mathbf{H}^H\mathbf{H} + \frac{\sigma_N^2}{\sigma_S^2}\mathbf{I}_{N_T}\right)^{-1} \quad (5.102)$$

From Equations (5.101) and (5.102) we see that the MMSE filter approaches the zero-forcing solution if the signal-to-noise ratio tends to infinity.

5.5.4 Original BLAST Detection

We recognized from Subsection 5.5.3 that the optimum APP preprocessor becomes quickly infeasible owing to its computational complexity. Moreover, linear detectors do not exploit the finite alphabet of digital modulation schemes. Therefore, alternative solutions have to be found that achieve a close-to-optimum performance at moderate implementation costs. Originally, a procedure consisting of a linear interference suppression stage and a subsequent detection and interference cancellation stage was proposed (Foschini, 1996; Foschini and Gans, 1998; Golden et al., 1998; Wolniansky et al., 1998). It detects the layers successively as shown in Figure 5.44, i.e. it uses already detected layers to cancel their interference onto remaining layers.

In order to get a deeper look inside, we start with the mathematical model of our MIMO system

$$\mathbf{r} = \mathbf{H}\mathbf{x} + \mathbf{n} \quad \text{with} \quad \mathbf{H} = \begin{pmatrix} \mathbf{h}_1 & \cdots & \mathbf{h}_{N_T} \end{pmatrix}$$

and express the channel matrix $\mathbf{H}$ by its column vectors $\mathbf{h}_\nu$. A linear suppression of the interference can be performed by applying a Zero Forcing (ZF) filter introduced on page 272. Since the ZF filter perfectly separates all layers, it totally suppresses the interference, and the only disturbance that remains is noise (Kühn, 2006).

$$\tilde{\mathbf{s}}_{\text{ZF}} = \mathbf{W}_{\text{ZF}}^H \mathbf{r} = \mathbf{s} + \mathbf{W}_{\text{ZF}}^H \mathbf{n}$$

However, the Moore-Penrose inverse incorporates the inverse $(\mathbf{H}^H\mathbf{H})^{-1}$ which may contain large values if $\mathbf{H}$ is poorly conditioned. They would lead to an amplification of the noise $\mathbf{n}$ and, thus, to small signal-to-noise ratios.

Owing to the successive detection of different layers, this procedure suffers from the risk of error propagation. Hence, choosing a suited order of detection is crucial. Obviously, one should start with the layer having the smallest error probability because this minimises the probability of error propagation. Since no interference disturbs the decision after the ZF filter, the best layer is the one with the largest SNR. This measure can be determined by the error covariance matrix defined in Equation (5.104). Its diagonal

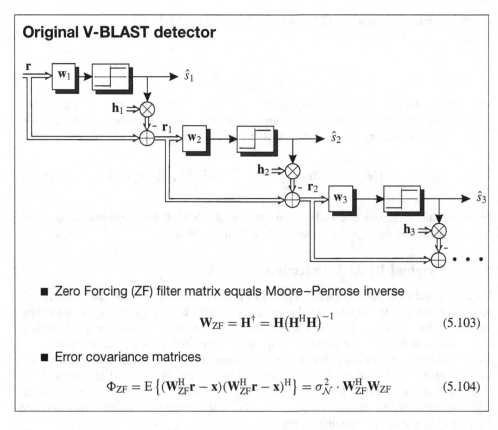

Figure 5.44: Structure of V-BLAST detector (Foschini, 1996; Foschini and Gans, 1998) with Zero Forcing (ZF) interference suppression

elements determine the squared Euclidean distances between the separated layers and the true symbols and equal the squared column norm of $\mathbf{W}_{ZF}$. Therefore, it suffices to determine $\mathbf{W}_{ZF}$; an explicit calculation of Φ_{ZF} is not required. We have to start with the layer that corresponds to the smallest column norm in $\mathbf{W}_{ZF}$. This leads to the algorithm presented in Figure 5.45.

The choice of the layer to be detected next in step 2 ensures that the risk of error propagation is minimised. Assuming that the detection in step 4 delivers the correct symbol $\hat{s}_{\lambda_\nu}$, the cancellation step 5 reduces the interference so that subsequent layers can be detected more reliably. Layers that have already been cancelled need not be suppressed by the ZF filter any more. Hence, the filter has more degrees of freedom and exploits more diversity (Kühn, 2006). Mathematically, this behaviour is achieved by removing columns from $\mathbf{H}$ which increases the null space for linear interference suppression.

In order to determine the linear filters in the different detection steps, the system matrices describing the reduced systems have to be inverted. This causes high implementation costs. However, a much more convenient way exists that avoids multiple matrix inversions. This approach leads to identical results and is presented in the next section.

ZF-V-BLAST detection algorithm

- Initialisation: set $\mathbf{r}_1 = \mathbf{r}$, $\mathbf{H}_1 = \mathbf{H}$
- The following steps are repeated for $\nu = 1, \ldots, N_T$ until all layers have been detected:
  1. Calculate the ZF filter matrix $\mathbf{W}_{ZF} = \mathbf{H}_\nu^\dagger$.
  2. Determine the filter $\mathbf{w}_\nu$ as the column in $\mathbf{W}_{ZF}$ with the smallest squared norm, e.g. the λ_νth column.
  3. Use $\mathbf{w}_\nu$ to extract the λ_νth layer by
  $$\tilde{s}_{\lambda_\nu} = \mathbf{w}_\nu^H \cdot \mathbf{r}_\nu.$$
  4. Detect the interference-free layer either by soft or hard decision
  $$\hat{s}_{\lambda_\nu} = \mathcal{Q}_\mathbb{S}(\tilde{s}_{\lambda_\nu}).$$
  5. Subtract the decided symbol $\hat{s}_{\lambda_\nu}$ from $\mathbf{r}_\nu$
  $$\mathbf{r}_{\nu+1} = \mathbf{r}_\nu - \mathbf{h}_{\lambda_\nu} \hat{s}_{\lambda_\nu}$$
  and delete the λ_νth column from $\mathbf{H}_\nu$, yielding $\mathbf{H}_{\nu+1}$.

Figure 5.45: V-BLAST detection algorithm (Foschini, 1996; Foschini and Gans, 1998) with Zero Forcing (ZF) interference suppression

MMSE Extension of V-BLAST

As mentioned above, the perfect interference suppression of the ZF filter comes at the expense of a severe noise amplification. Especially at low signal-to-noise ratios, it may happen that the ZF filter performs even worse than the simple matched filter. The noise amplification can be avoided by using the MMSE filter derived on page 274. It provides the smallest mean-squared error and represents a compromise between residual interference power that cannot be suppressed and noise power. Asymptotically, the MMSE filter approaches the matched filter for very low SNRs, and the ZF filter for very high SNRs.

From the optimisation approach

$$\mathbf{W}_{MMSE} = \underset{\mathbf{W} \in \mathbb{C}^{N_T \times N_R}}{\arg\min} \; \mathrm{E}\left\{ \|\mathbf{W}^H \mathbf{r} - \mathbf{s}\|^2 \right\}$$

we easily obtain the MMSE solution presented in Equation (5.105) in Figure 5.46. A convenient way to apply the MMSE filter to the BLAST detection problem without new derivation is provided in Equation (5.106). Extending the channel matrix $\mathbf{H}$ in the way

> **MMSE extension of V-BLAST detection**
>
> - Minimum Mean-Square Error (MMSE) filter matrix
>
> $$\mathbf{W}_{\text{MMSE}} = \mathbf{H}\left(\mathbf{H}^H\mathbf{H} + \frac{\sigma_\mathcal{N}^2}{\sigma_\mathcal{N}^2}\mathbf{I}_{N_T}\right)^{-1} \quad (5.105)$$
>
> - Relationship between ZF and MMSE filter
>
> $$\mathbf{W}_{\text{MMSE}}(\mathbf{H}) = \mathbf{W}_{\text{ZF}}(\underline{\mathbf{H}}) \quad \text{with} \quad \underline{\mathbf{H}} = \begin{pmatrix} \mathbf{H} \\ \frac{\sigma_\mathcal{N}}{\sigma_\mathcal{X}}\mathbf{I}_{N_T} \end{pmatrix} \quad (5.106)$$
>
> - Error covariance matrices
>
> $$\Phi_{\text{MMSE}} = \mathrm{E}\left\{(\mathbf{W}_{\text{MMSE}}\mathbf{r} - \mathbf{x})(\mathbf{W}_{\text{MMSE}}\mathbf{r} - \mathbf{x})^H\right\}$$
>
> $$= \sigma_\mathcal{N}^2 \cdot \left(\mathbf{H}^H\mathbf{H} + \frac{\sigma_\mathcal{N}^2}{\sigma_\mathcal{N}^2}\mathbf{I}_{N_T}\right)^{-1} \quad (5.107)$$

Figure 5.46: Procedure of V-BLAST detection (Foschini, 1996; Foschini and Gans, 1998) with Minimum Mean-Square Error (MMSE) interference suppression

presented yields the relation

$$\underline{\mathbf{H}}^H\underline{\mathbf{H}} = \begin{pmatrix} \mathbf{H}^H & \frac{\sigma_\mathcal{N}}{\sigma_\mathcal{X}}\mathbf{I}_{N_T} \end{pmatrix} \cdot \begin{pmatrix} \mathbf{H} \\ \frac{\sigma_\mathcal{N}}{\sigma_\mathcal{X}}\mathbf{I}_{N_T} \end{pmatrix} = \left(\mathbf{H}^H\mathbf{H} + \frac{\sigma_\mathcal{N}^2}{\sigma_\mathcal{X}^2}\mathbf{I}_{N_T}\right) \ .$$

Obviously, applying the ZF solution to the extended matrix $\underline{\mathbf{H}}$ delivers the MMSE solution of the original matrix $\mathbf{H}$ (Hassibi, 2000). Hence, the procedure described on page 275 can also be applied to the MMSE-BLAST detection. There is only a slight difference that has to be emphasised. The error covariance matrix Φ_{MMSE} given in Equation (5.107) cannot be expressed by the product of the MMSE filter $\mathbf{W}_{\text{MMSE}}$ and its Hermitian. Hence, the squared column norm of $\mathbf{W}_{\text{MMSE}}$ is not an appropriate measure to determine the optimum detection order. Instead, the error covariance matrix has to be explicitly determined.

5.5.5 QL Decomposition and Interference Cancellation

The last subsection made obvious that multiple matrix inversions have to be calculated, one for each detection step. Although the size of the channel matrix is reduced in each detection step, this procedure is computationally demanding. Kühn proposed a different implementation (Kühn, 2006). It is based on the QL decomposition of $\mathbf{H}$, the complexity of which is identical to a matrix inversion. Since it has to be computed only once, it

SPACE-TIME CODES

saves valuable computational resources. The QR decomposition of **H** is often used in the literature (Wübben *et al.*, 2001).

ZF-QL Decomposition by Modified Gram–Schmidt Procedure

We start with the derivation of the QL decomposition for the zero-forcing solution. An extension to the MMSE solution will be presented on page 284. In order to illuminate the basic principle of the QL decomposition, we will consider the modified Gram–Schmidt algorithm (Golub and van Loan, 1996). Nevertheless, different algorithms that are based on Householder reflections or Givens rotations (see Appendix B) may exhibit a better behaviour concerning numerical robustness. Basically, the QL decomposition decomposes an $N_R \times N_T$ matrix $\mathbf{H} = \mathbf{QL}$ into an $N_R \times N_T$ matrix

$$\mathbf{Q} = \begin{pmatrix} \mathbf{q}_1 & \cdots & \mathbf{q}_{N_T-1} & \mathbf{q}_{N_T} \end{pmatrix}$$

and a lower triangular $N_T \times N_T$ matrix

$$\mathbf{L} = \begin{pmatrix} L_{1,1} & & & \\ L_{2,1} & L_{2,2} & & \\ \vdots & & \ddots & \\ L_{N_T,1} & \cdots & & L_{N_T,N_T} \end{pmatrix}.$$

For $N_T \leq N_R$, which is generally true, we can imagine that the columns in **H** span an N_T-dimensional subspace from $\mathbb{C}^{N_R}$. The columns in **Q** span exactly the same subspace and represent an orthogonal basis. They are determined successively from $\mu = N_T$ to $\mu = 1$, as described by the pseudocode in Figure 5.47.

First, the matrices **L** and **Q** are initialised with the all-zero matrix and the channel matrix **H** respectively. In the first step, the vector $\mathbf{q}_{N_T}$ is chosen such that it points in the direction of $\mathbf{h}_{N_T}$ (remember the initialisation). After its normalisation to unit length by

$$L_{N_T,N_T} = \|\mathbf{h}_{N_T}\| \quad \Rightarrow \quad \mathbf{q}_{N_T} = \mathbf{h}_{N_T}/L_{N_T,N_T},$$

the projections of all remaining vectors $\mathbf{q}_{\nu < N_T}$ onto $\mathbf{q}_{N_T}$ are subtracted from $\mathbf{q}_{\nu < N_T}$. The resulting vectors

$$\mathbf{q}_\nu - L_{N_T,\nu}\mathbf{q}_{N_T}$$

point in directions that are perpendicular to $\mathbf{q}_{N_T}$ and build the basis for the next step. Continuing with $\mathbf{q}_{N_T-1}$, the procedure is repeated down to $\mathbf{q}_1$. The projection and subtraction ensure that the remaining vectors $\mathbf{q}_1$ to $\mathbf{q}_\mu$ are perpendicular to the hyperplane spanned by the already fixed vectors $\mathbf{q}_{\mu+1}$ to $\mathbf{q}_{N_T}$. Hence, the columns of **Q** are mutually orthogonal. After N_T steps, **Q** and **L** are totally determined.

Successive Interference Cancellation (SIC)

Inserting the QL decomposition into our system model yields

$$\mathbf{r} = \mathbf{QLx} + \mathbf{n} \tag{5.108}$$

Modified Gram–Schmidt algorithm

- Columns q_μ of $\mathbf{Q}$ are determined from right to left.
- Elements of triangular matrix $\mathbf{L}$ are determined from lower right corner to upper left corner

Step	Task
(1)	Initialisation: $\mathbf{L} = \mathbf{0}, \mathbf{Q} = \mathbf{H}$
(2)	for $\mu = N_T, \ldots, 1$
(3)	set diagonal element $L_{\mu,\mu} = \|\mathbf{q}_\mu\|$
(4)	normalise $\mathbf{q}_\mu = \mathbf{q}_\mu / L_{\mu,\mu}$ to unit length
(5)	for $\nu = 1, \ldots, \mu - 1$
(6)	calculate projections $L_{\mu,\nu} = \mathbf{q}_\mu^H \cdot \mathbf{q}_\nu$
(7)	$\mathbf{q}_\nu = \mathbf{q}_\nu - L_{\mu,\nu} \cdot \mathbf{q}_\mu$
(8)	end
(9)	end

Figure 5.47: Pseudocode of modified Gram–Schmidt algorithm for QL decomposition of channel matrix $\mathbf{H}$ (Golub and van Loan, 1996)

The advantage of this representation becomes obvious when it is multiplied from the left-hand side by $\mathbf{Q}^H$. On account of the to orthogonal columns in $\mathbf{Q}$, $\mathbf{Q}^H \mathbf{Q} = \mathbf{I}_{N_T}$ holds and the multiplication results in

$$\mathbf{y} = \mathbf{Q}^H \cdot \mathbf{r} = \mathbf{L}\mathbf{x} + \mathbf{Q}^H \mathbf{n} = \begin{pmatrix} L_{1,1} & 0 & & \\ L_{2,1} & L_{2,2} & & 0 \\ \vdots & \vdots & \ddots & \\ L_{N_T,1} & L_{N_T,2} & \cdots & L_{N_T,N_T} \end{pmatrix} \cdot \mathbf{x} + \tilde{\mathbf{n}} \qquad (5.109)$$

The modified noise vector $\tilde{\mathbf{n}}$ still represents white Gaussian noise because $\mathbf{Q}$ consists of orthogonal columns. However, the most important property is the triangular structure of $\mathbf{L}$. Multiplication with $\mathbf{Q}^H$ suppresses the interference partly, i.e. x_1 is totally free of interference, x_2 only suffers from x_1, x_3 from x_1 and x_2 and so on. This property allows a successive detection strategy as depicted in Figure 5.48. It starts with the first layer, whose received sample of which

$$y_1 = L_{1,1} x_1 + \tilde{n}_1$$

can be characterised the signal-to-noise ratio $\gamma_1 = L_{1,1}^2 E_s / N_0$. After appropriate scaling, a hard decision $\mathcal{Q}(\cdot)$ delivers the estimate

$$\hat{s}_1 = \mathcal{Q}\left(L_{1,1}^{-1} \cdot y_1\right) \qquad (5.110)$$

SPACE–TIME CODES

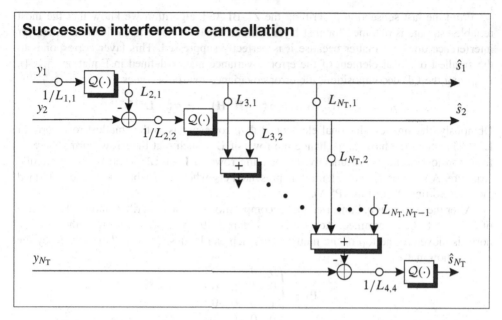

Figure 5.48: Illustration of interference cancellation after multiplying **r** with $\mathbf{Q}^H$. Reproduced by permission of John Wiley & Sons, Ltd

The obtained estimate can be used to remove the interference in the second layer which is then only perturbed by noise. Generally, the μth estimate is obtained by

$$\hat{s}_\mu = \mathcal{Q}\left(\frac{1}{L_{\mu,\mu}} \cdot \left[y_\mu - \sum_{\nu=1}^{\mu-1} L_{\mu,\nu} \cdot \hat{s}_\nu\right]\right) \tag{5.111}$$

Supposing that previous decisions have been correct, the signal-to-noise ratio amounts to $\gamma_\mu = L_{\mu,\mu}^2 E_s/N_0$. This entire procedure consists of the QL decomposition and a subsequent successive interference cancellation and is therefore termed 'QL-SIC'. Obviously, the matrix inversions of the original BLAST detection have been circumvented, and only a single QL decomposition has to be performed.

Optimum Post-Sorting Algorithm

Although the QL-SIC approach described above is an efficient way to suppress and cancel interference, it does not take into account the risk of error propagation. So far, the algorithm has simply started with the column on the right-hand side of **H** and continued to the left-hand side. Hence, the corresponding signal-to-noise ratios that each layer experiences are ordered randomly, and it might happen that the first layer to be detected is the worst one. Intuitively, we would like to start with the layer having the largest SNR. Assuming perfect interference cancellation, the layer-specific SNRs should decrease monotonically from the first to the last layer.

From the last subsection describing the ZF-BLAST algorithm we know that the most reliable estimate is obtained for that layer with the smallest noise amplification. Interlayer interference does not matter because it is perfectly suppressed. This layer corresponds to the smallest diagonal element of the error covariance matrix defined in Equation (5.104). Applying the QL decomposition, the error covariance matrix becomes

$$\Phi_{ZF} = \sigma_\mathcal{N}^2 \cdot \mathbf{W}_{ZF}^H \mathbf{W}_{ZF} = \sigma_\mathcal{N}^2 \cdot \left(\mathbf{H}^H \mathbf{H}\right)^{-1} = \sigma_\mathcal{N}^2 \cdot \mathbf{L}^{-1} \mathbf{L}^{-H}.$$

Obviously, the smallest diagonal element of Φ_{ZF} corresponds to the smallest row norm of $\mathbf{L}^{-1}$. Therefore, we have to exchange the rows of $\mathbf{L}^{-1}$ such that their row norms increase from top to bottom. Unfortunately, exchanging rows in $\mathbf{L}$ or $\mathbf{L}^{-1}$ destroys its triangular structure. A solution to this dilemma is presented elsewhere (Hassibi, 2000) and is termed the Post-Sorting Algorithm (PSA).

After the conventional unsorted QL decomposition of $\mathbf{H}$ as described above, the inverse of $\mathbf{L}^{-1}$ has to be determined. According to Figure 5.49, the row of $\mathbf{L}^{-1}$ with the smallest norm is moved to the top of the matrix. This step can be described mathematically by the permutation matrix

$$\mathbf{P}_1 = \begin{pmatrix} 0 & 0 & 1 & 0 \\ 0 & 1 & 0 & 0 \\ 1 & 0 & 0 & 0 \\ 0 & 0 & 0 & 1 \end{pmatrix}.$$

Since the first row should consist of only one non-zero element, Householder reflections or Givens rotations (Golub and van Loan, 1996) can be used to retrieve the triangular shape again. They are briefly described in Appendix B. In this book, we confine ourselves to Householder reflections denoted by unitary matrices Θ_μ. The multiplication of $\mathbf{P}_1 \mathbf{L}^{-1}$ with Θ_1 forces all elements of the first rows except the first one to zero without changing the row norm. Hence, the norm is now concentrated in a single non-zero element.

Now, the first row of the intermediate matrix $\mathbf{P}_1 \mathbf{L}^{-1} \Theta_1$ already has its final shape. Assuming a correct decision of the first layer in the Successive Interference Cancellation (SIC) procedure, x_1 does not influence other layers any more. Therefore, the next recursion is restricted to a 3×3 submatrix corresponding to the remaining three layers. Although the row norms to be compared are only taken from this submatrix, permutation and Householder matrices are constructed for the original size. We obtain the permutation matrix

$$\mathbf{P}_2 = \begin{pmatrix} 1 & 0 & 0 & 0 \\ 0 & 0 & 0 & 1 \\ 0 & 1 & 0 & 0 \\ 0 & 0 & 1 & 0 \end{pmatrix}$$

and the Householder matrix Θ_2. With the same argumentation as above, the relevant matrix is reduced again to a 2×2 matrix for which we obtain

$$\mathbf{P}_3 = \begin{pmatrix} 1 & 0 & 0 & 0 \\ 0 & 1 & 0 & 0 \\ 0 & 0 & 0 & 1 \\ 0 & 0 & 1 & 0 \end{pmatrix}$$

and Θ_3. The optimised inverse matrix has the form

$$\mathbf{L}_{opt}^{-1} = \mathbf{P}_{N_T-1} \cdots \mathbf{P}_1 \mathbf{L}^{-1} \cdot \Theta_1 \cdots \Theta_{N_T-1} \tag{5.112}$$

SPACE–TIME CODES

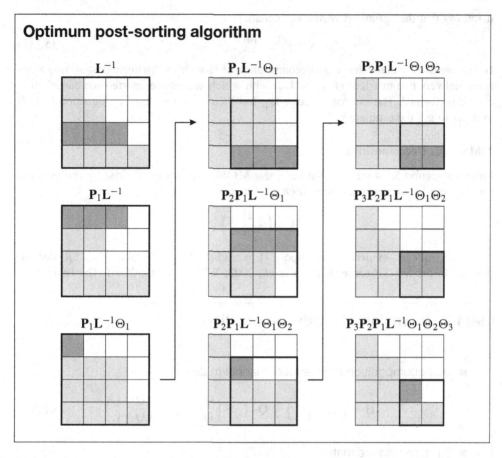

Figure 5.49: Illustration of the post-sorting algorithm (white squares indicate zeros, light-grey squares indicate non-zero elements, dark-grey squares indicate a row with a minimum norm) (Kühn, 2006). Reproduced by permission of John Wiley & Sons, Ltd

Since all permutation and Householder matrices are unitary with $\mathbf{PP}^H = \mathbf{P}^H\mathbf{P} = \mathbf{I}$, we obtain the final triangular matrix

$$\mathbf{L}_{\text{opt}} = \mathbf{\Theta}_{N_T-1}^H \cdots \mathbf{\Theta}_1^H \cdot \mathbf{L} \cdot \mathbf{P}_1^H \cdots \mathbf{P}_{N_T-1}^H \tag{5.113}$$

At most $N_T - 1$, permutations and Householder reflections have to be carried out. The optimised triangular matrix results in the QL decomposition of a modified channel matrix

$$\mathbf{H} = \mathbf{QL} \quad \Rightarrow \quad \mathbf{H}_{\text{opt}} = \mathbf{H} \cdot \mathbf{P}_1^H \cdots \mathbf{P}_{N_T-1}^H = \mathbf{Q}_{\text{opt}}\mathbf{L}_{\text{opt}} \tag{5.114}$$

with

$$\mathbf{Q}_{\text{opt}} = \mathbf{Q} \cdot \mathbf{\Theta}_1 \cdots \mathbf{\Theta}_{N_T-1} \tag{5.115}$$

Applying this result to the received vector $\mathbf{r}$ delivers

$$\mathbf{r} = \mathbf{H} \cdot \mathbf{x} + \mathbf{n} = \mathbf{H}_{\text{opt}} \cdot \mathbf{x}_{\text{opt}} + \mathbf{n} = \mathbf{H} \cdot \mathbf{P}_1^H \cdots \mathbf{P}_{N_T-1}^H \cdot \mathbf{x}_{\text{opt}} + \mathbf{n} \tag{5.116}$$

It follows that the optimised vector $\mathbf{x}_{opt}$ equals

$$\mathbf{x} = \mathbf{P}_1^H \cdots \mathbf{P}_{N_T-1}^H \cdot \mathbf{x}_{opt} \quad (5.117)$$

Hence, we can conclude that a QL decomposition of $\mathbf{H}$ with a subsequent post-sorting algorithm delivers two matrices $\mathbf{Q}_{opt}$ and $\mathbf{L}_{opt}$ with which successive interference cancellation can be performed. The estimated vector $\hat{\mathbf{x}}_{opt}$ has then to be reordered by Equation (5.117) in order to get the estimate $\hat{\mathbf{x}}$.

MMSE-QL Decomposition

From Subsection 5.5.4 we saw that the linear MMSE detector is obtained by applying the zero-forcing solution to the extended channel matrix

$$\underline{\mathbf{H}} = \begin{pmatrix} \mathbf{H} \\ \frac{\sigma_\mathcal{N}}{\sigma_\mathcal{X}} \mathbf{I}_{N_T} \end{pmatrix}.$$

Therefore, it is self-evident to decompose $\underline{\mathbf{H}}$ instead of $\mathbf{H}$ in order to obtain a QL decomposition in the MMSE sense (Böhnke et al., 2003; Wübben et al., 2003). The basic steps

---

MMSE extension of QL decomposition

- QL decomposition of extended channel matrix

$$\underline{\mathbf{H}} = \begin{pmatrix} \mathbf{H} \\ \frac{\sigma_\mathcal{N}}{\sigma_\mathcal{X}} \mathbf{I}_{N_T} \end{pmatrix} = \underline{\mathbf{Q}} \cdot \underline{\mathbf{L}} = \begin{pmatrix} \underline{\mathbf{Q}}_1 \\ \underline{\mathbf{Q}}_2 \end{pmatrix} \cdot \underline{\mathbf{L}} = \begin{pmatrix} \underline{\mathbf{Q}}_1 \cdot \underline{\mathbf{L}} \\ \underline{\mathbf{Q}}_2 \cdot \underline{\mathbf{L}} \end{pmatrix} \quad (5.118)$$

- Error covariance matrix

$$\Phi_{MMSE} = \underline{\Phi}_{ZF} = \sigma_\mathcal{N}^2 \cdot \left(\underline{\mathbf{H}}^H \underline{\mathbf{H}}\right)^{-1} = \sigma_\mathcal{N}^2 \cdot \underline{\mathbf{L}}^{-1} \underline{\mathbf{L}}^{-H} \quad (5.119)$$

- Inverse of $\underline{\mathbf{L}}$ as byproduct obtained

$$\frac{\sigma_\mathcal{N}}{\sigma_\mathcal{X}} \cdot \mathbf{I}_{N_T} = \underline{\mathbf{Q}}_2 \cdot \underline{\mathbf{L}} \quad \Rightarrow \quad \underline{\mathbf{L}}^{-1} = \frac{\sigma_\mathcal{X}}{\sigma_\mathcal{N}} \cdot \underline{\mathbf{Q}}_2, \quad (5.120)$$

- Received vector after filtering with $\underline{\mathbf{Q}}^H$

$$\mathbf{y} = \underline{\mathbf{Q}}^H \cdot \underline{\mathbf{r}} = \begin{pmatrix} \underline{\mathbf{Q}}_1^H & \underline{\mathbf{Q}}_2^H \end{pmatrix} \cdot \begin{pmatrix} \mathbf{r} \\ \mathbf{0}_{N_T \times 1} \end{pmatrix} = \underline{\mathbf{Q}}_1^H \cdot \mathbf{Hx} + \underline{\mathbf{Q}}_1^H \cdot \mathbf{n} \quad (5.121a)$$

$$= \underline{\mathbf{L}} \cdot \mathbf{x} - \frac{\sigma_\mathcal{N}}{\sigma_\mathcal{X}} \cdot \underline{\mathbf{Q}}_2^H \mathbf{x} + \underline{\mathbf{Q}}_1^H \mathbf{n} \quad (5.121b)$$

Figure 5.50: Summary of MMSE extension of QL decomposition

SPACE–TIME CODES

are summarised in Figure 5.50. The QL decomposition presented in Equation (5.118) now delivers a unitary $(N_R + N_T) \times N_T$ matrix $\underline{\mathbf{Q}}$ and a lower triangular $N_T \times N_T$ matrix $\underline{\mathbf{L}}$. Certainly, $\underline{\mathbf{Q}}$ and $\underline{\mathbf{L}}$ are not identical with $\mathbf{Q}$ and $\mathbf{L}$, e.g. $\underline{\mathbf{Q}}$ has N_T more rows than the original matrix $\mathbf{Q}$. It can be split into the submatrices $\underline{\mathbf{Q}}_1$ and $\underline{\mathbf{Q}}_2$ such that $\underline{\mathbf{Q}}_1$ has the same size as $\mathbf{Q}$ for the ZF approach.

Similarly to the V-BLAST algorithm, the order of detection has to be determined with respect to the error covariance matrix given in Equation (5.119). Please note that underlined vectors and matrices are associated with the extended system model. Since the only difference between ZF and MMSE is the use of $\underline{\mathbf{H}}$ instead of $\mathbf{H}$, it is not surprising that the row norm of the inverse extended triangular matrix $\underline{\mathbf{L}}$ determines the optimum sorting. However, in contrast to the ZF case, $\underline{\mathbf{L}}$ need not be explicitly inverted. Looking at Equation (5.118), we recognise that the lower part of the equation delivers the relation given in Equation (5.120). Hence, $\underline{\mathbf{L}}^{-1}$ is gained as a byproduct of the initial QL decomposition and the optimum post-sorting algorithm exploits the row norms of $\underline{\mathbf{Q}}_2$. This compensates for the higher computational costs due to QL decomposing a larger matrix $\underline{\mathbf{H}}$.

Since the MMSE approach represents a compromise between matched and zero-forcing filters, residual interference remains in its outputs. This effect will now be considered in more detail. Using the extended channel matrix requires modification of the received vector $\mathbf{r}$ as well. An appropriate way is to append N_T zeros. The detection starts by multiplying $\underline{\mathbf{r}}$ with $\underline{\mathbf{Q}}^H$, yielding the result in Equation (5.121a). In fact, only a multiplication with $\underline{\mathbf{Q}}_1^H$ is performed, having the same complexity as filtering with $\mathbf{Q}^H$ for the zero-forcing solution. However, in contrast to $\mathbf{Q}$ and $\underline{\mathbf{Q}}$, $\underline{\mathbf{Q}}_1$ does not contain orthogonal columns because it consists of only the first N_R rows of $\underline{\mathbf{Q}}$. Hence, the noise term $\underline{\mathbf{Q}}_1^H \mathbf{n}$ is coloured, i.e. its samples are correlated and a symbol-by-symbol detection as considered here is suboptimum.

Furthermore, the product $\underline{\mathbf{Q}}_1 \mathbf{Q}$ does not equal the identity matrix any more. In order to illuminate the consequence, we will take a deeper look at the product of the extended matrices $\underline{\mathbf{Q}}$ and $\underline{\mathbf{H}}$. Inserting their specific structures given in Equation (5.118) results in

$$\underline{\mathbf{Q}}^H \underline{\mathbf{H}} = \underline{\mathbf{Q}}_1^H \mathbf{H} + \underline{\mathbf{Q}}_2^H \cdot \frac{\sigma_\mathcal{N}}{\sigma_\mathcal{X}} \mathbf{I}_{N_T} \overset{!}{=} \underline{\mathbf{L}} \quad \Leftrightarrow \quad \underline{\mathbf{Q}}_1^H \mathbf{H} = \underline{\mathbf{L}} - \frac{\sigma_\mathcal{N}}{\sigma_\mathcal{X}} \cdot \underline{\mathbf{Q}}_2^H \qquad (5.122)$$

Replacing the term $\underline{\mathbf{Q}}_1^H \cdot \mathbf{H}$ in Equation (5.121a) delivers Equation (5.121b). We observe the desired term $\underline{\mathbf{L}}\mathbf{x}$, the coloured noise contribution $\underline{\mathbf{Q}}_1^H \mathbf{n}$ and a third term in the middle also depending on $\mathbf{x}$. It is this term that represents the residual interference after filtering with $\underline{\mathbf{Q}}_1^H$. If the noise power $\sigma_\mathcal{N}^2$ becomes small, the problem of noise amplification becomes less severe and the filter can concentrate on the interference suppression. For $\sigma_\mathcal{N}^2 \to 0$, the MMSE solution tends to the zero-forcing solution and no interference remains in the system.

Sorted QL Decomposition

We saw from the previous discussion that the order of detection is crucial in successive interference cancellation schemes. However, reordering the layers by the explained post-sorting algorithm requires rather large computational costs owing to several permutations and Householder reflections. They can be omitted by realising that the QL decomposition is performed column by column of $\mathbf{H}$. Therefore, it should be possible to change the order of columns during the modified Gram–Schmidt procedure in order directly to obtain the optimum detection order without post-processing.

Unfortunately, a computational conflict arises if we pursue this approach. The QL decomposition starts with the rightmost column of $\mathbf{H}$; the first element to be determined is L_{N_T,N_T} in the lower right corner. By contrast, the detection starts with $L_{1,1}$ in the upper left corner, since the corresponding layer does not suffer from interference. Remember that the risk of error propagation is minimised if the diagonal elements decrease with each iteration step, i.e. $L_{1,1} \geq L_{2,2} \geq \cdots \geq L_{N_T,N_T}$. Therefore, the QL decomposition should be performed such that $L_{1,1}$ is the largest among all diagonal elements. However, it is the last element to be determined.

A way out of this conflict is found by exploiting the property of triangular matrices that their determinants equal the product of their diagonal elements. Moreover, the determinant is invariant with respect to row or column permutations. Therefore, the following strategy (Wübben et al., 2001) can be pursued. If the diagonal elements are determined in ascending order, i.e. starting with the smallest for the lower right corner first, the last one for the upper left corner should be the desired largest value because the total product is constant.

The corresponding procedure is sketched in Figure 5.51. Equivalently to the modified Gram–Schmidt algorithm in Figure 5.47, the algorithm is initialised with $\mathbf{Q} = \mathbf{H}$ and $\mathbf{L} = \mathbf{0}$. Next, step 3 determines the column with the smallest norm and exchanges it with the rightmost unprocessed vector. After normalising its length to unity and, therefore, determining the corresponding diagonal element $L_{\mu,\mu}$ (steps 5 an 6), the projections of the remaining columns onto the new vector $\mathbf{q}_\mu$ provide the off-diagonal elements $L_{\mu,\nu}$.

Sorted QL decomposition

Step	Task
(1)	Initialisation: $\mathbf{L} = \mathbf{0}$, $\mathbf{Q} = \mathbf{H}$
(2)	for $\mu = N_T, \ldots, 1$
(3)	search for minimum norm among remaining columns in $\mathbf{Q}$ $k_\mu = \underset{\nu=1,\ldots,\mu}{\operatorname{argmin}} \|\mathbf{q}_\nu\|^2$
(4)	exchange columns μ and k_μ in $\mathbf{Q}$, and determine $\mathbf{P}_\mu$
(5)	set diagonal element $L_{\mu,\mu} = \|\mathbf{q}_\mu\|$
(6)	normalise $\mathbf{q}_\mu = \mathbf{q}_\mu / L_{\mu,\mu}$ to unit length
(7)	for $\nu = 1, \ldots, \mu - 1$
(8)	calculate projections $L_{\mu,\nu} = \mathbf{q}_\mu^H \cdot \mathbf{q}_\nu$
(9)	$\mathbf{q}_\nu = \mathbf{q}_\nu - L_{\mu,\nu} \cdot \mathbf{q}_\mu$
(10)	end
(11)	end

Figure 5.51: Pseudocode for sorted QL decomposition (Kühn, 2006)

SPACE–TIME CODES

At the end of the procedure, we obtain a orthonormal matrix $\mathbf{Q}$, a triangular matrix $\mathbf{L}$ as well as a set of permutation matrices $\mathbf{P}_\mu$ with $1 \leq \mu \leq N_T$. It has to be mentioned that the computational overheads compared with the conventional Gram–Schmidt procedure are negligible.

At this point, we have no wish to conceal that the strategy does not always achieve the optimum succession as is the case with the post-sorting algorithm. The algorithm especially fails in situations where two column vectors $\mathbf{q}_\mu$ and $\mathbf{q}_\nu$ have large norms but point in similar directions. Owing to their large norms, these vectors are among the latest columns to be processed. When the smaller one has been chosen as the next vector to be processed, its projection to the other vector is rather large on account of their similar directions. Hence, the remaining orthogonal component may become small, leading to a very small diagonal element in the upper left corner of $\mathbf{L}$.

Fortunately, simulations demonstrate that those pathological events occur very rarely, so that the SQLD performs nearly as well as the optimum post-sorting algorithm. Moreover, it is possible to concatenate the sorted QL decomposition and the post-sorting algorithm. This combination always ensures the best detection order. The supplemental costs are rather small because only very few additional permutations are required owing to the presorting of the SQLD. Hence, this algorithm is suited to perform the QL decomposition and to provide a close-to-optimal order of detection.

5.5.6 Performance of Multilayer Detection Schemes

This section analyses the performance of the different detection schemes described above. The comparison is drawn for a multiple-antenna system with $N_T = 4$ transmit and $N_R = 4$ receive antennas. The channel matrix $\mathbf{H}$ consists of independent identically distributed complex channel coefficients $h_{\mu,\nu}$ whose real and imaginary parts are independent and Gaussian distributed with zero mean and variance $1/2$. Moreover, the channel is constant during one coded frame and $\mathbf{H}$ is assumed to be perfectly known to the receiver, while the transmitter has no channel knowledge at all. QPSK was chosen as a modulation scheme. It has to be emphasised that all derived techniques work as well with other modulation schemes. However, the computational costs increase exponentially with M in the case of the optimal APP detector, while the complexity of the QL decomposition based approach is independent of the modulation alphabet's size.

Turbo Multilayer Detection

We start with the turbo detection approach from Subsection 5.5.2. For the simulations, we used a simple half-rate convolutional code with memory $m = 2$ and generator polynomials $g_1(D) = 1 + D^2$ and $g_2(D) = 1 + D + D^2$. In all cases, a max-log MAP decoder was deployed. Figure 5.52 shows the obtained results, where solid lines correspond to perfectly interleaved channels, i.e. each transmitted vector $\mathbf{x}[k]$ experiences a different channel matrix $\mathbf{H}[k]$. Dashed lines indicate a block fading channel with only a single channel matrix for the entire coded frame. In the left-hand diagram, the bit error rates of the max-log MAP solution are depicted versus the signal-to-noise ratio E_b/N_0. We observe that the performance increases from iteration to iteration. However, the largest gains are achieved in the first iterations, while additional runs lead to only minor improvements. This coincides with the observations made in the context of turbo decoding in Chapter 4. Moreover, the

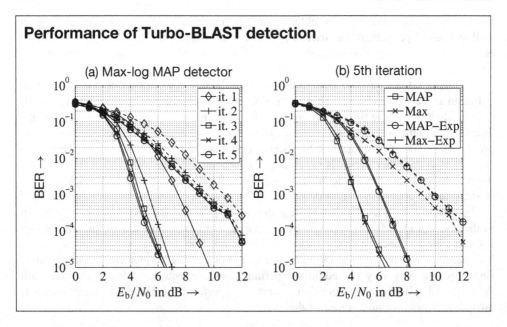

Figure 5.52: Error rate performance of V-BLAST system with $N_T = N_R = 4$, a convolutional code with $g_1(D) = 1 + D^2$ and $g_2(D) = 1 + D + D^2$ and turbo detection (solid lines: perfectly interleaved channel; dashed lines: block fading channel)

perfectly interleaved channel leads to a much better performance because the FEC decoders exploit temporal diversity and, therefore, provide more reliable estimates of interfering symbols. Hence, the total diversity degree amounts to $N_T N_R d_{\min}$ instead of $N_T N_R$.

The right-hand diagram illustrates the performance of different algorithms for the multilayer detector. Obviously, MAP and max-log MAP detectors show a similar behavior. The performance loss of the max-log MAP approximation compared with the optimal MAP solution is very small and less than 0.2 dB. However, the algorithms incorporating expected interfering symbols ('MAP-EXP' and 'Max-Exp') perform worse. The loss compared with MAP and max-log MAP amounts to 1 dB for the block fading channel and to nearly 2 dB for the perfectly interleaved channel. It has to be mentioned that the expected interfering symbols $\bar{s}_\mu$ have been obtained by applying the optimum MAP or max-log MAP solutions only in the first iteration owing to the lack of a-priori information. In all subsequent iteration steps, expected symbols have been used, reducing the computational costs markedly.

QL-Based Successive Interference Cancellation

Next, we will focus on the QL decomposition based approaches. In contrast to the turbo detector discussed before, we consider now an uncoded system. The reason is that we compare QL-SIC strategies with the optimum maximum likelihood detection. Besides the brute-force approach, which considers all possible hypotheses, Maximum Likelihood

SPACE-TIME CODES

Decoding (MLD) can also be accomplished by means of sphere detection (Agrell et al., 2002; Fincke and Pohst, 1985; Schnoor and Euchner, 1994) with lower computational costs. Since these detection algorithms originally deliver hard-decision outputs, a subsequent channel decoder would suffer from the missing reliability information and a direct comparison with the turbo detector would be unfair. Although there exist soft-output extensions for sphere detectors, their derivation is outside the scope of this book.

Figure 5.53 shows the results for different detection schemes, diagram (a) for the ZF solution and diagram (b) for the MMSE solution. Performing only a linear detection according to Section 5.5.3 has obviously the worst performance. As expected, the MMSE filter performs slightly better than the ZF approach because the background noise is not amplified. An additional QL-based successive interference cancellation without appropriate sorting ('QLD') already leads to improvements of roughly 3 dB for the ZF filter and 2 dB for the MMSE filter. However, the loss compared with the optimum maximum likelihood detector is still very high and amounts to more than 10 dB. The reason is that the average error rate is dominated by the layer detected first owing to error propagation. If this layer has incidentally a low SNR, its error rate, and hence the interference in subsequent layers, becomes quite high. Consequently, the average error rate is rather high. This effect will be illustrated in more detail later on and holds for both ZF and MMSE solutions.

Optimal post-sorting with the algorithm proposed in Figure 5.49 leads to remarkable improvements at least for the MMSE filter. It gains 8 dB compared with the unsorted detection at an error rate of $2 \cdot 10^{-3}$ and shows a loss of only 2 dB to the maximum likelihood detector. This behaviour emphasises the severe influence of error propagation.

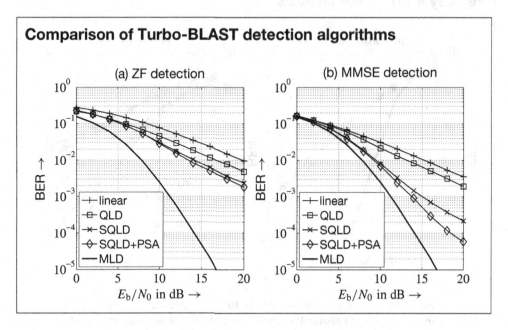

Figure 5.53: Error rate performance of an uncoded V-BLAST system with $N_T = N_R = 4$, QPSK and different detection methods (Kühn, 2006)

Regarding the ZF filter, the amplification of the background noise is too high to approach the MLD performance. The suboptimum SQLD with its negligible complexity overheads comes very close to the optimum sorting algorithm. Only at high signal-to-noise ratios does the loss increase for the MMSE solution.

Error Propagation and Diversity

The effects of error propagation and layer-specific diversity gains will be further examined in Figure 5.54 for a system with $N_T = N_R = 4$ and QPSK. Diagram (a) showing layer-specific error rates for the turbo detection scheme shows that all layers have the same error probability and error propagation is not an issue. Moreover, the slope of the curves indicates that the full diversity degree is obtained for all layers. This holds for both the perfectly interleaved and the block fading channel, although the overall performance of the former is much better.

By Contrast, the zero-forcing SQLD with successive interference cancellation shows a diverging behavior as illustrated in diagram (b). The solid lines represent the real world where error propagation from one layer to the others occurs. Hence, all error rates are dominated by the layer detected first. If the first detection generates many errors, subsequent layers cannot be reliably separated. Error propagation could be avoided by a genie-aided detector which cancels the interference always, perfectly although the layer-wise detection

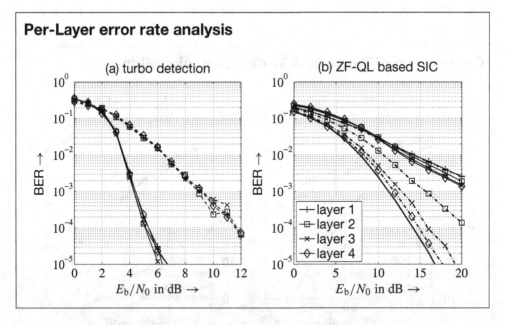

Figure 5.54: Per-layer error rate analysis of **(a)** turbo-decoded system (solid lines: perfectly interleaved channel; dashed lines: block fading channel) and **(b)** ZF-SQLD-based interference cancellation for $N_T = N_R = 4$ antennas (dashed-dotted line: genie-aided detector)

SPACE-TIME CODES

may be erroneous. This detector is analysed by the dashed-dotted lines. While the first detected layer naturally has the same error rate as in the real-world scenario, substantial improvements are achieved with each perfect cancellation step. Besides the absence of error propagation, the main observation is that the slope of the curves varies. Layer 4 comes very close to the performance of the maximum likelihood detection (bold solid line). Since the slope of the error rate curves depends on the diversity gain, we can conclude that the MLD detector and the last detected layer have the same diversity gain. In our example, the diversity degree amounts to $N_R = 4$.

However, all other layers seem to have a lower diversity degree, and the first layer has no diversity at all. This effect is explained by the zero-forcing filter $\mathbf{Q}$. In the first detection step, three interfering layers have to be suppressed using four receive antennas. Hence, the null space of the interference contains only $N_R - (N_T - 1) = 1$ 'dimension' and there is no degree of freedom left. Assuming perfect interference cancellation as for the genie-aided detector, the linear filter has to suppress only $N_T - 2 = 2$ interferers for the next layer. With $N_R = 4$ receive antennas, there is one more degree of freedom, and diversity of order $N_R - (N_T - 2) = 2$ is obtained. Continuing this thought, we obtain the full diversity degree N_R for the last layer because no interference has to be suppressed any more. Distinguishing perfectly interleaved and block fading channels makes no sense for uncoded transmissions because the detection is performed symbol by symbol and temporal diversity cannot be exploited.

5.5.7 Unified Description by Linear Dispersion Codes

Comparing orthogonal space–time block codes and multilayer transmission schemes such as the BLAST system, we recognise that they have been introduced for totally different purposes. While orthogonal space–time codes have been designed to achieve the highest possible diversity gain, they generally suffer from a rate loss due to the orthogonality constraint. By Contrast, BLAST-like systems aim to multiply the data rate without looking at the diversity degree per layer. Hence, one may receive the impression that both design goals exclude each other.

However, it is possible to achieve a trade-off between diversity and multiplexing gains (Heath and Paulraj, 2002). In order to reach this target, a unified description of both techniques would be helpful and can be obtained by Linear Dispersion (LD) codes (Hassibi and Hochwald, 2000, 2001, 2002). As the name suggests, the information is distributed or spread in several dimensions similarly to the physical phenomenon dispersion. In this context, we consider the dimensions space and time. Obviously, the dimension frequency can be added as well.

Taking into account that space–time code words are generally made up of K symbols s_μ and their conjugate complex counterparts s_μ^*, the matrix $\mathbf{X}$ describing a linear dispersion code word can be constructed

$$\mathbf{X} = \sum_{\mu=1}^{K} \mathbf{B}_{1,\mu} \cdot s_\mu + \mathbf{B}_{2,\mu} \cdot s_\mu^* \quad (5.123)$$

The $N_T \times L$ dispersion matrices $\mathbf{B}_{1,\mu}$ and $\mathbf{B}_{2,\mu}$ distribute the information into N_T spatial and L temporal directions. Hence, the resulting code word has a length of L time instants. In order to illustrate this general description, we apply Equation (5.123) to Alamouti's space–time code.

LD Description of Alamouti's Scheme

We remember from Section 5.4.1 that a code word $\mathbf{X}_2$ consists of $K = 2$ symbols s_1 and s_2 that are transmitted over two antennas within two time slots. The matrix has the form

$$\mathbf{X}_2 = \frac{1}{\sqrt{2}} \cdot \begin{pmatrix} s_1 & -s_2^* \\ s_2 & s_1^* \end{pmatrix}.$$

Comparing the last equation with Equation (5.123), the four matrices describing the linear dispersion code are

$$\mathbf{B}_{1,1} = \frac{1}{\sqrt{2}} \cdot \begin{pmatrix} 1 & 0 \\ 0 & 0 \end{pmatrix}, \quad \mathbf{B}_{1,2} = \frac{1}{\sqrt{2}} \cdot \begin{pmatrix} 0 & 0 \\ 1 & 0 \end{pmatrix},$$

$$\mathbf{B}_{2,1} = \frac{1}{\sqrt{2}} \cdot \begin{pmatrix} 0 & -1 \\ 0 & 0 \end{pmatrix}, \quad \mathbf{B}_{2,2} = \frac{1}{\sqrt{2}} \cdot \begin{pmatrix} 0 & 0 \\ 0 & 1 \end{pmatrix}.$$

For different space–time block codes, an equivalent description is obtained in the same way. Relaxing the orthogonality constraint, one can design arbitrary codes with specific properties. However, the benefit of low decoding complexity gets lost in this case. +

LD Description of Multilayer Transmissions

The next step will be to find an LD description for multilayer transmission schemes. As we know from the last section, BLAST-like systems transmit N_T independent symbols at each time instant. Hence, the code word length equals $L = 1$ and the dispersion matrices reduce to column vectors. Moreover, no complex-values symbols are used, so that the vectors $\mathbf{B}_{2,\mu}$ contain only zeros. Finally, each symbol is transmitted over a single antenna, resulting in vectors $\mathbf{B}_{1,\mu}$ that contain only a single 1 at the μth row. For the special case of $N_T = 4$ transmit antennas, we obtain

$$\mathbf{B}_{1,1} = \begin{pmatrix} 1 \\ 0 \\ 0 \\ 0 \end{pmatrix}, \quad \mathbf{B}_{1,2} = \begin{pmatrix} 0 \\ 1 \\ 0 \\ 0 \end{pmatrix}, \quad \mathbf{B}_{1,3} = \begin{pmatrix} 0 \\ 0 \\ 1 \\ 0 \end{pmatrix}, \quad \mathbf{B}_{1,4} = \begin{pmatrix} 0 \\ 0 \\ 0 \\ 1 \end{pmatrix},$$

$$\mathbf{B}_{2,1} = \begin{pmatrix} 0 \\ 0 \\ 0 \\ 0 \end{pmatrix}, \quad \mathbf{B}_{2,2} = \begin{pmatrix} 0 \\ 0 \\ 0 \\ 0 \end{pmatrix}, \quad \mathbf{B}_{2,3} = \begin{pmatrix} 0 \\ 0 \\ 0 \\ 0 \end{pmatrix}, \quad \mathbf{B}_{2,4} = \begin{pmatrix} 0 \\ 0 \\ 0 \\ 0 \end{pmatrix}.$$

Detection of Linear Dispersion Codes

Unless linear dispersion codes can be reduced to special cases such as orthogonal space–time block codes, the detection requires the multilayer detection philosophies introduced in this section. In order to be able to apply the discussed algorithms such as QL-based interference cancellation or multilayer turbo detection, we need an appropriate system model. Using the transmitted code word defined in Equation (5.123), the received word has the form

$$\mathbf{R} = \mathbf{H}\mathbf{X} + \mathbf{N} = \mathbf{H} \cdot \sum_{\mu=1}^{K} \mathbf{B}_{1,\mu} \cdot s_\mu + \mathbf{B}_{2,\mu} \cdot s_\mu^* + \mathbf{N} \quad (5.124)$$

SPACE–TIME CODES

and consists of N_R rows according to the number of receive antennas and L columns denoting the duration of a space–time code word. The operator

$$\text{vec}\{\mathbf{A}\} = \text{vec}\{(\mathbf{a}_1 \cdots \mathbf{a}_n)\} = \begin{pmatrix} \mathbf{a}_1 \\ \vdots \\ \mathbf{a}_n \end{pmatrix}$$

stacks the columns of matrix $\mathbf{A}$ on top of each other. Applying it to $\mathbf{B}_{1,\mu}$ and $\mathbf{B}_{2,\mu}$ transforms the sum in Equation (5.124) into

$$\sum_{\mu=1}^{K} \text{vec}\{\mathbf{B}_{1,\mu}\} \cdot s_\mu + \text{vec}\{\mathbf{B}_{2,\mu}\} \cdot s_\mu^* = \mathbf{B}_1 \cdot \mathbf{s} + \mathbf{B}_2 \cdot \mathbf{s}^* \qquad (5.125)$$

The vectors $\mathbf{s}$ and $\mathbf{s}^*$ comprise all data symbols s_μ and their complex conjugates s_μ^* respectively. Equivalently, the matrices $\mathbf{B}_\nu$ are made up of the vectors $\text{vec}\{\mathbf{B}_{\nu,\mu}\}$. Both terms of the sum can be merged into the expression

$$\mathbf{B}_1 \cdot \mathbf{s} + \mathbf{B}_2 \cdot \mathbf{s}^* = (\mathbf{B}_1 \quad \mathbf{B}_2) \cdot = \begin{pmatrix} \mathbf{s} \\ \mathbf{s}^* \end{pmatrix} = \mathbf{B} \cdot \underline{\mathbf{s}}.$$

Owing to the arrangement of the time axis along a single column, the channel matrix $\mathbf{H}$ has to be enlarged by repeating it L times. This can be accomplished by the Kronecker product which is generally defined as

$$\mathbf{A} \otimes \mathbf{B} = \begin{pmatrix} A_{1,1} \cdot \mathbf{B} & \cdots & A_{1,N} \cdot \mathbf{B} \\ \vdots & & \vdots \\ A_{M,1} \cdot \mathbf{B} & \cdots & A_{M,N} \cdot \mathbf{B} \end{pmatrix}.$$

Finally, the application of the vec-operator to the matrices $\mathbf{R}$ and $\mathbf{N}$ results in

$$\mathbf{r} = \text{vec}\{\mathbf{R}\} = (\mathbf{I}_L \otimes \mathbf{H}) \cdot \mathbf{B} \cdot \underline{\mathbf{s}} + \text{vec}\{\mathbf{N}\} = \mathbf{H}^{\text{ld}} \cdot \underline{\mathbf{s}} + \text{vec}\{\mathbf{N}\} \qquad (5.126)$$

We have derived an equivalent system description that resembles the structure of a BLAST system. Hence, the same detection algorithms can be applied for LD codes as well. However, we have to be aware that $\underline{\mathbf{s}}$ may not only contain independent data symbols but conjugate complex versions of them as well. In that case, an appropriate combination of the estimates for s_μ and s_μ^* is required, and an independent detection suboptimum. This effect can be circumvented by using a real-valued description (Kühn, 2006).

Optimising Linear Dispersion Codes

Optimising linear dispersion codes can be done with respect to different goals. A maximisation of the ergodic capacity according to Section 5.3.1 would result in

$$\mathbf{B} = \underset{\tilde{\mathbf{B}}}{\text{argmax}} \log_2 \det \left(\mathbf{I}_{LN_R} + \frac{\sigma_N^2}{\sigma_X^2} \cdot (\mathbf{I}_L \otimes \mathbf{H}) \tilde{\mathbf{B}} \tilde{\mathbf{B}}^H (\mathbf{I}_L \otimes \mathbf{H}) \right) \qquad (5.127a)$$

subject to a power constraint, e.g.

$$\text{tr}\left\{\sum_{\mu=1}^{K} \mathbf{B}_{1,\mu}(\mathbf{B}_{1,\mu})^{\text{H}} + \mathbf{B}_{2,\mu}(\mathbf{B}_{2,\mu})^{\text{H}}\right\} = K \ . \tag{5.127b}$$

Such an optimisation was performed elsewhere (Hassibi and Hochwald, 2000, 2001, 2002). A different approach also considering the error rate performance was taken by (Heath and Paulraj, 2002). Generally, the obtained LD codes do not solely pursue diversity or multiplexing gains but can achieve a trade-off between the two aspects.

5.6 Summary

This chapter introduced some examples for space–time signal processing. Based on the description of the MIMO channel and some evaluation criteria, the principle of orthogonal space–time block codes was explained. All presented coding schemes have achieved the full diversity degree, and a simple linear processing at the receiver was sufficient for data detection. However, Alamouti's scheme with $N_T = 2$ transmit antennas is the only code with rate $R = 1$. Keeping the orthogonality constraint for more transmit antennas directly leads to a loss of spectral efficiency that has to be compensated for by choosing modulation schemes with $M > 2$. At high signal-to-noise ratios, the diversity effect is dominating and more transmit antennas are beneficial. By contrast, only two transmit antennas and a more robust modulation scheme is an appropriate choice at medium and low SNRs.

In contrast to diversity achieving space–time codes, spatial multiplexing increases the data rate. In the absence of channel knowledge at the transmitter, the main complexity of this approach has to be spent at the receiver. For coded systems, we discussed turbo detectors whose structure is similar to that of the turbo decoder explained in Chapter 4. Since the complexity grows exponentially with the number of layers (transmit antennas) and the modulation alphabet size, it becomes quickly infeasible for a practical implementation. A suitable detection strategy has been proposed that is based on the QL decomposition of the channel matrix $\mathbf{H}$. It consists of a linear interference suppression and a non-linear interference cancellation step. Using the MMSE solution and an appropriate sorting algorithm, this approach performs almost as well to the maximum likelihood solution. Moreover, its complexity grows only polynomially with N_T and is independent of the alphabet size $\$$. Finally, we showed that space–time block codes and spatial multiplexing can be uniquely described by linear dispersion codes. They also offer a way to obtain a trade-off between diversity and multiplexing gains.

As the discussed MIMO techniques offer the potential of high spectral efficiencies, they are also discussed for the standardisation of UMTS Terrestrial Radio Access (UTRA) extensions. In the context of HSDPA, spatial multiplexing and space–time coding concepts, as well as the combination of the two, are considered. A multitude of proposals from different companies is currently being evaluated in actual standardisation bodies of 3GPP (3GPP, 2007). Furthermore, the upcoming standards Worldwide Interoperability for Microwave Access (WIMAX) (IEEE, 2004) and IEEE 802.11n will also incorporate space–time coding concepts.

A

Algebraic Structures

In this appendix we will give a brief overview of those algebraic basics that we need throughout the book. We start with the definition of some algebraic structures (Lin and Costello, 2004; McEliece, 1987; Neubauer, 2006b).

A.1 Groups, Rings and Finite Fields

The most important algebraic structures in the context of algebraic coding theory are groups, rings and finite fields.

A.1.1 Groups

A non-empty set $\mathbb{G}$ together with a binary operation '·' is called a *group* if for all elements $a, b, c \in \mathbb{G}$ the following properties hold:

(G1) $a \cdot b \in \mathbb{G}$,

(G2) $a \cdot (b \cdot c) = (a \cdot b) \cdot c$,

(G3) $\exists e \in \mathbb{G} : \forall a \in \mathbb{G} : a \cdot e = e \cdot a = a$,

(G4) $\forall a \in \mathbb{G} : \exists a' \in \mathbb{G} : a \cdot a' = e$.

The element e is the identity element, and a' is the inverse element of a. If, additionally, the commutativity property

(G5) $a \cdot b = b \cdot a$

holds, then the group is called commutative.

The number of elements of a group $\mathbb{G}$ is given by the order $\mathrm{ord}(\mathbb{G})$. If the number of elements is finite $\mathrm{ord}(\mathbb{G}) < \infty$, we have a *finite group*. A *cyclic group* is a group where all elements $\gamma \in \mathbb{G}$ are obtained from the powers α^i of one element $\alpha \in \mathbb{G}$. The powers are defined according to

$$\alpha^0 = e, \alpha^1 = \alpha, \alpha^2 = \alpha \cdot \alpha, \ldots$$

This particular element α is the *primitive element* of the group $\mathbb{G}$. All elements $\gamma \in \mathbb{G}$ of a finite group $\mathbb{G}$ of order $\text{ord}(\mathbb{G})$ fulfil the condition

$$\gamma^{\text{ord}(\mathbb{G})} = e.$$

A.1.2 Rings

If we define two operations – multiplication '$\cdot$' and addition '+' – over the set $\mathbb{S}$, then $\mathbb{S}$ is called a *ring* if the following properties are fulfilled for all elements $a, b, c \in \mathbb{S}$:

(R1) $a + b \in \mathbb{S}$,

(R2) $a + (b + c) = (a + b) + c$,

(R3) $\exists 0 \in \mathbb{S} : \forall a \in \mathbb{S} : a + 0 = 0 + a = a$,

(R4) $\forall a \in \mathbb{S} : \exists -a \in \mathbb{S} : a + (-a) = 0$,

(R5) $a + b = b + a$,

(R6) $a \cdot b \in \mathbb{S}$,

(R7) $a \cdot (b \cdot c) = (a \cdot b) \cdot c$,

(R8) $a \cdot (b + c) = a \cdot b + a \cdot c$.

The element 0 is the zero element. The inverse element of a with respect to addition is given by $-a$. If, additionally, the following two properties

(R9) $\exists 1 \in \mathbb{S} : \forall a \in \mathbb{S} : a \cdot 1 = 1 \cdot a = a$,

(R10) $a \cdot b = b \cdot a$

are met, then the ring is called a *commutative ring with identity*. The element 1 denotes the identity element. In the following, we will use the common notation $a\,b$ for the multiplication $a \cdot b$. The set of integers $\mathbb{Z}$ with ordinary addition '+' and multiplication '$\cdot$' forms a commutative ring with identity. Another example of a commutative ring with identity is the *residue class ring* defined in Figure A.1.

An important operation for rings is the division with remainder. As a well-known example, we will consider the set of integers $\mathbb{Z}$. For two integers $a, b \in \mathbb{Z}$ there exist numbers $q, r \in \mathbb{Z}$ such that

$$a = q\,b + r$$

with quotient q, remainder r and $0 \leq r < b$. Written with congruences, this reads

$$a \equiv r \mod b.$$

If the remainder r is zero according to $a \equiv 0$ modulo b, the number b divides a which we will write as $b | a$. b is called a divisor of a. A prime number $p > 1$ is defined as a number that is only divisible by 1 and itself.

The greatest common divisor $\gcd(a, b)$ of two numbers a and b can be calculated with the help of *Euclid's algorithm*. This algorithm is based on the observation that $\gcd(a, b) = \gcd(a, b - c\,a)$ for each integer $c \in \mathbb{Z}$. Therefore, the greatest common divisor $\gcd(a, b)$

Residue class ring

- As an example we consider the set of integers

$$\mathbb{Z}_m = \{0, 1, 2, \ldots, m-1\}$$

- This set with addition and multiplication modulo m fulfils the ring properties. $\mathbb{Z}_m$ is called the residue class ring. In order to denote the calculation modulo m, we will call this ring $\mathbb{Z}/(m)$.

- The element $a \in \mathbb{Z}_m$ has a multiplicative inverse if the greatest common divisor $\gcd(a, m) = 1$.

- If $m = p$ is a prime number p, then $\mathbb{Z}_p$ yields the finite field $\mathbb{F}_p$.

Figure A.1: Residue class ring $\mathbb{Z}_m$

can be obtained from successive subtractions. A faster implementation uses successive divisions according to the following well-known iterative algorithm.

After the initialisation $r_{-1} = a$ and $r_0 = b$ in each iteration we divide the remainder r_{i-2} by r_{i-1} which yields the new remainder r_i according to

$$r_i \equiv r_{i-2} \mod r_{i-1}.$$

Because $0 \leq r_i < r_{i-1}$, the algorithm finally converges with $r_{n+1} = 0$. The greatest common divisor $\gcd(a, b)$ of a and b is then given by

$$\gcd(a, b) = r_n.$$

In detail, Euclid's algorithm reads

$$\begin{aligned}
r_{-1} &\equiv r_1 \mod r_0 &\quad &\text{with } r_{-1} = a \text{ and } r_0 = b \\
r_0 &\equiv r_2 \mod r_1 \\
r_1 &\equiv r_3 \mod r_2 \\
r_2 &\equiv r_4 \mod r_3 \\
&\vdots \\
r_{n-2} &\equiv r_n \mod r_{n-1} \\
r_{n-1} &\equiv 0 \mod r_n &\quad &\text{with } r_{n+1} = 0
\end{aligned}$$

> **Euclid's algorithm for integers**
>
> ■ Initialisation
> $$r_{-1} = a \quad \text{and} \quad r_0 = b$$
> $$f_{-1} = 1 \quad \text{and} \quad f_0 = 0$$
> $$g_{-1} = 0 \quad \text{and} \quad g_0 = 1$$
>
> ■ Iterations until $r_{n+1} = 0$
> $$r_i \equiv r_{i-2} \mod r_{i-1}$$
> $$f_i = f_{i-2} - q_i f_{i-1}$$
> $$g_i = g_{i-2} - q_i g_{i-1}$$
>
> ■ Greatest common divisor
> $$\gcd(a, b) = r_n = f_n a + g_n b \qquad (A.1)$$

Figure A.2: Euclid's algorithm for integers

It can be shown that each remainder r_i can be written as a linear combination of a and b according to
$$r_i = f_i a + g_i b.$$
The numbers f_i and g_i can be obtained with the initialisations
$$f_{-1} = 1 \quad \text{and} \quad f_0 = 0,$$
$$g_{-1} = 0 \quad \text{and} \quad g_0 = 1$$
and by successively calculating the iterations
$$f_i = f_{i-2} - q_i f_{i-1},$$
$$g_i = g_{i-2} - q_i g_{i-1}.$$
Euclid's algorithm for calculating the greatest common divisor of two integers is summarised in Figure A.2.

A.1.3 Finite Fields

If it is possible to divide two arbitrary non-zero numbers we obtain the algebraic structure of a *field*. A field $\mathbb{F}$ with addition '+' and multiplication '·' is characterised by the following properties for all elements $a, b, c \in \mathbb{F}$:

ALGEBRAIC STRUCTURES

(F1) $a + b \in \mathbb{F}$,

(F2) $a + (b + c) = (a + b) + c$,

(F3) $\exists 0 \in \mathbb{F} : \forall a \in \mathbb{F} : a + 0 = 0 + a = a$,

(F4) $\forall a \in \mathbb{F} : \exists -a \in \mathbb{F} : a + (-a) = 0$,

(F5) $a + b = b + a$,

(F6) $a \cdot b \in \mathbb{F}$,

(F7) $a \cdot (b \cdot c) = (a \cdot b) \cdot c$,

(F8) $\exists 1 \in \mathbb{F} : \forall a \in \mathbb{F} : a \cdot 1 = 1 \cdot a = a$,

(F9) $\forall a \in \mathbb{F} \setminus \{0\} : \exists a^{-1} \in \mathbb{F} : a \cdot a^{-1} = 1$,

(F10) $a \cdot b = b \cdot a$,

(F11) $a \cdot (b + c) = a \cdot b + a \cdot c$.

The element a^{-1} is called the multiplicative inverse if $a \neq 0$. If the number of elements of $\mathbb{F}$ is finite, i.e.

$$|\mathbb{F}| = q,$$

then $\mathbb{F}$ is called a *finite field*. We will write a finite field of cardinality q as $\mathbb{F}_q$. A deep algebraic result is the finding that every finite field $\mathbb{F}_q$ has a prime power of elements, i.e.

$$q = p^l$$

with the prime number p. Finite fields are also called *Galois fields*.

A.2 Vector Spaces

For linear block codes, which we will discuss in Chapter 2, code words are represented by n-dimensional vectors over the finite field $\mathbb{F}_q$. A vector $\mathbf{a}$ is defined as the n-tuple

$$\mathbf{a} = (a_0, a_1, \ldots, a_{n-1})$$

with $a_i \in \mathbb{F}_q$. The set of all n-dimensional vectors is the n-dimensional space $\mathbb{F}_q^n$ with q^n elements. We define the vector addition of two vectors $\mathbf{a} = (a_0, a_1, \ldots, a_{n-1})$ and $\mathbf{b} = (b_0, b_1, \ldots, b_{n-1})$ according to

$$\mathbf{a} + \mathbf{b} = (a_0, a_1, \ldots, a_{n-1}) + (b_0, b_1, \ldots, b_{n-1})$$
$$= (a_0 + b_0, a_1 + b_1, \ldots, a_{n-1} + b_{n-1})$$

as well as the scalar multiplication

$$\beta \mathbf{a} = \beta (a_0, a_1, \ldots, a_{n-1})$$
$$= (\beta a_0, \beta a_1, \ldots, \beta a_{n-1})$$

with the scalar $\beta \in \mathbb{F}_q$. The set $\mathbb{F}_q^n$ is called the *vector space* over the finite field $\mathbb{F}_q$ if for two vectors $\mathbf{a}$ and $\mathbf{b}$ in $\mathbb{F}_q^n$ and two scalars α and β in $\mathbb{F}_q$ the following properties hold:

(V1) $\mathbf{a} + (\mathbf{b} + \mathbf{c}) = (\mathbf{a} + \mathbf{b}) + \mathbf{c}$,

(V2) $\mathbf{a} + \mathbf{b} = \mathbf{b} + \mathbf{a}$,

(V3) $\exists \mathbf{0} \in \mathbb{F}_q^n : \forall \mathbf{a} \in \mathbb{F}_q^n : \mathbf{a} + \mathbf{0} = \mathbf{a}$,

(V4) $\forall \mathbf{a} \in \mathbb{F}_q^n : \exists -\mathbf{a} \in \mathbb{F}_q^n : \mathbf{a} + (-\mathbf{a}) = \mathbf{0}$,

(V5) $\alpha \cdot (\mathbf{a} + \mathbf{b}) = \alpha \cdot \mathbf{a} + \alpha \cdot \mathbf{b}$,

(V6) $(\alpha + \beta) \cdot \mathbf{a} = \alpha \cdot \mathbf{a} + \beta \cdot \mathbf{a}$,

(V7) $(\alpha \cdot \beta) \cdot \mathbf{a} = \alpha \cdot (\beta \cdot \mathbf{a})$,

(V8) $1 \cdot \mathbf{a} = \mathbf{a}$.

For $\mathbf{a} + (-\mathbf{b})$ we will also write $\mathbf{a} - \mathbf{b}$. Because of property V3 the zero vector $\mathbf{0} = (0, 0, \ldots, 0)$ is always an element of the vector space $\mathbb{F}_q^n$.

A non-empty subset $\mathbb{B} \subseteq \mathbb{F}_q^n$ is called a *subspace* of $\mathbb{F}_q^n$ if the addition and the scalar multiplication of elements from $\mathbb{B}$ lead to elements in $\mathbb{B}$, i.e. the set $\mathbb{B}$ is closed under addition and scalar multiplication.

Another important concept is the concept of linear independency. A finite number of vectors $\mathbf{a}_1, \mathbf{a}_2, \ldots, \mathbf{a}_k$ is called linearly independent if

$$\beta_1 \cdot \mathbf{a}_1 + \beta_2 \cdot \mathbf{a}_2 + \cdots + \beta_k \cdot \mathbf{a}_k = \mathbf{0}$$

implies

$$\beta_1 = \beta_2 = \cdots = \beta_k = 0.$$

Otherwise, the vectors $\mathbf{a}_1, \mathbf{a}_2, \ldots, \mathbf{a}_k$ are said to be linearly dependent. With the help of this term we can define the dimension of a subspace. The dimension of the subspace $\mathbb{B} \subseteq \mathbb{F}^n$ is equal to $\dim \mathbb{B} = k$ if there exist k linearly independent vectors in $\mathbb{B}$ but $k+1$ vectors are always linearly dependent.

A.3 Polynomials and Extension Fields

We have already noted that finite fields $\mathbb{F}_q$ always have a prime power $q = p^l$ of elements. In the case of $l = 1$, the residue class ring $\mathbb{Z}_p$ in Figure A.1 yields the finite field $\mathbb{F}_p$, i.e.

$$\mathbb{F}_p \cong \mathbb{Z}_p.$$

We now turn to the question of how finite fields $\mathbb{F}_{p^l}$ with $l > 1$ can be constructed. To this end, we consider the set $\mathbb{F}_p[z]$ of all polynomials

$$a(z) = a_0 + a_1 z + a_2 z^2 + \cdots + a_{n-1} z^{n-1}$$

over the finite field $\mathbb{Z}_p$. The degree $\deg(a(z)) = n - 1$ of the polynomial $a(z)$ corresponds to the highest power of z with $a_{n-1} \neq 0$. Since $\mathbb{F}_p[z]$ fulfils the ring properties, it is called the *polynomial ring* over the finite field $\mathbb{F}_p$.

ALGEBRAIC STRUCTURES

Euclid's algorithm for polynomials

- ■ Initialisation

$$r_{-1}(z) = a(z) \quad \text{and} \quad r_0(z) = b(z)$$
$$f_{-1}(z) = 1 \quad \text{and} \quad f_0(z) = 0$$
$$g_{-1}(z) = 0 \quad \text{and} \quad g_0(z) = 1$$

- ■ Iterations until $r_{n+1}(z) = 0$

$$r_i(z) \equiv r_{i-2}(z) \mod r_{i-1}(z)$$
$$f_i(z) = f_{i-2}(z) - q_i(z) f_{i-1}(z)$$
$$g_i(z) = g_{i-2}(z) - q_i(z) g_{i-1}(z)$$

- ■ Greatest common divisor

$$\gcd(a(z), b(z)) = r_n(z) = f_n(z) a(z) + g_n(z) b(z) \tag{A.2}$$

Figure A.3: Euclid's algorithm for polynomials

Similar to the ring $\mathbb{Z}$ of integers, two polynomials $a(z), b(z) \in \mathbb{F}_p[z]$ can be divided with remainder according to

$$a(z) = q(z) b(z) + r(z)$$

with quotient $q(z)$ and remainder $r(z) = 0$ or $0 \leq \deg(r(z)) < \deg(b(z))$. With congruences this can be written as

$$a(z) \equiv r(z) \mod b(z).$$

If $b(z) \in \mathbb{F}_p[z] \setminus \{0\}$ divides $a(z)$ according to $b(z) \,|\, a(z)$, then there exists a polynomial $q(z) \in \mathbb{F}_p[z]$ such that $a(z) = q(z) \cdot b(z)$. On account of the ring properties, Euclid's algorithm can also be formulated for polynomials in $\mathbb{F}_p[z]$, as shown in Figure A.3.

Similarly to the definition of the residue class ring $\mathbb{Z}/(m)$, we define the so-called *factorial ring* $\mathbb{F}_p[z]/(m(z))$ by carrying out the calculations on the polynomials modulo $m(z)$. This factorial ring fulfils the ring properties. If the polynomial

$$m(z) = m_0 + m_1 z + m_2 z^2 + \cdots + m_l z^l$$

of degree $\deg(m(z)) = l$ with coefficients $m_i \in \mathbb{F}_p$ in the finite field $\mathbb{F}_p$ is irreducible, i.e. it cannot be written as the product of two polynomials of smaller degree, then

$$\mathbb{F}_p[z]/(m(z)) \cong \mathbb{F}_{p^l}$$

Finite field $\mathbb{F}_{2^4}$

- Let $m(z)$ be the irreducible polynomial $m(z) = 1 + z + z^4$ in the polynomial ring $\mathbb{F}_2[z]$.

- Each element $a(z) = a_0 + a_1 z + a_2 z^2 + a_3 z^3$ of the extension field $\mathbb{F}_2[z]/(1 + z + z^4)$ corresponds to one element **a** of the finite field $\mathbb{F}_{2^4}$.

- Addition and multiplication are carried out on the polynomials modulo $m(z)$.

Index	Element **a**	$a(z) \in \mathbb{F}_{2^4}$			
		a_3	a_2	a_1	a_0
0	0000	0	0	0	0
1	0001	0	0	0	1
2	0010	0	0	1	0
3	0011	0	0	1	1
4	0100	0	1	0	0
5	0101	0	1	0	1
6	0110	0	1	1	0
7	0111	0	1	1	1
8	1000	1	0	0	0
9	1001	1	0	0	1
10	1010	1	0	1	0
11	1011	1	0	1	1
12	1100	1	1	0	0
13	1101	1	1	0	1
14	1110	1	1	1	0
15	1111	1	1	1	1

Figure A.4: Finite field $\mathbb{F}_{2^4}$. Reproduced by permission of J. Schlembach Fachverlag

yields a finite field with p^l elements. Therefore, the finite field $\mathbb{F}_q$ with cardinality $q = p^l$ can be generated by an irreducible polynomial $m(z)$ of degree $\deg(m(z)) = l$ over the finite field $\mathbb{F}_p$. This field is the so-called *extension field* of $\mathbb{F}_p$. In Figure A.4 the finite field $\mathbb{F}_{2^4}$ is constructed with the help of the irreducible polynomial $m(z) = 1 + z + z^4$.

In order to simplify the multiplication of elements in the finite field $\mathbb{F}_{p^l}$, the modular polynomial $m(z)$ can be chosen appropriately. To this end, we define the root α of the irreducible polynomial $m(z)$ according to $m(\alpha) = 0$. Since the polynomial is irreducible over the finite field $\mathbb{F}_p$, this root is an element of the extension field $\mathbb{F}_p[z]/(m(z))$ or equivalently $\alpha \in \mathbb{F}_{p^l}$.

ALGEBRAIC STRUCTURES

The powers of α are defined by

$$\alpha^0 \equiv 1 \mod m(\alpha),$$
$$\alpha^1 \equiv \alpha \mod m(\alpha),$$
$$\alpha^2 \equiv \alpha \cdot \alpha \mod m(\alpha),$$
$$\alpha^3 \equiv \alpha \cdot \alpha \cdot \alpha \mod m(\alpha),$$
$$\vdots$$

or equivalently in the notation of the finite field $\mathbb{F}_{p^l}$ by

$$\alpha^0 = 1,$$
$$\alpha^1 = \alpha,$$
$$\alpha^2 = \alpha \cdot \alpha,$$
$$\alpha^3 = \alpha \cdot \alpha \cdot \alpha,$$
$$\vdots$$

If these powers of α run through all $p^l - 1$ non-zero elements of the extension field $\mathbb{F}_p[z]/(m(z))$ or the finite field $\mathbb{F}_{p^l}$ respectively, the element α is called a *primitive root*. Each irreducible polynomial $m(z)$ with a primitive root is itself called a primitive polynomial. With the help of the primitive root α, all non-zero elements of the finite field $\mathbb{F}_{p^l}$ can be generated. Formally, the element 0 is denoted by the power $\alpha^{-\infty}$ (see Figure A.5). The multiplication of two elements of the extension field $\mathbb{F}_{p^l}$ can now be carried out using the respective powers α^i and α^j. In Figure A.6 some primitive polynomials over the finite field $\mathbb{F}_2$ are listed.

The operations of addition and multiplication within the finite field $\mathbb{F}_q = \mathbb{F}_{p^l}$ can be carried out with the help of the respective polynomials or the primitive root α in the case of a primitive polynomial $m(z)$. As an example, the addition table and multiplication table for the finite field $\mathbb{F}_{2^4}$ are given in Figure A.7 and Figure A.8 using the primitive polynomial $m(z) = 1 + z + z^4$. Figure A.9 illustrates the arithmetics in the finite field $\mathbb{F}_{2^4}$.

The order $\mathrm{ord}(\gamma)$ of an arbitrary non-zero element $\gamma \in \mathbb{F}_{p^l}$ is defined as the smallest number for which $\gamma^{\mathrm{ord}(\gamma)} \equiv 1$ modulo $m(\gamma)$ or $\gamma^{\mathrm{ord}(\gamma)} = 1$ in the notation of the finite field $\mathbb{F}_{p^l}$. Thus, the order of the primitive root α is equal to $\mathrm{ord}(\alpha) = p^l - 1$. For each non-zero element $\gamma \in \mathbb{F}_{p^l}$ we have

$$\gamma^{p^l} = \gamma \quad \Leftrightarrow \quad \gamma^{p^l-1} = 1.$$

Furthermore, the order $\mathrm{ord}(\gamma)$ divides the number of non-zero elements $p^l - 1$ in the finite field $\mathbb{F}_{p^l}$, i.e.

$$\mathrm{ord}(\gamma) \mid p^l - 1.$$

There is a close relationship between the roots of an irreducible polynomial $m(z)$ of degree $\deg(m(z)) = l$ over the finite field $\mathbb{F}_p$. If α is a root of $m(z)$ with $m(\alpha) = 0$, the powers

$$\alpha^p, \alpha^{p^2}, \ldots, \alpha^{p^{l-1}}$$

Primitive root in the finite field $\mathbb{F}_{2^4}$

- Let $m(z)$ be the irreducible polynomial $m(z) = 1 + z + z^4$ in the polynomial ring $\mathbb{F}_2[z]$.
- We calculate the powers of the root α using $1 + \alpha + \alpha^4 = 0$.
- Since α yields all $2^4 - 1 = 15$ non-zero elements of $\mathbb{F}_{2^4}$, α is a primitive root of the primitive polynomial $m(z) = 1 + z + z^4$.

α^j					a
$\alpha^{-\infty}$				0	0000
α^0				1	0001
α^1			α		0010
α^2		α^2			0100
α^3	α^3				1000
α^4			$\alpha\ +$	1	0011
α^5		$\alpha^2\ +$	α		0110
α^6	$\alpha^3\ +$	α^2			1100
α^7	α^3		$+\ \alpha\ +$	1	1011
α^8		α^2	$+$	1	0101
α^9	α^3		$+\ \alpha$		1010
α^{10}		$\alpha^2\ +$	$\alpha\ +$	1	0111
α^{11}	$\alpha^3\ +$	$\alpha^2\ +$	α		1110
α^{12}	$\alpha^3\ +$	$\alpha^2\ +$	$\alpha\ +$	1	1111
α^{13}	$\alpha^3\ +$	α^2	$+$	1	1101
α^{14}	α^3		$+$	1	1001
α^{15}				1	0001

Figure A.5: Primitive root in the finite field $\mathbb{F}_{2^4}$. Reproduced by permission of J. Schlembach Fachverlag

are also roots of $m(z)$. These elements are called the *conjugate roots* of $m(z)$. For a primitive root α of a primitive polynomial $m(z)$, the powers $\alpha, \alpha^p, \alpha^{p^2}, \ldots, \alpha^{p^{l-1}}$ are all different. The primitive polynomial $m(z)$ can thus be written as a product of linear factors

$$m(z) = (z - \alpha)(z - \alpha^p)(z - \alpha^{p^2}) \cdots (z - \alpha^{p^{l-1}}).$$

Correspondingly, the product over all different linear factors $z - \alpha^{i\,p^j}$ for a given power α^i yields the so-called *minimal polynomial* $m_i(z)$, which we will encounter in Section 2.3.6.

ALGEBRAIC STRUCTURES

Primitive polynomials over the finite field $\mathbb{F}_2$

■ Primitive polynomials $m(z)$ of degree $l = \deg(m(z))$ for the construction of the finite field $\mathbb{F}_{2^l}$

$l = \deg(m(z))$	$m(z)$
1	$1 + z$
2	$1 + z + z^2$
3	$1 + z + z^3$
4	$1 + z + z^4$
5	$1 + z^2 + z^5$
6	$1 + z + z^6$
7	$1 + z + z^7$
8	$1 + z^4 + z^5 + z^6 + z^8$
9	$1 + z^4 + z^9$
10	$1 + z^3 + z^{10}$
11	$1 + z^2 + z^{11}$
12	$1 + z^3 + z^4 + z^7 + z^{12}$
13	$1 + z + z^3 + z^4 + z^{13}$
14	$1 + z + z^6 + z^8 + z^{14}$
15	$1 + z + z^{15}$
16	$1 + z + z^3 + z^{12} + z^{16}$

Figure A.6: Primitive polynomials over the finite field $\mathbb{F}_2$

A.4 Discrete Fourier Transform

As is wellknown from signal theory, a finite complex sequence $\{x[0], x[1], \ldots, x[n-1]\}$ of samples $x[k] \in \mathbb{C}$ can be transformed into the spectral sequence $\{X[0], X[1], \ldots, X[n-1]\}$ with the help of the Discrete Fourier Transform (DFT) according to

$$X[l] = \sum_{k=0}^{n-1} x[k]\, w_n^{kl}$$

$$x[k] = \frac{1}{n} \sum_{l=0}^{n-1} X[l]\, w_n^{-kl}$$

using the so-called twiddle factor

$$w_n = e^{-j\,2\pi/n}.$$

Addition table of the finite field $\mathbb{F}_{2^4}$

+	0000	0001	0010	0011	0100	0101	0110	0111	1000	1001	1010	1011	1100	1101	1110	1111
0000	0000	0001	0010	0011	0100	0101	0110	0111	1000	1001	1010	1011	1100	1101	1110	1111
0001	0001	0000	0011	0010	0101	0100	0111	0110	1001	1000	1011	1010	1101	1100	1111	1110
0010	0010	0011	0000	0001	0110	0111	0100	0101	1010	1011	1000	1001	1110	1111	1100	1101
0011	0011	0010	0001	0000	0111	0110	0101	0100	1011	1010	1001	1000	1111	1110	1101	1100
0100	0100	0101	0110	0111	0000	0001	0010	0011	1100	1101	1110	1111	1000	1001	1010	1011
0101	0101	0100	0111	0110	0001	0000	0011	0010	1101	1100	1111	1110	1001	1000	1011	1010
0110	0110	0111	0100	0101	0010	0011	0000	0001	1110	1111	1100	1101	1010	1011	1000	1001
0111	0111	0110	0101	0100	0011	0010	0001	0000	1111	1110	1101	1100	1011	1010	1001	1000
1000	1000	1001	1010	1011	1100	1101	1110	1111	0000	0001	0010	0011	0100	0101	0110	0111
1001	1001	1000	1011	1010	1101	1100	1111	1110	0001	0000	0011	0010	0101	0100	0111	0110
1010	1010	1011	1000	1001	1110	1111	1100	1101	0010	0011	0000	0001	0110	0111	0100	0101
1011	1011	1010	1001	1000	1111	1110	1101	1100	0011	0010	0001	0000	0111	0110	0101	0100
1100	1100	1101	1110	1111	1000	1001	1010	1011	0100	0101	0110	0111	0000	0001	0010	0011
1101	1101	1100	1111	1110	1001	1000	1011	1010	0101	0100	0111	0110	0001	0000	0011	0010
1110	1110	1111	1100	1101	1010	1011	1000	1001	0110	0111	0100	0101	0010	0011	0000	0001
1111	1111	1110	1101	1100	1011	1010	1001	1000	0111	0110	0101	0100	0011	0010	0001	0000

Figure A.7: Addition table of the finite field $\mathbb{F}_{2^4}$. Reproduced by permission of J. Schlembach Fachverlag

This twiddle factor corresponds to an nth root of unity in the field $\mathbb{C}$ of complex numbers, i.e. $w_n^n = 1$. The discrete Fourier transform can be written in matrix form as follows

$$\begin{pmatrix} X[0] \\ X[1] \\ \vdots \\ X[n-1] \end{pmatrix} = \begin{pmatrix} 1 & 1 & \cdots & 1 \\ 1 & w_n^{1 \cdot 1} & \cdots & w_n^{1 \cdot (n-1)} \\ 1 & w_n^{2 \cdot 1} & \cdots & w_n^{2 \cdot (n-1)} \\ \vdots & \vdots & \ddots & \vdots \\ 1 & w_n^{(n-1) \cdot 1} & \cdots & w_n^{(n-1) \cdot (n-1)} \end{pmatrix} \cdot \begin{pmatrix} x[0] \\ x[1] \\ \vdots \\ x[n-1] \end{pmatrix}.$$

Correspondingly, the inverse discrete Fourier transform reads

$$\begin{pmatrix} x[0] \\ x[1] \\ \vdots \\ x[n-1] \end{pmatrix} = \frac{1}{n} \begin{pmatrix} 1 & 1 & \cdots & 1 \\ 1 & w_n^{-1 \cdot 1} & \cdots & w_n^{-1 \cdot (n-1)} \\ 1 & w_n^{-2 \cdot 1} & \cdots & w_n^{-2 \cdot (n-1)} \\ \vdots & \vdots & \ddots & \vdots \\ 1 & w_n^{-(n-1) \cdot 1} & \cdots & w_n^{-(n-1) \cdot (n-1)} \end{pmatrix} \cdot \begin{pmatrix} X[0] \\ X[1] \\ \vdots \\ X[n-1] \end{pmatrix}.$$

With the Fast Fourier Transform (FFT) there exist fast algorithms for calculating the DFT.

The discrete Fourier transform can also be defined over finite fields $\mathbb{F}_{q^l}$. To this end, we consider the vector $\mathbf{a} = (a_0, a_1, a_2, \ldots, a_{n-1})$ over the finite field $\mathbb{F}_{q^l}$ which can also be represented by the polynomial

$$a(z) = a_0 + a_1 z + a_2 z^2 + \cdots + a_{n-1} z^{n-1}$$

ALGEBRAIC STRUCTURES

Multiplication table of the finite field $\mathbb{F}_{2^4}$

·	0000	0001	0010	0011	0100	0101	0110	0111	1000	1001	1010	1011	1100	1101	1110	1111
0000	0000	0000	0000	0000	0000	0000	0000	0000	0000	0000	0000	0000	0000	0000	0000	0000
0001	0000	0001	0010	0011	0100	0101	0110	0111	1000	1001	1010	1011	1100	1101	1110	1111
0010	0000	0010	0100	0110	1000	1010	1100	1110	0011	0001	0111	0101	1011	1001	1111	1101
0011	0000	0011	0110	0101	1100	1111	1010	1001	1011	1000	1101	1110	0111	0100	0001	0010
0100	0000	0100	1000	1100	0011	0111	1011	1111	0110	0010	1110	1010	0101	0001	1101	1001
0101	0000	0101	1010	1111	0111	0010	1101	1000	1110	1011	0100	0001	1001	1100	0011	0110
0110	0000	0110	1100	1010	1011	1101	0111	0001	0101	0011	1001	1111	1110	1000	0010	0100
0111	0000	0111	1110	1001	1111	1000	0001	0110	1101	1010	0011	0100	0010	0101	1100	1011
1000	0000	1000	0011	1011	0110	1110	0101	1101	1100	0100	1111	0111	1010	0010	1001	0001
1001	0000	1001	0001	1000	0010	1011	0011	1010	0100	1101	0101	1100	0110	1111	0111	1110
1010	0000	1010	0111	1101	1110	0100	1001	0011	1111	0101	1000	0010	0001	1011	0110	1100
1011	0000	1011	0101	1110	1010	0001	1111	0100	0111	1100	0010	1001	1101	0110	1000	0011
1100	0000	1100	1011	0111	0101	1001	1110	0010	1010	0110	0001	1101	1111	0011	0100	1000
1101	0000	1101	1001	0100	0001	1100	1000	0101	0010	1111	1011	0110	0011	1110	1010	0111
1110	0000	1110	1111	0001	1101	0011	0010	1100	1001	0111	0110	1000	0100	1010	1011	0101
1111	0000	1111	1101	0010	1001	0110	0100	1011	0001	1110	1100	0011	1000	0111	0101	1010

Figure A.8: Multiplication table of the finite field $\mathbb{F}_{2^4}$. Reproduced by permission of J.Schlembach Fachverlag

---

Arithmetics in the finite field $\mathbb{F}_{2^4}$

- Let $a_5 = 0101$ and $a_6 = 0110$.
- Addition

$$a_5 + a_6 = 0101 + 0110 = 0011 = a_3.$$

- Multiplication

$$a_5 \cdot a_6 = 0101 \cdot 0110 = 1101 = a_{13}$$

corresponding to the product

$$(\alpha^2 + 1) \cdot (\alpha^2 + \alpha) = \alpha^8 \cdot \alpha^5 = \alpha^{13} = \alpha^3 + \alpha^2 + 1.$$

Figure A.9: Arithmetics in the finite field $\mathbb{F}_{2^4}$

---

in the factorial ring $\mathbb{F}_{q^l}[z]/(z^n - 1)$. With the help of the nth root of unity $\alpha \in \mathbb{F}_{q^l}$ in the extension field $\mathbb{F}_{q^l}$ with $\alpha^n = 1$, the discrete Fourier transform over the factorial ring

Discrete Fourier transform over finite fields

- Let α be an nth root of unity in the extension field $\mathbb{F}_{q^l}$ with $\alpha^n = 1$.
- DFT equations

$$A_j = a(\alpha^j) = \sum_{i=0}^{n-1} a_i \alpha^{ij} \qquad (A.3)$$

$$a_i = n^{-1} A(\alpha^{-i}) = n^{-1} \sum_{j=0}^{n-1} A_j \alpha^{-ij} \qquad (A.4)$$

Figure A.10: Discrete Fourier transform over the finite field $\mathbb{F}_{q^l}$

$\mathbb{F}_{q^l}[z]/(z^n - 1)$ is defined by

$$A(z) = A_0 + A_1 z + A_2 z^2 + \cdots + A_{n-1} z^{n-1}$$

$$a(z) = a_0 + a_1 z + a_2 z^2 + \cdots + a_{n-1} z^{n-1}$$

with the DFT formulas given in Figure A.10.

The discrete Fourier transform is a mapping of a polynomial $a(z)$ in the factorial ring $\mathbb{F}_{q^l}[z]/(z^n - 1)$ onto a polynomial $A(z)$ in $\mathbb{F}_{q^l}[z]/(z^n - 1)$, i.e.

$$\text{DFT}: \quad \mathbb{F}_{q^l}[z]/(z^n - 1) \mapsto \mathbb{F}_{q^l}[z]/(z^n - 1).$$

In matrix form the transform equations read

$$\begin{pmatrix} A_0 \\ A_1 \\ \vdots \\ A_{n-1} \end{pmatrix} = \begin{pmatrix} 1 & 1 & \cdots & 1 \\ 1 & \alpha^{1 \cdot 1} & \cdots & \alpha^{1 \cdot (n-1)} \\ 1 & \alpha^{2 \cdot 1} & \cdots & \alpha^{2 \cdot (n-1)} \\ \vdots & \vdots & \ddots & \vdots \\ 1 & \alpha^{(n-1) \cdot 1} & \cdots & \alpha^{(n-1) \cdot (n-1)} \end{pmatrix} \begin{pmatrix} a_0 \\ a_1 \\ \vdots \\ a_{n-1} \end{pmatrix}$$

and

$$\begin{pmatrix} a_0 \\ a_1 \\ \vdots \\ a_{n-1} \end{pmatrix} = n^{-1} \begin{pmatrix} 1 & 1 & \cdots & 1 \\ 1 & \alpha^{-1 \cdot 1} & \cdots & \alpha^{-1 \cdot (n-1)} \\ 1 & \alpha^{-2 \cdot 1} & \cdots & \alpha^{-2 \cdot (n-1)} \\ \vdots & \vdots & \ddots & \vdots \\ 1 & \alpha^{-(n-1) \cdot 1} & \cdots & \alpha^{-(n-1) \cdot (n-1)} \end{pmatrix} \begin{pmatrix} A_0 \\ A_1 \\ \vdots \\ A_{n-1} \end{pmatrix}.$$

ALGEBRAIC STRUCTURES

The polynomial $A(z)$ – possibly in different order – is also called the Mattson–Solomon polynomial. Similarly to the discrete Fourier transform over the field $\mathbb{C}$ of complex numbers there also exist fast transforms for calculating the discrete Fourier transform over finite fields $\mathbb{F}_{q^l}$. The discrete Fourier transform is, for example, used in the context of Reed–Solomon codes, as we will see in Chapter 2.

B

Linear Algebra

This appendix aims to provide some basics of linear algebra as far as it is necessary for the topics of this book. For further information the reader is referred to the rich literature that is available for linear algebra. Generally, we denote an $N \times N$ identity matrix by $\mathbf{I}_N$, $\mathbf{0}_{N \times M}$ is an $N \times M$ matrix containing only 0s and $\mathbf{1}_{N \times M}$ is a matrix of the same size consisting only of 1s.

Definition B.0.1 (Hermitian Matrix) *A Hermitian matrix $\mathbf{A}$ is a square matrix whose complex conjugate transposed version is identical to itself*

$$\mathbf{A} = \mathbf{A}^{\mathrm{H}} \tag{B.1}$$

The real part $\mathbf{A}'$ of a Hermitian matrix is symmetric while the imaginary part $\mathbf{A}''$ is asymmetric, i.e.

$$\mathbf{A}' = (\mathbf{A}')^{\mathrm{T}} \quad \text{and} \quad \mathbf{A}'' = -(\mathbf{A}'')^{\mathrm{T}} \tag{B.2}$$

holds. Obviously, the properties symmetric and Hermitian are identical for real matrices.

According to strang (Strang, 1988), Hermitian matrices have the following properties:

- Its diagonal elements $A_{i,i}$ are real.
- For each element, $A_{i,j} = A_{j,i}^*$ holds.
- For all complex vectors x, the number $\mathbf{x}^{\mathrm{H}} \mathbf{A} \mathbf{x}$ is real.
- From (B.1), it follows directly that $\mathbf{A}\mathbf{A}^{\mathrm{H}} = \mathbf{A}^{\mathrm{H}}\mathbf{A}$.
- The eigenvalues λ_i of a Hermitian matrix are real.
- The eigenvectors $\mathbf{x}_i$ belonging to different eigenvalues λ_i of a real symmetric or Hermitian matrix are mutually orthogonal.

Coding Theory – Algorithms, Architectures, and Applications   André Neubauer, Jürgen Freudenberger, Volker Kühn
© 2007 John Wiley & Sons, Ltd

Definition B.0.2 (Spectral Norm) *Following Golub and van Loan, (Golub and van Loan, 1996), the spectral norm or ℓ_2 norm of an arbitrary $N \times M$ matrix $\mathbf{A}$ is defined as*

$$\|\mathbf{A}\|_2 = \sup_{x \neq 0} \frac{\|\mathbf{A}x\|}{\|x\|} \tag{B.3}$$

The ℓ_2 norm describes the maximal amplification of a vector x that experiences a linear transformation by $\mathbf{A}$. It has the following basic properties:

- The spectral norm equals its largest singular value σ_{max}

$$\|\mathbf{A}\|_2 = \sigma_{max}(\mathbf{A}) \tag{B.4}$$

- The spectral norm of the inverse $\mathbf{A}^{-1}$ is identical to the reciprocal of the smallest singular value σ_{min} of $\mathbf{A}$

$$\|\mathbf{A}^{-1}\|_2 = \frac{1}{\sigma_{min}(\mathbf{A})} \tag{B.5}$$

Definition B.0.3 (Frobenius Norm) *The Frobenius norm of an arbitrary $N \times M$ matrix $\mathbf{A}$ resembles the norm of a vector and is defined as the sum over the squared magnitudes of all matrix elements (Golub and van Loan, 1996)*

$$\|\mathbf{A}\|_F = \sqrt{\sum_{i=1}^{N}\sum_{j=1}^{M}|A_{i,j}|^2} = \sqrt{\mathrm{tr}\{\mathbf{A}\mathbf{A}^H\}} \tag{B.6}$$

Consequently, the squared Frobenius norm is $\|\mathbf{A}\|_F^2 = \mathrm{tr}\{\mathbf{A}\mathbf{A}^H\}$.

Definition B.0.4 (Rank) *The rank $r = \mathrm{rank}(\mathbf{A})$ of an arbitrary matrix $\mathbf{A}$ equals the largest number of linear independent columns or rows.*

According to this definition, we can conclude that the rank of an $N \times M$ matrix is always upper bounded by the minimum of N and M

$$r = \mathrm{rank}(\mathbf{A}) \leq \min(N, M) \tag{B.7}$$

The following properties hold:

- An $N \times N$ matrix $\mathbf{A}$ is called regular if its determinant is non-zero and, therefore, $r = \mathrm{rank}(\mathbf{A}) = N$ holds. For regular matrices, the inverse $\mathbf{A}^{-1}$ with $\mathbf{A}^{-1}\mathbf{A} = \mathbf{I}_{N \times N}$ exists.

- If the determinant is zero, the rank r is smaller than N and the matrix is singular. The inverse does not exist for singular matrices.

- For each $N \times N$ matrix $\mathbf{A}$ of rank r there exist at least one $r \times r$ submatrix whose determinant is non-zero. The determinants of all $(r+1) \times (r+1)$ submatrices of $\mathbf{A}$ are zero.

- The rank of the product $\mathbf{A}\mathbf{A}^H$ equals

$$\mathrm{rank}(\mathbf{A}\mathbf{A}^H) = \mathrm{rank}(\mathbf{A}) \tag{B.8}$$

LINEAR ALGEBRA

Definition B.0.5 (Orthogonality) *An orthogonal real-valued matrix consists of columns that are orthogonal to each other, i.e. the inner product between different columns equals $\mathbf{q}_i^T \mathbf{q}_j = 0$. A matrix is termed orthonormal if its columns are orthogonal and additionally have unit length*

$$\mathbf{q}_i^T \mathbf{q}_j = \delta(i, j) \tag{B.9}$$

Orthonormal matrices have the properties

$$\mathbf{Q}^T \mathbf{Q} = \mathbf{I}_N \quad \Leftrightarrow \quad \mathbf{Q}^T = \mathbf{Q}^{-1} \tag{B.10}$$

Definition B.0.6 (Unitary Matrix) *An $N \times N$ matrix $\mathbf{U}$ with orthonormal columns and complex elements is called unitary. The Hermitian of a unitary matrix is also its inverse*

$$\mathbf{U}^H \mathbf{U} = \mathbf{U}\mathbf{U}^H = \mathbf{I}_N \quad \Leftrightarrow \quad \mathbf{U}^H = \mathbf{U}^{-1} \tag{B.11}$$

The columns of $\mathbf{U}$ span an N-dimensional orthonormal vector space.

Unitary matrices $\mathbf{U}$ have the following properties:

- The eigenvalues of $\mathbf{U}$ have unit magnitude ($|\lambda_i| = 1$).

- The eigenvectors belonging to different eigenvalues are orthogonal to each other.

- The inner product $\mathbf{x}^H \mathbf{y}$ between two vectors is invariant to multiplications with a unitary matrix because $(\mathbf{U}\mathbf{x})^H(\mathbf{U}\mathbf{y}) = \mathbf{x}^H \mathbf{U}^H \mathbf{U} \mathbf{y} = \mathbf{x}^H \mathbf{y}$.

- The norm of a vector is invariant to the multiplication with a unitary matrix, $\|\mathbf{U}\mathbf{x}\| = \|\mathbf{x}\|$.

- A random matrix $\mathbf{B}$ has the same statistical properties as the matrices $\mathbf{B}\mathbf{U}$ and $\mathbf{U}\mathbf{B}$.

- The determinant of a unitary matrix equals $\det(\mathbf{U}) = 1$ (Blum, 2000).

Definition B.0.7 (Eigenvalue Problem) *The calculation of the eigenvalues λ_i and the associated eigenvectors x_i of a square $N \times N$ matrix $\mathbf{A}$ is called the eigenvalue problem. The basic problem is to find a vector $\mathbf{x}$ proportional to the product $\mathbf{A}\mathbf{x}$. The corresponding equation*

$$\mathbf{A} \cdot x = \lambda \cdot x \tag{B.12}$$

can be rewritten as $(\mathbf{A} - \lambda \mathbf{I}_N) x = 0$. For the non-trivial solution $x \neq \mathbf{0}$, the matrix $(\mathbf{A} - \lambda \mathbf{I}_N)$ has to be singular, i.e. its columns are linear dependent. This results in $\det(\mathbf{A} - \lambda \mathbf{I}_N) = 0$ and the eigenvalues λ_i represent the zeros of the characteristic polynomial $p_N(\lambda) = \det(\mathbf{A} - \lambda \mathbf{I}_N)$ of rank N. Each $N \times N$ matrix has exactly N eigenvalues that need not be different.

In order to determine the associated eigenvectors x_i, the equation $(\mathbf{A} - \lambda_i \mathbf{I}_N) \mathbf{x}_i = 0$ has to be solved for each eigenvalue λ_i. Since $\mathbf{x}_i$ as well as $c \cdot \mathbf{x}_i$ fulfil the above equation, a unique eigenvector is only obtained by normalising it to unit length. The eigenvectors $x_1, \ldots, x_k$ belonging to different eigenvalues $\lambda_1, \ldots, \lambda_k$ are linear independent of each other (Horn and Johnson, 1985; Strang, 1988).

The following relationships exist between the matrix $\mathbf{A}$ and its eigenvalues:

- The sum over all eigenvalues equals the trace of a square matrix **A**

$$\text{tr}(\mathbf{A}) = \sum_{i=1}^{r=N} A_{i,i} = \sum_{i=1}^{r=N} \lambda_i \tag{B.13}$$

- The product of the eigenvalues of a square matrix **A** with full rank equals the determinant of **A**

$$\prod_{i=1}^{r=N} \lambda_i = \det(\mathbf{A}) \tag{B.14}$$

- If eigenvalues $\lambda_i = 0$ exist, the matrix is singular, i.e. $\det(\mathbf{A}) = 0$ holds.

Definition B.0.8 (Eigenvalue Decomposition) *An $N \times N$ matrix **A** with N linear independent eigenvectors $\mathbf{x}_i$ can be transformed into a diagonal matrix according to*

$$\mathbf{U}^{-1}\mathbf{A}\mathbf{U} = \Lambda = \begin{pmatrix} \lambda_1 & & & \\ & \lambda_2 & & \\ & & \ddots & \\ & & & \lambda_N \end{pmatrix} \tag{B.15}$$

*if $\mathbf{U} = (x_1, x_2, \ldots, x_N)$ contains the N independent eigenvectors of **A** (Horn and Johnson, 1985). The resulting diagonal matrix Λ contains the eigenvalues λ_i of **A**.*

Since **U** is unitary, it follows from definition B.0.8 that each matrix **A** can be decomposed as $\mathbf{A} = \mathbf{U}\Lambda\mathbf{U}^{-1} = \mathbf{U}\Lambda\mathbf{U}^H$.

Definition B.0.9 (Singular Value Decomposition) *A generalisation of definition (B.0.8) for arbitrary $N \times M$ matrices **A** is called singular value decomposition (SVD). A matrix **A** can be expressed by*

$$\mathbf{A} = \mathbf{U}\Sigma\mathbf{V}^H \tag{B.16}$$

*with the unitary $N \times N$ matrix **U** and the unitary $M \times M$ matrix **V**. The columns of **U** contain the eigenvectors of $\mathbf{A}\mathbf{A}^H$ and the columns of **V** the eigenvectors of $\mathbf{A}^H\mathbf{A}$. The matrix Σ is an $N \times M$ diagonal matrix with non-negative, real-valued elements σ_k on its diagonal. Denoting the eigenvalues of $\mathbf{A}\mathbf{A}^H$ and, therefore, also of $\mathbf{A}^H\mathbf{A}$ with λ_k, the diagonal elements σ_k are the positive square roots of λ_k*

$$\sigma_k = \sqrt{\lambda_k} \tag{B.17}$$

*They are called singular values of **A**. For a matrix **A** with rank r, we obtain*

$$\Sigma = \left(\begin{array}{cccc|ccc} \sigma_1 & 0 & \ldots & 0 & 0 & \ldots & 0 \\ 0 & \sigma_2 & & 0 & 0 & & 0 \\ \vdots & & \ddots & \vdots & \vdots & & \vdots \\ 0 & 0 & \ldots & \sigma_r & 0 & \ldots & 0 \\ \hline +0 & 0 & \ldots & 0 & 0 & \ldots & 0 \\ \vdots & & \ddots & \vdots & \vdots & & \vdots \\ 0 & 0 & \ldots & 0 & 0 & \ldots & 0 \end{array}\right) \begin{array}{l} \left.\begin{array}{l} \\ \\ \\ \\ \end{array}\right\} r \text{ rows} \\ \left.\begin{array}{l} \\ \\ \\ \end{array}\right\} M\text{-}r \text{ rows} \end{array} \tag{B.18}$$

$\underbrace{}_{r \text{ columns}} \underbrace{}_{N\text{-}r \text{ columns}}$

LINEAR ALGEBRA

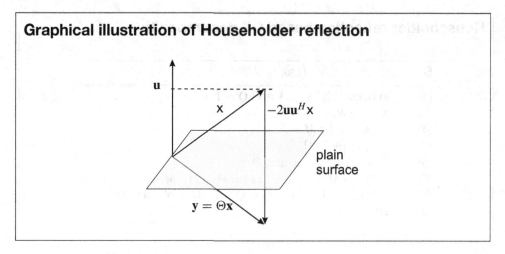

Figure B.1: Graphical illustration of Householder reflection

Definition B.0.10 (vec-Operator) *The application of the vec-operator onto a matrix* $\mathbf{A}$ *stacks its columns on top of each other. For* $\mathbf{A} = [\mathbf{a}_1, \ldots, \mathbf{a}_n]$, *we obtain*

$$\mathrm{vec}(\mathbf{A}) = \begin{bmatrix} \mathbf{a}_1 \\ \vdots \\ \mathbf{a}_n \end{bmatrix} \qquad (B.19)$$

Definition B.0.11 (Householder Reflections) *A vector* x *can be reflected at a plane or line onto a new vector* $\mathbf{y}$ *of the same length by Householder reflections. The reflection is performed by a multiplication of* $\mathbf{x}$ *with a unitary matrix* Θ. *Since we generally consider column vectors, we obtain* $\mathbf{y} = \Theta \cdot \mathbf{x}$ *with the unitary matrix*

$$\Theta = \mathbf{I}_N - (1+w) \cdot \mathbf{u}\mathbf{u}^H \qquad (B.20)$$

The vector $\mathbf{u}$ *and the scalar* w *are defined as*

$$\mathbf{u} = \frac{\mathbf{x} - \mathbf{y}}{\|\mathbf{x} - \mathbf{y}\|} \quad \text{and} \quad w = \frac{\mathbf{x}^H \mathbf{u}}{\mathbf{u}^H \mathbf{x}} \qquad (B.21)$$

If $\mathbf{x}$ *contains only real-valued elements,* $w = 1$ *holds and Equation (B.20) becomes*

$$\Theta = \mathbf{I}_N - 2 \cdot \mathbf{u}\mathbf{u}^H \qquad (B.22)$$

The reflection is graphically illustrated in Figure B.1 for a real-valued vector $\mathbf{x}$. The plane $\mathbf{x}$ has to be reflected at the plain surface being represented by the vector $\mathbf{u}$ that is perpendicular to the plane. The projection $\mathbf{u}\mathbf{u}^H\mathbf{x}$ of $\mathbf{x}$ onto $\mathbf{u}$ has to be subtracted twice from the vector $\mathbf{x}$ in order to obtain a reflection at the line perpendicular to $\mathbf{u}$.

Householder reflection based QL decomposition

Step	Task
(1)	Initialise with $\mathbf{L} = \mathbf{A}$ and $\mathbf{Q} = \mathbf{I}_M$
(2)	for $k = N, \ldots, 1$
(3)	$\quad \mathbf{x} = \mathbf{L}[1 : M - N + k, k]$
(4)	$\quad \mathbf{y} = \begin{bmatrix} \mathbf{0} & \|\mathbf{x}\| \end{bmatrix}^T$
(5)	$\quad$ calculate $\mathbf{u}$, w and Θ
(6)	$\quad \mathbf{L}[1 : M - N + k, 1 : k] = \Theta \cdot \mathbf{L}[1 : M - N + k, 1 : k]$
(7)	$\quad \mathbf{Q}[:, 1 : M - N + k] = \mathbf{Q}[:, 1 : M - N + k] \cdot \Theta^H$
(8)	end

Figure B.2: Pseudocode for QL decomposition via Householder reflections

If a row vector $\underline{\mathbf{x}}$ instead of a column vector has to be reflected, w and Θ have the form

$$\Theta = \mathbf{I}_N - (1 + w) \cdot \underline{\mathbf{u}}^H \underline{\mathbf{u}} \quad \text{and} \quad w = \frac{\underline{\mathbf{u}} \, \underline{\mathbf{x}}^H}{\underline{\mathbf{x}} \, \underline{\mathbf{u}}^H} \tag{B.23}$$

The reflection is performed by $\underline{\mathbf{y}} = \underline{\mathbf{x}} \cdot \Theta$.

Householder reflections have been used for the Post-Sorting Algorithm (PSA) in Section 5.5 to force certain elements of a matrix to zero and thus restore the triangular structure after permutations. For this special case, the target vector has only one non-zero element and becomes

$$\mathbf{y} = \begin{bmatrix} \mathbf{0} & \|\mathbf{x}\| \end{bmatrix}^T .$$

Similarly, Householder reflections can be used to decompose an $M \times N$ matrix $\mathbf{A}$ with $M \geq N$ into the matrices $\mathbf{Q}$ and $at L$. The algorithm for this QL decomposition is shown as a pseudocode in Figure B.2.

Definition B.0.12 (Givens Rotation) *Let $\mathbf{G}(i, k, \theta)$ be an $N \times N$ identity matrix except for the elements $G_{i,i}^* = G_{k,k} = \cos\theta = \alpha$ and $-G_{i,k}^* = G_{k,i} = \sin\theta = \beta$, i.e. it has the form*

$$\mathbf{G}(i, k, \theta) = \begin{bmatrix} 1 & & & & & & \\ & \ddots & & & & & \\ & & \alpha & \cdots & -\beta & & \\ & & \vdots & \ddots & \vdots & & \\ & & \beta^* & \cdots & \alpha^* & & \\ & & & & & \ddots & \\ & & & & & & 1 \end{bmatrix} \tag{B.24}$$

LINEAR ALGEBRA

Hence, $\mathbf{G}(i, k, \theta)$ is unitary and describes a rotation by the angle θ in the N-dimensional vector space. If θ is chosen appropriately, the rotation can force the ith element of a vector to zero.

As an example, the ith element of an arbitrary column vector $\mathbf{x} = [x_1, \ldots, x_N]^T$ should equal zero. By choosing

$$\alpha = \cos\theta = \frac{x_k}{\sqrt{|x_i|^2 + |x_k|^2}} \quad \text{and} \quad \beta = \sin\theta = \frac{x_i}{\sqrt{|x_i|^2 + |x_k|^2}} \qquad (B.25)$$

we obtain the new vector $\mathbf{y} = \mathbf{G}(i, k, \theta)\mathbf{x}$ with

$$\begin{bmatrix} x_1 \\ \vdots \\ 0 \\ \vdots \\ \sqrt{|x_i|^2 + |x_k|^2} \\ \vdots \\ x_N \end{bmatrix} = \begin{bmatrix} 1 & & & & & & \\ & \ddots & & & & & \\ & & \frac{x_k}{\sqrt{|x_i|^2+|x_k|^2}} & \cdots & \frac{-x_i}{\sqrt{|x_i|^2+|x_k|^2}} & & \\ & & \vdots & \ddots & \vdots & & \\ & & \frac{x_i^*}{\sqrt{|x_i|^2+|x_k|^2}} & \cdots & \frac{x_k^*}{\sqrt{|x_i|^2+|x_k|^2}} & & \\ & & & & & \ddots & \\ & & & & & & 1 \end{bmatrix} \cdot \begin{bmatrix} x_1 \\ \vdots \\ x_i \\ \vdots \\ x_k \\ \vdots \\ x_N \end{bmatrix}$$

It can be recognised that only two elements of $\mathbf{x}$ and $\mathbf{y}$ differ: $y_i = 0$ and $y_k = \sqrt{|x_i|^2 + |x_k|^2}$. The Givens rotation can also be used to perform QL decompositions similarly to the application of Householder reflections.

C

Acronyms

3GPP Third-Generation Partnership Project

4PSK Four-point Phase Shift Keying

8PSK Eight-point Phase Shift Keying

16QAM 16-Point Quadrature Amplitude Modulation

ACK Acknowledgement

ACSU Add Compare Select Unit

AGC Automatic Gain Control

AMR Adaptive MultiRate

AP Acknowledgement Period

APP A-Posteriori Probability

ARQ Automatic Repeat Request

ASK Amplitude Shift Keying

AWGN Additive White Gaussian Noise

BCJR Bahl, Cocke, Jelinek, Raviv

BCH Bose, Chaudhuri, Hocquenghem

BCS Block Check Sequence

BEC Binary Erasure Channel

BER Bit Error Rate

Coding Theory – Algorithms, Architectures, and Applications   André Neubauer, Jürgen Freudenberger, Volker Kühn
© 2007 John Wiley & Sons, Ltd

BLAST Bell Labs Layered Space–Time

BPSK Binary Phase Shift Keying

BSC Binary Symmetric Channel

CDMA Code Division Multiple Access

COST European Cooperation in the field of Scientific and Technical Research

CRC Cyclic Redundancy Check

CSI Channel State Information

DAB Digital Audio Broadcast

D-BLAST diagonal BLAST

DFT Discrete Fourier Transform

DoA Direction of Arrival

DoD Direction of Departure

DRAM Dynamic Random Access Memory

DS-CDMA direct-sequence CDMA

DSL Digital Subscriber Line

DSP Digital Signal Processor

DVB Digital Video Broadcast

EDGE Enhanced Data rates for GSM Evolution

EFR Enhanced Full Rate

EGPRS Enhanced General Packet Radio Service

EXIT EXtrinsic Information Transfer

FDD Frequency Division Duplex

FDMA Frequency Division Multiple Access

FEC Forward Error Correction

FER Frame Error Rate

FFT Fast Fourier Transform

FHT Fast Hadamard Transform

FR Full-Rate

ACRONYMS

GF Galois Field

GMSK Gaussian Minimum Key Shifting

GPRS General Packet Radio Service

GSM Global System for Mobile communications

HCS Header Check Sequence

HSCSD High-Speed Circuit Switched Data

HSDPA High-Speed Downlink Packet Access

IEEE Institute of Electrical and Electronics Engineers

i.i.d. independent identically distributed

IOPEF Input–Output Path Enumerator Function

IOWEF Input–Output Weight Enumerating Function

IR Incremental Redundancy

ISI Inter Symbol Interference

LA Link Adaptation

LAN Local Area Network

LD Linear Dispersion

LDPC Low-Density Parity Check

LFSR Linear Feedback Shift Register

LLL Lenstra, Lenstra and Lovász

LLR Log-Likelihood Ratio

LoS Line of Sight

LR lattice reduction

LTI Linear Time-Invariant

MAC Medium Access Control

MAP Maximum A-Posteriori

MDS Maximum Distance Separable

MCS Modulation and Coding Scheme

MED Minimum Error Probability Decoding

MIMO Multiple-Input Multiple-Output

MISO Multiple-Input Single-Output

ML Maximum Likelihood

MLD Maximum Likelihood Decoding

MMSE Minimum Mean-Square Error

MRC Maximum Ratio Combining

NACK Not Acknowledgement

NLoS Non-Line of Sight

OFDM Orthogonal Frequency Division Multiplexing

PDF Probability Density Function

PEF Path Enumerator Function

PSA Post-Sorting Algorithm

PSK Phase Shift Keying

QAM Quadrature Amplitude Modulation

QLD QL Decomposition

QoS Quality of Service

QPSK Quaternary Phase Shift Keying

RLC Radio Link Control

SCC Serially Concatenated Code

SCM Spatial Channel Model

SDM Space Division Multiplexing

SDMA Space Division Multiple Access

SIC Successive Interference Cancellation

SIMO Single-input Multiple-Output

SINR Signal to Interference plus Noise Ratio

SISO Soft-Input Soft-Output

SMU Survivor Memory Unit

SNR Signal-to-Noise Ratio

ACRONYMS

SOVA Soft-Output Viterbi Algorithm

SQLD Sorted QL Decomposition

SR-ARQ Selective Repeat ARQ

STBC Space-Time Block Code

STC Space-Time Code

STTC Space-Time Trellis Code

SVD Singular Value Decomposition

TB Tail Bits

TC Turbo Code

TDD Time Division Duplex

TDMA Time Division Multiple Access

TMU Transition Metric Unit

TFCI Transport Format Combination Indicator

TU Typical Urban

UEP Unequal Error Protection

UMTS Universal Mobile Telecommunications System

UTRA UMTS Terrestrial Radio Access

V-BLAST vertical BLAST

VLSI Very Large-Scale Integration

WCC Woven Convolutional Code

WEF Weight Enumerating Function

WER Word Error Rate

WIMAX Worldwide Interoperability for Microwave Access

WSSUS wide sense stationary uncorrelated scattering

WLAN Wireless Local Area Network

WTC Woven Turbo Code

ZF Zero Forcing

Bibliography

3GPP (1999) *Physical Channels and Mapping of Transport Channels onto Physical Channels (FDD)*. 3rd Generation Partnership Project, Technical Specification Group Radio Access Network, TS 25.211, http://www.3gpp.org/ftp/Specs/html-info/25-series.htm.

3GPP (2003) *Spatial Channel Model for Multiple Input Multiple Output (MIMO) Simulations (Release 6)*. 3rd Generation Partnership Project, Technical Specification Group Radio Access Network, TR25.996, http://www.3gpp.org/ftp/Specs/html-info/25-series.htm.

3GPP (2007) *Multiple-Input Multiple-Output in UTRA*. 3rd Generation Partnership Project, Technical Specification Group Radio Access Network, TS25.876, http://www.3gpp.org/ftp/Specs/html-info/25-series.htm.

Agrell, E., Eriksson, T., Vardy, A. and Zeger, K. (2002) Closest point search in lattices. *IEEE Transactions on Information Theory*, **48** (8), 2201–2214.

Alamouti, S. (1998) A simple transmit diversity technique for wireless communications. *IEEE Journal on Selected Areas in Communications*, **16** (8), 1451–1458.

Bahl, L., Cocke, J., Jelinek, F. and Raviv, J. (1974) Optimal decoding of linear codes for minimum symbol error rate. *IEEE Transactions on Information Theory*, **20**, 284–287.

Ball, C., Ivanov, K., Bugl, L. and Stöckl, P. (2004a) Improving GPRS/EDGE end-to-end performance by optimization of the RLC protocol and parameters. IEEE Proceedings of Vehicular Technology Conference, Los angeles, CA, USA, pp. 4521–4527.

Ball, C., Ivanov, K., Stöckl P., Masseroni, C., Parolari, S. and Trivisonno, R. (2004b) Link quality control benefits from a combined incremental redundancy and link adaptation in EDGE networks. IEEE Proceedings of Vehicular Technology Conference, Milan, Italy, pp. 1004–1008.

Bäro, S., Bauch, G. and Hansmann, A. (2000a) Improved codes for space–time trellis coded modulation. *IEEE Communications Letters*, **4** (1), 20–22.

Bäro, S., Bauch, G. and Hansmann, A. (2000b) New trellis codes for Space–time coded modulation. ITG Conference on Source and Channel Coding, Munich, Germany, pp. 147–150.

Benedetto, S. and Biglieri, E. (1999) *Principles of Digital Transmission with Wireless Applications*, Kluwer Academic/Plenum Publishers, New York.

Benedetto, S. and Montorsi, G. (1996) Unveiling turbo codes: some results on parallel concatenated coding schemes. *IEEE Transactions on Information Theory*, **42**, 409–429.

Benedetto, S. and Montorsi, G. (1998) Serial concatenation of interleaved codes: performance analysis, design, and iterative decoding. *IEEE Transactions on Information Theory*, **44**, 909–926.

Berlekamp, E. (1984) *Algebraic Coding Theory*, revised 1984, Aegean Park Press, Laguna Hills, CA, USA.

Berrou, C., Glavieux, A. and Thitimasjshima, P. (1993) Near shannon limit error-correcting coding and decoding: turbo-codes (1). IEEE Proceedings of International Conference on Communications, Geneva, Switzerland, pp. 1064–1070.

Bluetooth (2004) *Specification of the Bluetooth System*, Bluetooth SIG. http://www.bluetooth.com.

Blum, R. (2000) Analytical tools for the design of space–time convolutional codes. Conference on Information Sciences and Systems, Princeton, NJ, USA.

Bocharova, I., Kudryashov, B., Handlery, M. and Johannesson, R. (2002) Convolutional codes with large slopes yield better tailbiting codes. IEEE Proceedings of International Symposium on Information Theory, Lausanne, Switzerland.

Bose, R., Ray-Chaudhuri, D. and Hocquenghem, A. (1960) On a class of error-correcting binary group codes. *Information and Control* **3**, 68–79, 279–290.

Bossert, M. (1999) *Channel Coding for Telecommunications*, John Wiley & Sons, Ltd, Chichester, UK.

Bossert, M., Gabidulin, E. and Lusina, P. (2000) Space–time codes based on Hadamard matrices. IEEE Proceedings of International Symposium on Information Theory, Sorrento, Italy, p. 283.

Bossert, M., Gabidulin, E. and Lusina, P. (2002) Space–time codes based on Gaussian integers IEEE Proceedings of International Symposium on Information Theory, Lausanne, Switzerland, p. 273.

Böhnke, R., Kühn, V. and Kammeyer, K.D. (2004a) Efficient near maximum-likelihood decoding of multistratum space-time codes. IEEE Semiannual Vehicular Technology Conference (VTC2004-Fall), Los Angeles, CA, USA.

Böhnke, R., Kühn, V. and Kammeyer, K.D. (2004b) Multistratum space-time codes for the asynchronous uplink of MIMO-CDMA systems. International Symposium on Spread Spectrum Techniques and Applications (ISSSTA'04), Sydney, Australia.

Böhnke R., Kühn, V. and Kammeyer, K.D. (2004c) Quasi-orthogonal multistratum space-time codes. IEEE Global Conference on Communications (Globecom'04), Dallas, TX, USA.

Böhnke R., Wübben, D., Kühn, V. and Kammeyer, K.D. (2003) Reduced complexity MMSE detection for BLAST architectures. IEEE Proceedings of Global Conference on Telecommunications, San Francisco, CA, USA.

Bronstein, I., Semendjajew, K., Musiol, G. and Mühlig, H. (2000) *Taschenbuch der Mathematik*, 5th edn, Verlag Harri Deutsch, Frankfurt, Germany.

Clark, G.C. and Cain, J.B. (1988) *Error-correcting Coding for Digital Communications*, Plenum Press, New York, NY, USA.

Costello, D., Hagenauer, J., Imai, H. and Wicker, S. (1998) Applications of error-control coding. *IEEE Transactions on Information Theory*, **44** (6), 2531–2560.

Cover, T. and Thomas, J. (1991) *Elements of Information Theory*, John Wiley & Sons, Inc., New York, NY, USA.

Di, C., Proietti, D., Telatar, E., Richardson, T. and Urbanke, R. (2002) Finite-length analysis of low-density parity-check codes on the binary erasure channel. *IEEE Transactions on Information Theory*, **48**, 1570–1579.

Divsalar, D. and McEliece, R.J. (1998) *On the Design of Generalized Concatenated Coding Systems with Interleavers*. TMO Progress Report 42-134, Jet Propulsion Laboratory, California Institute of Technology, Pasadena, CA, USA.

Divsalar, D. and Pollara, F. (1995) *Multiple Turbo Codes for Deep-Space Communications*. TDA Progress Report 42-121, Jet Propulsion Laboratory, California Institute of Technology, Pasadena, CA, USA.

Dornstetter, J. (1987) On the equivalence between Berlekamp's and Euclid's algorithms. *IEEE Transactions on Information Theory*, **33** (3), 428–431.

Elias, P. (1954) Error-free coding. *IEEE Transactions on Information Theory*, **4**, 29–37.

Elias, P. (1955) Coding for noisy channels. IRE Convention Record 4, pp. 37–46.

ETSI (2001) *Broadband Radio Access Networks (BRAN); HIPERLAN type 2; Physical (PHY) Layer*. Norme ETSI, Document RTS0023003-R2, Sophia-Antipolis, France.

BIBLIOGRAPHY

ETSI (2006) *Digital Video Broadcasting (DVB), Framing Structure, Channel Coding and Modulation for Digital Terrestrial Television*, European Telecommunications Standards Institute, Sophia Antipolis, France, http://www.etsi.org/about_etsi/5_minutes/home.htm.

Fano, R.M. (1963) A heuristic discussion of probabilistic decoding. *IEEE Transactions on Information Theory*, **9**, 64–73.

Fincke, U. and Pohst, M. (1985) Improved methods for calculating vectors of short length in a lattice, including a complexity analysis. *Mathematics of Computation*, **44**, 463–471.

Fischer, R. (2002) *Precoding and Signal Shaping for Digital Transmission*, John Wiley & Sons, Inc., New York, NY, USA.

Fischer, R. and Huber, J. (1996) A New loading algorithm for discrete multitone transmission. IEEE Global Conference on Telecommunications (Globecom'96), London, UK, pp. 724, 728.

Forney, Jr, G. (1966) *Concatenated Codes*, MIT Press, Cambridge, MA, USA.

Forney, Jr, G.D. (1970) Convolutional codes I: algebraic structure. *IEEE Transactions on Information Theory*, **16**, 720–738.

Forney, Jr, G.D. (1973a) Structural analyses of convolutional codes via dual codes. *IEEE Transactions on Information Theory*, **19**, 512–518.

Forney, Jr, G.D. (1973b) The Viterbi algorithm. *Proceedings of the IEEE*, **61**, 268–278.

Forney, Jr, G.D. (1974) Convolutional codes II: maximum likelihood decoding. *Information Control*, **25**, 222–266.

Forney, Jr, G. (1991) Algebraic structure of convolutional codes, and algebraic system theory, In *Mathematical System Theory* (ed. A. C. Antoulas), Springer-Verlag, Berlin, Germany, pp. 527–558.

Forney, Jr, G., Johannesson, R. and Wan, Z.X. (1996) Minimal and canonical rational generator matrices for convolutional codes. *IEEE Transactions, on Information Theory*, **42**, 1865–1880.

Foschini, G. (1996) Layered space–time architecture for wireless communication in a fading environment when using multiple antennas. *Bell Labs Technical Journal*, **1** (2), 41–59.

Foschini, G. and Gans, M. (1998) On limits of wireless communications in a fading environment when using multiple antennas. *Wireless Personal Communications*, **6** (3), 311–335.

Foschini, G., Golden, G., Valencuela, A. and Wolniansky, P. (1999) Simplified processing for high spectral efficiency wireless communications emplying multi-element arrays. *IEEE Journal on Selected Areas in Communications*, **17** (11), 1841–1852.

Freudenberger, J. and Stender, B. (2004) An algorithm for detecting unreliable code sequence segments and its applications. *IEEE Transactions on Communications*, **COM-52**, 1–7.

Freudenberger, J., Bossert, M. and Shavgulidze, S. (2004) Partially concatenated convolutional codes. *IEEE Transactions on Communications*, **COM-52**, 1–5.

Freudenberger, J., Bossert, M., Shavgulidze, S. and Zyablov, V. (2000a) Woven turbo codes. Proceedings of 7th International Workshop on Algebraic and Combinatorial Coding Theory, Bansko, Bulgaria, pp. 145–150.

Freudenberger, J., Bossert, M., Shavgulidze, S. and Zyablov, V. (2001) Woven codes with outer warp: variations, design, and distance properties. IEEE Journal on Selected Areas in Communications, **19**, 813–824.

Freudenberger, J., Jordan, R., Bossert, M. and Shavgulidze, S. (2000b) Serially concatenated convolutional codes with product distance Proceedings of 2nd International Symposium on Turbo Codes and Related Topics, Brest, France, pp. 81–84.

Gabidulin, E., Bossert, M. and Lusina, P. (2000) Space–time codes based on rank codes. IEEE Proceedings of International Symposium on Information Theory, Sorrento, Italy, p. 284.

Gallager, R. (1963) *Low-Density Parity-Check Codes*, MIT Press, Cambridge, MA, USA.

Gesbert, D., Shafi, M., Shiu, D., Smith, P. and Naguib, A. (2003) From theory to practice: an overview of MIMO Space–time coded wireless systems. *IJSAC*, **21** (3), 281–302.

Gibson, D., Berger, T., Lookabaugh, T., Lindbergh, D. and Baker, R. (1998) *Digital Compression for Multimedia–Principles and Standards*, Morgan Kaufmann, San Franciso, CA, USA.

Golden, G., Foschini, G., Wolniansky, P. and Valenzuela, R. (1998) V-BLAST: a high capacity space–time architecture for the rich-scattering wireless channel. Proceedings of International Symposium on Advanced Radio Technologies, Boulder, CO, USA.

Golub, G. and van Loan, C. (1996) *Matrix Computations*, 3rd edn, The John Hopkins University Press, London, UK.

Gorenstein, D. and Zierler, N. (1961) A class of error-correcting codes in p^m symbols. *Journal of the Society for Industrial and Applied Mathematics*, 9 (3), 207–214.

Hagenauer, J. (1988) Rate-compatible punctured convolutional codes (RCPC codes) and their applications. *IEEE Transactions on Communications*, **COM-36**, 389–400.

Hagenauer, J. and Hoeher, P. (1989) A Viterbi algorithm with soft-decision outputs and its applications. IEEE Proceedings of International Conference on Communications, Dallas, TX, USA, pp. 1680–1686.

Hagenauer, J. Offer, E. and Papke, L. (1996) Iterative decoding of binary block and convolutional codes. *IEEE Transactions on Information Theory*, 42, 429–445.

Hamming, R. (1950) Error detecting and error correcting codes. *The Bell System Technical Journal*, 29, 147–160.

Hamming, R. (1986) *Information und Codierung*.

Hanzo, L., Webb, W. and Keller, T. (2000) *Single- and Multi-carrier Quadrature Amplitude Modulation–Principles and Applications for Personal Communications WLANs Broadcasting* 2d edn, IEEE-Press, John Wiley & Sons, Ltd, Chichester, UK.

Hanzo, L., Wong, C. and Yee, M. (2002) *Adaptive Wireless Transceivers: Turbo-Coded, Space–Time Coded TDMA, CDMA and OFDM Systems*, IEEE-Press, John Wiley & Sons, Ltd, Chichester, UK.

Hassibi, B. (2000) An efficient square-root algorithm for BLAST. IEEE Proceedings of International Conference on Acoustics, Speech and Signal Processing, Istanbul, Jurkey, pp. 5–9.

Hassibi, B. and Hochwald, B. (2000) High rate codes that are linear in space and time. Allerton Conference on Communication, Control and Computing, University of Jllinois at Urbana-Champaign, Urbana, IL, USA.

Hassibi, B. and Hochwald, B. (2001) Linear dispersion codes. *IEEE Proceedings of International Symposium on Information Theory*, Washington, DC, USA.

Hassibi, B. and Hochwald, B. (2002) High rate codes that are linear in space and time. *IEEE Transactions on Information Theory*, 48 (7), 1804–1824.

Heath, R. and Paulraj, A. (2002) Linear dispersion codes for MIMO systems based on frame theory. *IEEE Transactions on Signal Processing*, 50 (10), 2429–2441.

Hochwald, B. and Marzetta, T. (2000) Unitary space–time modulation for multiple-antenna communications in Rayleigh flat fading. *IEEE Transactions on Information Theory*, 46 (2), 543–564.

Hochwald, B. and Sweldens, W. (2000) Differential unitary space–time modulation. *IEEE Transactions on Communications Technology*, 48 (12), 2041–2052.

Hochwald, B., Marzetta, T., Richardson, T., Sweldens, W. and Urbanke, R. (2000) Systematic design of unitary Space–time constellations. *IEEE Transactions on Information Theory*, 46 (6), 1962–1973.

Hoeve, H., Timmermans, J. and Vries, L. (1982) Error correction and concealment in the compact disc system. *Philips Technical Review*, 40 (6), 166–172.

Holma, H. and Toskala, A. (2004) *WCDMA for UMTS*, 3rd edn, John Wiley & Sons, Ltd, Chichester, UK.

Honig, M. and Tsatsanis, M. (2000) Multiuser CDMA receivers. *IEEE Signal Processing Magazine*, (3), 49–61.

Horn, R. and Johnson, C. (1985) *Matrix Analysis*, Cambridge University Press, Cambridge, UK.

Höst, S. (1999) *On Woven Convolutional Codes*. PhD thesis, Lund University, ISBN 91-7167-016-5.

Höst, S., Johannesson, R. and Zyablov, V. (1997) A first encounter with binary woven convolutional codes Proceedings of International Symposium on Communication Theory and Applications, Ambleside, Lake District, UK.

Höst, S., Johannesson, R. and Zyablov, V. (2002) Woven convolutional codes I: encoder properties. *IEEE Transactions on Information Theory*, **48**, 149–161.

Höst, S., Johannesson, R., Zigangirov, K. and Zyablov, V. (1999) Active distances for convolutional codes. *IEEE Transactions on Information Theory*, **45**, 658–669.

Höst, S., Johannesson, R., Sidorenko, V., Zigangirov, K. and Zyablov, V. (1998) Cascaded convolutional codes, in *Communication and Coding (Eds M. Darnell and B. Honary)*, Research Studies Press Ltd and John Wiley & Sons, Ltd, Chichester, UK, pp. 10–29.

Hughes, B. (2000) Differential space–time modulation. *IEEE Transactions on Information Theory*, **46** (7), 2567–2578.

Hughes-Hartogs, D. (1989) Ensemble modem structure for imperfect transmission media. US Patents 4,679,227 and 4,731,816 and 4,833,706, Orlando, FL, USA.

Hübner, A. and Jordan, R. (2006) On higher order permutors for serially concatenated convolutional codes. *IEEE Transactions on Information Theory*, **52**, 1238–1248.

Hübner, A. and Richter, G. (2006) On the design of woven convolutional encoders with outer warp row permutors. *IEEE Transactions on Communications*, **COM-54**, 438–444.

Hübner, A., Truhachev, D.V. and Zigangirov, K.S. (2004) On permutor designs based on cycles for serially concatenated convolutional codes. *IEEE Transactions on Communications*, **COM-52**, 1494–1503.

IEEE (2004) *Part 16: Air Interface for Fixed Broadband Wireless Access Systems*, IEEE, New York, NY, USA.

Jelinek, F. (1969) A fast sequential decoding algorithm using a stack. *IBM Journal of Research and Development*, **13**, 675–685.

Johannesson, R. and Zigangirov, K.S. (1999) *Fundamentals of Convolutional Coding*, IEEE Press, Piscataway, NJ, USA.

Jordan, R., Pavlushkov, V., and Zyablov, V. (2004b) Maximum slope convolutional codes. *IEEE Transactions on Information Theory*, **50**, 2511–2521.

Jordan, R., Freudenberger, J., Dieterich, H., Bossert, M. and Shavgulidze, S. (1999) Simulation results for woven codes with outer warp. *Proceedings of 4th. ITG Conference on Mobile Communication, Munich, Germany*, pp. 439–444.

Jordan, R., Freudenberger, J., Pavlouchkov, V., Bossert, M. and Zyablov, V. (2000) Optimum slope convolutional codes. IEEE Proceedings of International Symposium on Information Theory, Sorrento, Italy.

Jordan, R., Höst, S., Johannesson, R., Bossert, M. and Zyablov, V. (2004a) Woven convolutional codes II: decoding aspects. *IEEE Transactions on Information Theory*, **50**, 2522–2531.

Jungnickel, D. (1995) *Codierungstheorie*, Spektrum Akademischer Verlag, Heidelberg, Germany.

Justesen. J., Thommensen, C. and Zyablov, V. (1988) Concatenated codes with convolutional inner codes. *IEEE Transactions on Information Theory*, **34**, 1217–1225.

Kallel, S. (1992) Sequential decoding with an efficient incremental redundancy ARQ strategy. *IEEE Transactions on Communications*, **40**, 1588–1593.

Kammeyer, K.D. (2004) *Nachrichtenübertragung*, 3rd edn, Teubner, Stuttgart, Germany.

Krongold, B., Ramchandran, K. and Jones, D. (1999) An efficient algorithm for optimum margin maximization in multicarrier communication systems. *IEEE Proceedings of Global Conference on Telecommunications*, Rio de Janeiro, Brazil, pp. 899–903.

Kühn, V. (2006) *Wireless Communications over MIMO Channels – Applications to CDMA and Multiple Antenna Systems*, John Wiley & Sons, Ltd, Chichester, UK.

Kühn, V. and Kammeyer, K.D. (2004) Multiple antennas in mobile communications: concepts and algorithms. International Symposium on Electromagnetic Theory (URSI 2004), Pisa, Italy.

Lee, H. (2003) An area-efficient Euclidean algorithm block for Reed–Solomon decoders. IEEE Computer Society Annual Symposium on VLSI, Tampa, FL, USA.

Lee, H. (2005) A high speed low-complexity Reed-Solomon decoder for optical communications. *IEEE Transactions on Circuits and Systems II*, **52** (8), 461–465.

Lee, L.H.C. (1997) *Convolutional Coding: Fundamentals and Applications*, Artech House, Boston, MA, USA.

Liew, T. and Hanzo, L. (2002) Space–time codes and concatenated channel codes for wireless communications. *IEEE Proceedings*, **90** (2), 187–219.

Lin, S. and Costello, D. (2004) *Error Control Coding–Fundamentals and Applications*, Pearson Prentice Hall, Upper Saddle River, NJ. USA.

Ling, S. and Xing, C. (2004) *Coding Theory–A First Course*, Cambridge University Press, Cambridge, UK.

Luby, M., Mitzenmacher, M. and Shokrollahi, A. (1998) Analysis of random processes via and-or tree evaluation. Proceedings of the 9th Annual ACM-SIAM Symposium on Discrete Algorithms, San Francisco, CA, USA, pp. 364–373.

Luby, M., Mitzenmacher, M., Shokrollahi, A. and Spielman, D. (2001) Analysis of low density codes and improved designs using irregular graphs. *IEEE Transactions on Information Theory*, **47**, 585–598.

Lusina, P., Gabidulin, E. and Bossert, M. (2001) Efficient decoding of space–time Hadamard codes using the Hadamard transform IEEE Proceedings of International Symposium on Information Theory, Washington, DC, USA, p. 243.

Lusina, P., Gabidulin, E. and Bossert, M. (2003) Maximum rank distance codes as space–time codes. *IEEE Transactions on Information Theory*, **49** (10), 2757–2760.

Lusina, P., Shavgulidze, S. and Bossert, M. (2002) Space time block code construction based on cyclotomic coset factorization. IEEE Proceedings of International Symposium on Information Theory, Lausanne, Switzerland, p. 135.

MacKay, D. (1999) Good error correcting codes based on very sparse matrices. *IEEE Transactions on Information Theory*, **45**, 399–431.

MacWilliams, F. and Sloane, N. (1998) *The Theory of Error-Correcting Codes*, North-Holland, Amsterdam, The Netherlands.

Massey, J. (1969) Shift register synthesis and bch decoding. *IEEE Transactions on Information Theory*, **15**, 122–127.

Massey, J. (1974) Coding and modulation in digital communications Proceedings of International Zürich Seminar on Digital Communications, Zürich Switzerland. pp. E2(1)–E2(4).

Massey, J. and Sain, M. (1967) Codes, automata, and continuous systems: explicit interconnections. *IEEE Transactions on Automatic Control*, **12**, 644–650.

Massey, J. and Sain, M. (1968) Inverses of linear sequential circuits. *IEEE Transactions on Computers*, **17**, 330–337.

McAdam, P.L., Welch, L.R. and Weber, C.L. (1972) MAP bit decoding of convolutional codes. IEEE Proceedings of International Symposium on Information Theory, p. 91.

McEliece, R. (1987) *Finite Fields for Computer Scientists and Engineers*, Kluwer Academic, Boston, MA, USA.

McEliece, R.J. (1998) The algebraic theory of convolutional codes, in Handbook of Coding Theory (eds V. Pless and W. Huffman), Elsevier Science, Amsterdam, The Netherlands, vol. 1, pp. 1065–1138.

McEliece, R. (2002) *The Theory of Information and Coding*, 2nd edn, Cambridge University Press, Cambridge, UK.

Moshavi, S. (1996) Multi-user detection for DS-CDMA communications. *IEEE Communications Magazine*, **34** (10), 124–136.

Muller, D. (1954) Applications of boolean algebra to switching circuits design and to error detection. *IRE Transactions on Electronic Computation*, **EC-3**, 6–12.

Mutti, C. and Dahlhaus, D. (2004) Adaptive loading procedures for multiple-input multiple-output OFDM systems with perfect channel state information. Joint COST Workshop on Antennas and Related System Aspects in Wireless Communications, Chalmers University of Technology, Gothenburg, Sweden.

Naguib, A., Tarokh, V., Seshadri, N. and Calderbank, A. (1997) Space–time coded modulation for high data rate wireless communications. IEEE Proceedings of Global Conference on Telecommunications, Phoenix, AZ, USA, vol. 1, pp. 102–109.

Naguib, A., Tarokh, V., Seshadri, N. and Calderbank, A. (1998) A space-time coding modem for high-data-rate wireless communications. *IEEE Journal on Selected Areas in Communications*, **16** (8), 1459–1478.

Neubauer, A. (2006a) *Informationstheorie und Quellencodierung–Eine Einführung für Ingenieure, Informatiker und Naturwissenschaftler*, J. Schlembach Fachverlag, Wilburgstetten, Germany.

Neubauer, A. (2006b) *Kanalcodierung–Eine Einführung für Ingenieure, Informatiker und Naturwissenschaftler*, J. Schlembach Fachverlag, Wilburgstetten, Germany.

Neubauer, A. (2007) *Digitale Signalübertragung–Eine Einführung in die Signal- und Systemtheorie*, J. Schlembach Fachverlag, Wilburgstetten, Germany.

Olofsson, H. and Furuskär, A. (1998) Aspects of introducing EDGE in existing GSM networks. IEEE Proceedings of International Conference on Universal Personal Communications, Florence, Italy, pp. 421–426.

Orten, P. (1999) Sequential decoding of tailbiting convolutional codes for hybrid ARQ on wireless channels. IEEE Proceedings of Vehicular Technology Conference, Houston, TX, USA, pp. 279–284.

Paulraj. A., Nabar, R. and Gore, D. (2003) *Introduction to Space–Time Wireless Communications*, Cambridge University Press, Cambridge, UK.

Pearl, J. (1988) *Probabilistic Reasoning in Intelligent Systems: Networks of Plausible Inference*, Morgan Kaufmann, San Francisco, CA, USA.

Perez, L., Seghers, J. and Costello, Jr, D.J. (1996) A distance spectrum interpretation of turbo codes. *IEEE Transactions on Information Theory*, **42**, 1698–1709.

Peterson, W. (1960) Encoding and error-correcting procedures for Bose–Chaudhuri codes. *IRE Transactions on Information Theory*, **IT-6**, 459–470.

Proakis, J. (2001) *Digital Communications*, 4th edn, McGraw-Hill, New York, NY, USA.

Reed, I. and Solomon, G. (1960) Polynomial codes over certain finite fields. *Journal of the Society for Industrial and Applied Mathematics*, **8**, 300–304.

Robertson, P., Hoeher, P. and Villebrun, E. (1997) Optimal and sub-optimal maximum a posteriori algorithms suitable for turbo decoding. *European Transactions on Telecommunications*, **8**, 119–125.

Richardson, T., Shokrollahi, A. and Urbanke, R. (2001) Design of capacity-approaching irregular low-density parity-check codes. *IEEE Transactions on Information Theory*, **47**, 619–637.

Richardson, T. and Urbanke, R. (2001) The capacity of low-density parity-check codes under message-passing decoding. *IEEE Transactions on Information Theory*, **47**, 599–618.

Sayood, K. (2000) *Introduction to Data Compression*, 2nd edn, Morgan Kaufmann, San Francisco, CA, USA.

Sayood, K. (2003) *Lossless Compression Handbook*, Academic Press, Amsterdam, The Netherlands.

Schnoor, C. and Euchner, M. (1994) Lattice basis reduction: improved practical algorithms and solving subset sum problems. *Mathematical Programming*, **66**, 181–191.

Schnug, W. (2002) *On Generalized Woven Codes*, VDI-Verlag, Düsseldorf, Germany.

Schober, R. and Lampe, L. (2002) Noncoherent Receivers for differential space–time modulation. *ITCT*, **50** (5), 768–777.

Schramm, P., Andreasson, H., Edholm, C., Edvardsson, N., Höök, M., Jäverbring, S., Müller, F. and Sköld, J. (1998) radio interface performance of EDGE, a proposal for enhanced data rates in existing digital cellular systems. IEEE Proceedings of Vehicular Technology Conference, Ottawa, Canada, pp. 1064–1068.

Schulze, H. and Lüders, C. (2005) *Theory and Applications of OFDM and CDMA*, John Wiley & Sons, Ltd, Chichester, UK.

Seshadri, N. and Winters, J. (1994) Two signaling schemes for improving the error performance of frequency division (FDD) transmission systems using transmitter antenna diversity. *International Journal of Wireless Networks*, **1** (1), 49–60.

Seshadri, N., Tarokh, V. and Calderbank, A. (1997) Space–time codes for wireless communication: code construction. IEEE Proceedings of Vehicular Technology Conference, Phoenix, AZ, USA, vol. 2-A, pp. 637–641.

Sezgin, A., Wübben, D. and Kühn, V. (2003) Analysis of mapping strategies for turbo coded space–time block codes. IEEE Information Theory Workshop (ITW03), Paris, France.

Shannon, C. (1948) A mathematical theory of communication. *The Bell System Technical Journal*, **27**, 379–424, 623–656.

Shung, C., Siegel, P., Ungerböck, G. and Thapar, H. (1990) Vlsi architectures for metric normalization in the Viterbi algorithm. IEEE Proceedings of International Conference on Communications, Atlanta, GA, USA. pp. 1723–1728.

Simon, M. and Alouini, M.S. (2000) *Digital Communication over Fading Channels*, John Wiley & Sons, Inc., New York, NY, USA.

Sklar, B. (2003) *Digital Communications: Fundamentals and Applications*. Prentice Hall PTR, Upper Saddle River, NJ, USA.

Strang, G. (1988) *Linear Algebra and its Applications*, 3rd edn, Harcout Brace Jovanovich College Publishers, Orlando, FL, USA.

Tanner, M. (1981) A recursive approach to low complexity codes. *IEEE Transactions on Information Theory*, **27**, 533–547.

Tarokh, V., Jafarkhani, H. and Calderbank, A. (1999a) Space–time block codes from orthogonal designs. *IEEE Transactions on Information Theory*, **45** (5), 1456–1467.

Tarokh, V., Jafarkhani, H. and Calderbank, A. (1999b) Space–time block coding for wireless communications: performance results. *IEEE Journal on Selected Areas in Communications*, **17** (3), 451–460.

Tarokh, V., Seshadri, N. and Calderbank, A. (1997) Space–time codes for high data rate wireless communication: performance criteria. IEEE Proceedings of International Conference on Communications, Montreal, Canada, vol. 1, pp. 299–303.

Tarokh, V., Seshadri, N. and Calderbank, A. (1998) Space–time codes for high data rate wireless communication: performance criterion and code construction. *IEEE Transactions on Information Theory*, **44** (2), 744–765.

Telatar, E. (1995) Capacity of multi-antenna Gaussian channels. ATT-Bell Labs Internal Tech. Memo.

ten Brink, S. (2000) Design of serially concatenated codes based on iterative decoding convergence. Proceedings 2nd International Symposium on Turbo Codes and Related Topics, Brest, France, pp. 319–322.

ten Brink, S. (2001) Convergence behavior of iteratively decoded parallel concatenated codes. *IEEE Transactions on Communications*, **COM-49**, 1727–1737.

van Lint, J. (1999) *Introduction to Coding Theory*, 3rd edn, Springer Verlag, Berlin, Germany.

Viterbi, A.J. (1967) Error bounds for convolutional codes and an asymptotically optimum decoding algorithm. *IEEE Transactions on Information Theory*, **13**, 260–269.

Viterbi, A.J. (1971) Convolutional codes and their performance in communication systems. *IEEE Transactions on Communication Technology*, **19**, 751–772.

Weiss, C., Bettstetter, C., Riedel, S. and Costello, D.J. (1998) Turbo decoding with tail-biting trellises. Proceedings of URSI International Symposium on Signals, Systems and Electronics, Pisa, Italy, pp. 343–348.

Windpassinger, C. and Fischer, R. (2003a) Low-complexity near-maximum-likelihood detection and precoding for MIMO systems using lattice reduction. IEEE Information Theory Workshop (ITW 2003), Paris, France.

Windpassinger, C. and Fischer, R. (2003b) Optimum and sub-optimum lattice-reduction-aided detection and precoding for MIMO communications. Canadian Workshop on Information Theory, Waterloo, Ontario, Canada, pp. 88–91.

Wittneben, A. (1991) Basestation modulation diversity for digital SIMUL-CAST. IEEE Proceedings of Vehicular Technology Conference, pp. 848–853, St Louis, MO, USA.

Wolniansky, P., Foschini, G., Golden, G. and Valenzuela, R. (1998) V-BLAST: an architecture for realizing very high data rates over the rich-scattering wireless channel. Invited paper, Proceedings of International Symposium on Signals, Systems and Electronics, Pisa, Italy.

Wozencraft, J.M. (1957) Sequential decoding for reliable communications. *IRE National Convention Record*, **5**, 2–11.

Wozencraft, J.M. and Reiffen, B. (1961) *Sequential Decoding*, MIT Press, Cambridge, MA, USA.

Wübben, D. (2006) *Effiziente Detektionsverfahren für Multilayer-MIMO-Systeme*. PhD thesis, Universität Bremen, Bremen, Germany.

Wübben, D, Böhnke, R, Kühn, V. and Kammeyer, K.D. (2003) MMSE extension of V-BLAST based on sorted QR decomposition. IEEE Semiannual Vehicular Technology Conference (VTC2003-Fall), Orlando, FL, USA.

Wübben, D., Böhnke, R, Kühn, V. and Kammeyer, K.D. (2004a) Near-maximum-likelihood detection of MIMO systems using MMSE-based lattice reduction. IEEE International Conference on Communications (ICC'2004), Paris, France.

Wübben, D., Böhnke, R., Rinas, J., Kühn, V. and Kammeyer, K.D. (2001) Efficient algorithm for decoding layered space–time codes. *IEE Electronic Letters*, **37** (22), 1348–1350.

Wübben, D., Kühn, V. and Kammeyer, K.D. (2004b) On the robustness of lattice-reduction aided detectors in correlated MIMO systems. IEEE Semiannual Vehicular Technology Conference (VTC2004-Fall), Los Angeles, CA, USA.

Zheng, L. and Tse, D. (2003) Diversity and multiplexing: a fundamental tradeoff in multiple antenna channels. *IEEE Transactions on Information Theory*, **49** (5), 1073–1096.

Zigangirov, K.S. (1966) Some sequential decoding procedures. *Problemy Peredachi Informatsii*, **2**, 13–25.

Zyablov, V., Shavgulidze, S., Skopintsev, O., Höst, S. and Johannesson, R. (1999b) On the error exponent for woven convolutional codes with outer warp. *IEEE Transactions on Information Theory*, **45**, 1649–1653.

Zyablov, V., Shavgulidze, S. and Johannesson, R. (2001) On the error exponent for woven convolutional codes with inner warp. *IEEE Transactions on Information Theory*, **47**, 1195–1199.

Zyablov. V., Johannesson, R., Skopintsev, O. and Höst, S. (1999a) Asymptotic distance capabilities of binary woven convolutional codes. *Problemy Peredachi Informatsii*, **35**, 29–46.

Index

(u,v) code construction, 46

a-posteriori probability, 22
acknowledgement, 150
active distances, 122, 202
    active burst distance, 123, 202
    active column distance, 124
    active reverse column distance, 124
    active segment distance, 124
addition table, 303
Alamouti, 257
algebraic decoding algorithm, 84
Amplitude Shift Keying, *see* ASK
angular spread, 231
APP decoding, 140, 145, 186
ARQ, 68
ASK, 218
automatic repeat request, 68
AWGN channel, 6, 57

Bahl, 140
basis, 27
Bayes' rule, 22
BCH bound, 78
BCH code, 80
BCJR algorithm, 140, 181, 186
beamforming, 257
BEC, binary erasure channel, 168
belief propagation algorithm, 168, 174
Benedetto, 184, 196, 212
Berger's channel diagram, 4
Berlekamp–Massey algorithm, 89
Berrou, 163
Bhattacharyya bound, 132
Bhattacharyya parameter, 132

binary symmetric channel, 5
binomial coefficient, 18
bit energy, 7
bit error probability, 5, 134
bit interleaving, 149
BLAST detection, 275
block check sequence, 153
block code, 13
    (n,k), 14
Bluetooth, 47, 51
Boolean functions, 58
bottleneck region, 194
bound, 37
    asymptotic, 40
    Bhattacharyya, 132, 134
    Gilbert–Varshamov, 40
    Griesmer, 40
    Hamming, 37
    Plotkin, 40
    Singleton, 37
    sphere packing, 37
    Viterbi, 134, 135
boxplus operation, 172, 173
BSC, 5, 114
burst, 202
burst error probability, 134

cardinality, 299
catastrophic generator matrix, 110
channel
    capacity, 4, 248
    code, 2
    coding, 1
    decoder, 2
    decoding, 2
    encoder, 2

Coding Theory – Algorithms, Architectures, and Applications   André Neubauer, Jürgen Freudenberger, Volker Kühn
© 2007 John Wiley & Sons, Ltd

check nodes, 165
CIRC, 83
Cocke, 140
code
    block, 13
    extension, 43
    interleaving, 44
    parameter, 16
    polynomial, 63
    puncturing, 43
    rate, 17
    shortening, 42
    space, 18
    space–time, 215
    termination, 104, 148, 153
    word, 14
commutative group, 295
commutative ring with identity, 296
commutativity, 295
compact disc, 83
complementary error function, 7
concatenated codes, 177, 182
concatenated convolutional codes, 182
conditional entropy, 4
conditional probability, 4
conjugate root, 76, 304
constraint length, 100
controller canonical form, 100
convolutional codes, 98–101, 121, 147, 182, 198
convolutional encoder, 98–101, 103, 106
correction ball, 25
correlated channels, 236
correlation matrix, 238
coset, 34
coset leader, 35
CRC, 66
cross-interleaved Reed–Solomon code, 83
cyclic code, 62
cyclic group, 295
cyclic redundancy check, 66
cyclotomic coset, 77

decision region, 20
decoding strategy, 19

demodulation, 2
demodulator, 2
design distance, 80
DFT, 82, 305
differential entropy, 6
digital video broadcasting, 83
dimension, 300
discrete convolution, 69
discrete Fourier transform, 82, 305
distance, 17
diversity, 256
    frequency diversity, 223
    polarization diversity, 223
    space diversity, 223
    time diversity, 223
diversity combining, 158
division with remainder, 296
divisor, 296
DoA, direction of arrival, 230
DoD, direction of departure, 230, 237
Doppler delay-angle scattering function, 230
DRAM, 29
dual code, 36
dual cyclic code, 71
DVB, 83
dynamic programming, 117

ECSD, 152
EDGE, 152
effective length, 204
EGPRS, 152
eigenvalue decomposition, 314
elementary symmetric polynomial, 89
Elias, 160
encoder state, 103
encoder state space, 103
entropy, 4
equidistant code, 40
equivalent encoders, 110
equivalent generator matrices, 110
erasure, 93
error
    burst, 44, 66
    covariance matrix, 275
    detection, 34
    evaluator polynomial, 89

INDEX

locator, 86
locator polynomial, 88
polynomial, 66, 86
position, 86
value, 86
error event, 121, 134
Euclid's algorithm, 89, 296
Euclidean metric, 136
EXIT charts, 188, 190, 192–196
EXIT, extrinsic information transfer, 188
expected weight distribution, 196
extended code, 43
extended Hamming code, 51
extension field, 75, 302
extrinsic log-likelihood ratio, 173

factorial ring, 63, 301
Fano, 160
fast Fourier transform, 82
fast Hadamard transform, 57
FDD, 239
FFT, 82
FHT, 57
field, 298
finite field, 299
finite geometric series, 91
finite group, 295
first event error probability, 134
Forney, 160
Forney's formula, 93
fractional rate loss, 105
free distance, 121, 122
frequency division duplex, see FDD
frequency hopping, 149
Frobenius norm, 312

Gallager, 163, 165, 167, 168
Galois field, 299
generating length, 203
generating tuples, 203
generator matrix, 29
generator polynomial, 64
girth, 177
Glavieux, 163
GMSK, 152
Golay code, 50
GPRS, 152

group, 295
GSM, 147, 152, 163

Hadamard matrix, 53
Hagenauer, 140, 169
Hamming code, 29, 49
Hamming distance, 17
hard output, 140
hard-decision decoding, 7
header check sequence, 153
Höher, 140
Höst, 212
HSCSD, 152
hybrid ARQ, 150
    type-I, 151, 156
    type-II, 151, 157

identity element, 295
incremental redundancy, 152, 158
independence assumption, 176
information polynomial, 65
information source, 3
information word, 14
inner code, 163
input–output path enumerator, 136
interleaved code, 44
interleaving, 153
interleaving depth, 44
inverse element, 295
IOPEF, 131
IOWEF, 128
irregular code, 167

Jacobian logarithm, 146, 172
Jelinek, 140, 160
Johannesson, 212

key equation, 89, 91

L-value, 169
latency, 46
LDPC codes, 163, 165, 168, 174
linear block code, 27
linear dispersion codes, 265
linear feedback shift register, 72
linear independency, 300

linear multilayer detection, 272
linearity, 27
link adaptation, 152, 156
log-likelihood algebra, 169
log-likelihood ratio, 141, 169, 171, 188
LoS, line-of-sight, 230

macrocell, 231
MacWilliams identity, 37
majority decision, 10
majority decoding, 47
majority logic decoding, 56
MAP, 22
MAP decoding, 112
Mariner, 58
Massey, 161
matched filter, 6
Mattson–Solomon polynomial, 309
max-log approximation, 147, 172
maximum a-posteriori, 22
maximum distance separable, 37
maximum likelihood decoding, 23, 112, 113, 116, 134, 196
maximum ratio combining, 159, 224
McAdam, 140
McEliece, 161
MCS, 152–155
MDS code, 37, 81
MED, 22
memory, 100
memoryless channel, 5
message nodes, 165
message passing, 174
message-passing algorithms, 168
MIMO channel
    frequency-selective, 234
    modelling, 237
minimal polynomial, 76, 304
minimum constraint length, 100
minimum distance decoding, 24, 113–115, 136
minimum error probability decoding, 22
minimum Hamming distance, 17
minimum length, 210, 211
minimum weight, 17
MISO channel, Multiple-Input Single-Output channel, 235

MLD, 23
modulation, 1
modulator, 2
Montorsi, 196, 212
Moore–Penrose inverse, 276
multilayer transmission, 266
multiplication table, 303
multiplicative inverse, 299
mutual information, 4

narrow-sense BCH code, 80, 84
NLoS, 230
node metric, 117
noise power, 6
noise vector, 57
normal burst, 149
not acknowledgement, 150

octal notation, 109
odd/even interleaver, 210
OFD codes, 122
optimal code, 37
optimum free distance codes, 122
order, 295, 303
orthogonal
    matrix, 313
    space–time block codes, 257
orthogonal code, 36
orthogonality, 30, 56
outer code, 163
overall constraint length, 100

pairwise error probability, 131, 134
parallel concatenation, 182, 185
parity
    bit, 28
    check code, 27, 48
    check condition, 31
    check equations, 31
    check matrix, 30
    check polynomial, 67
    check symbol, 29, 48
    frequency, 83
partial concatenation, 185
partial distance, 205
partial rate, 185
partial weights, 196

INDEX

path enumerator, 129
path enumerator function, 134
PEF, 131
perfect code, 39
Phase shift keying, *see* PSK
pinch-off region, 194
pit, 85
polling, 155
polynomial, 63, 300
    irreducible, 301
polynomial channel model, 72, 86
polynomial generator matrix, 110
polynomial ring, 300
power spectral density, 7
prime number, 296
primitive BCH code, 80
primitive element, 296
primitive polynomial, 303
primitive root, 303
primitive root of unity, 75
product code, 163, 177–179
PSK, 219
punctured code, 43
puncturing, 106, 137, 153–155, 159
pure code combining, 159

QAM, 218
QL decomposition, 278
Quadrature Amplitude Modulation, *see* QAM

rank criterion, 256
rank of a matrix, 312
rate-compatible punctured convolutional codes, 160
Raviv, 140
received polynomial, 66
received word, 14
Reed–Solomon code, 81
Reed–Muller code, 55
regular code, 167
repetition code, 47
residue class ring, 297
ring, 296
RLC, 152, 153
root, 302

SCCC, 184
SDMA, 216
selective repeat ARQ, 155
self-dual code, 36
sequence estimation, 112, 116
serial concatenation, 184, 185
Shannon, 163
Shannon function, 4
shortened code, 42
signal, 2
signal power, 6
signal-to-noise ratio, 6
SIMO channel, Single-Input Multiple-Output channel, 235
simplex code, 51
singular value decomposition, 314
SISO, 192
SISO decoder, 181
SISO decoding, 186
SISO, soft-in/soft-out, 181
soft output, 140
soft-decision decoding, 7, 57
soft-input, 136
soft-output decoding, 140
sorted QL decomposition, *see* SQLD
source coding, 1
source decoder, 2
source decoding, 2
source encoder, 2
space–time code, 215
sparse graphs, 165
spatial multiplexing, 265
spectral encoding, 82
spectral norm, 312
spectral polynomial, 82
spectral sequence, 305
sphere detection, 289
SQLD, 285
squared Euclidean distance, 137
state diagram, 103
subspace, 27, 300
superposition, 27
survivor, 118
symbol error probability, 19

symbol-by-symbol decoding, 112
syndrome, 33, 86
syndrome polynomial, 72, 89
systematic block code, 29
systematic encoding, 70, 111
systematic generator matrix, 111

tail-biting, 104, 153
tail-biting code, 104
Tanner graph, 164, 165
TDD, 239
termination of a convolutional code, 104
TFCI, 58
Thitimasjshima, 163
time division duplex, *see* TDD
transmission, 1
trellis diagram, 95, 112, 115
triple repetition code, 9
truncation of a convolutional code, 104
turbo codes, 163, 182, 183
turbo decoding, 163, 186, 188
turbo decoding algorithm, 180
twiddle factor, 305

UART, 49
UMTS, 58, 147, 183
unitary matrix, 313

V-BLAST, 266
Vandermonde matrix, 79
vector, 299
vector space, 18, 299
Viterbi, 97, 160
Viterbi algorithm, 112, 116, 136
Viterbi bound, 134
Voyager, 50

waterfall region, 193
Weber, 140
WEF, 126
weight, 17
weight distribution, 17, 126, 196
weight enumerator, 126
Welch, 140
wide-open region, 195
Wiener solution, 274
word error probability, 19
woven convolutional codes, 177, 198–200, 202, 203, 205
woven turbo codes, 201
Wozencraft, 160

zero, 75
zero element, 296
zero padding, 46
zero-forcing, 273
Zigangirov, 160
Zyablov, 212